W9-AWS-020

ALSO BY JAMES D. WATSON

The Molecular Biology of the Gene (1965, 1970, 1976, coauthor: 1987)

The Double Helix: A Personal Account of the Discovery of the Structure of DNA (1968)

The DNA Story: A Documentary History of Gene Cloning (coauthor: 1981)

Recombinant DNA (coauthor: 1983, 1992)

The Molecular Biology of the Cell (coauthor: 1983, 1989, 1994)

A Passion for DNA: Genes, Genomes, and Society (2000)

Genes, Girls, and Gamow: After the Double Helix (2002)

DNA

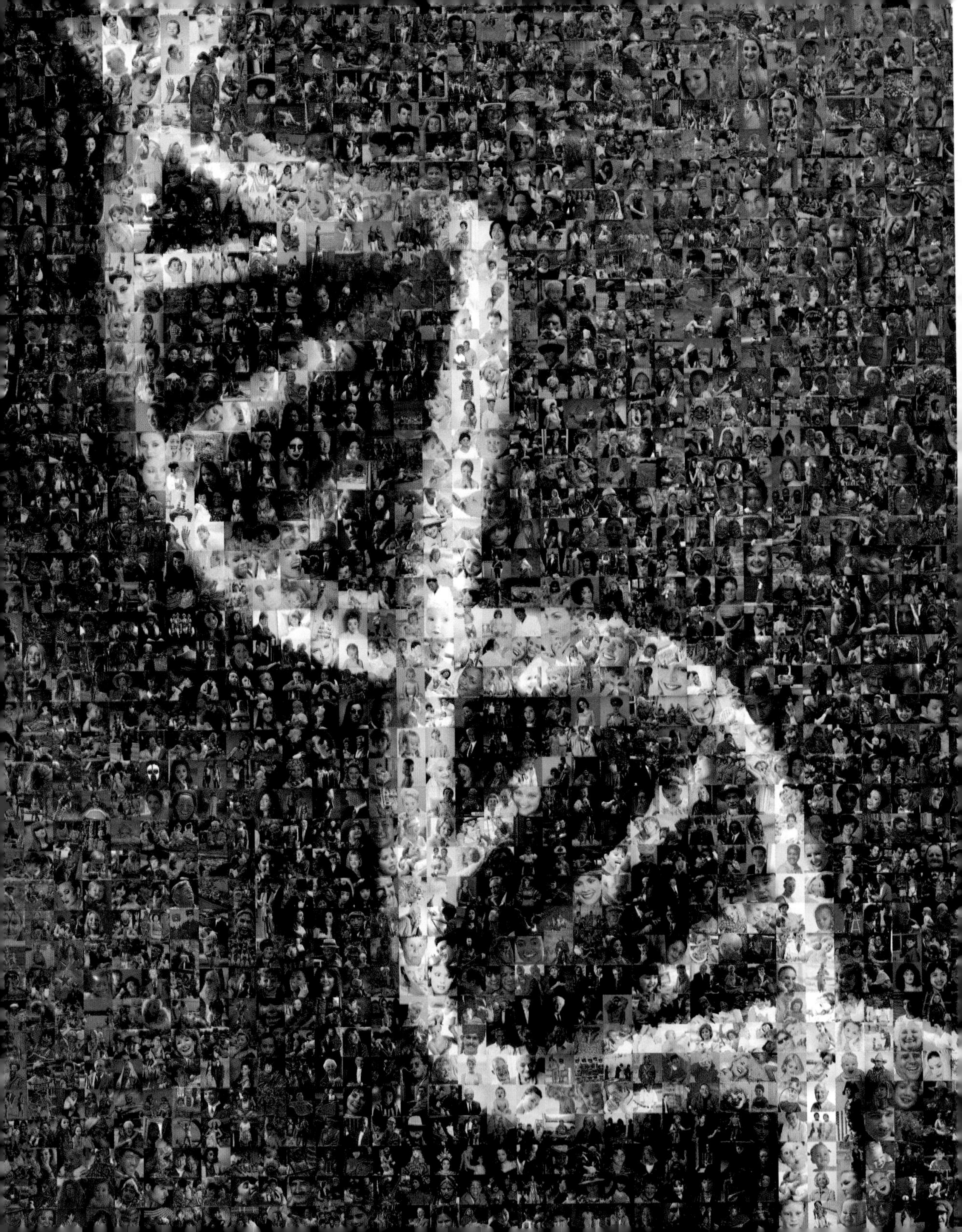

DNA

THE SECRET OF LIFE

JAMES D. WATSON

WITH ANDREW BERRY

ALFRED A. KNOPF
NEW YORK
2004

THIS IS A BORZOI BOOK
PUBLISHED BY ALFRED A. KNOPF

 Published in the United States by Alfred A. Knopf, a division of Random House, Inc., New York, and simultaneously in Canada by Alfred A. Knopf Canada, Limited, Toronto. Distributed by Random House, Inc., New York.
www.aaknopf.com

Library of Congress Cataloging-in-Publication Data

Watson, James D., 1928–
DNA: the secret of life / James D. Watson, with Andrew Berry.
p. cm.
Includes bibliographical references and index.
ISBN 0-375-71007-8 (pbk)
1. Genetics—Popular works. 2. DNA—Popular works.
I. Berry, Andrew. II. Title.
QH437.W387 2003
576.5—dc21 2002190725

Manufactured in the United States of America
Published April 7, 2003
First Paperback Edition, August 2004

For Francis Crick

CONTENTS

1953: Francis Crick (right) and me with our model of the double helix

AUTHORS' NOTE

DNA: *The Secret of Life* was conceived over dinner in 1999. Under discussion was how best to mark the fiftieth anniversary of the discovery the double helix. Publisher Neil Patterson joined one of us, James D. Watson, in dreaming up a multifaceted venture including this book, a television series, and additional more avowedly educational projects. Neil's presence was no accident: he published JDW's first book, *The Molecular Biology of the Gene,* in 1965, and ever since has lurked genielike behind JDW's writing projects. Doron Weber at the Alfred P. Sloan Foundation then secured seed money to ensure that the idea would turn into something more concrete. Andrew Berry was recruited in 2000 to hammer out a detailed outline for the TV series and has since become a regular commuter between his base in Cambridge, Massachusetts, and JDW's at Cold Spring Harbor Laboratory on the north coast of Long Island, close to New York City.

From the start, our goal was to go beyond merely recounting the events of the past fifty years. DNA has moved from being an esoteric molecule only of interest to a handful of specialists to being the heart of a technology that is transforming many aspects of the way we all live. With that transformation has come a host of difficult questions about its impact—practical, social, and ethical. Taking the fiftieth anniversary as an opportunity to pause and take stock of where we are, we give an unabashedly personal view both of the history and of the issues. Moreover, it is JDW's personal view and is accordingly written in the first-person singular. The double helix was already ten years old when DNA was working its *in utero* magic on a fetal AB.

Authors' Note

We have tried to write for a general audience, intending that someone with zero biological knowledge should be able to understand the book's every word. Every technical term is explained when first introduced. Should you need to refresh your memory about a term when you come across one of its later appearances, you can refer to the index, where such words are printed in bold to make locating them easy; a number also in bold will take you to the page on which the term is defined. We have inevitably skimped on many of the technical details and recommend that readers interested in learning more go to DNAi.org, the Web site of the multimedia companion project, DNA Interactive, aimed at high-schoolers and entry-level college students. Here you will find animations explaining basic processes and an extensive archive of interviews with the scientists involved. In addition, the Further Reading section lists books relevant to each chapter. Where possible we have avoided the technical literature, but the titles listed nevertheless provide a more in-depth exploration of particular topics than we supply.

We thank the many people who contributed generously to this project in one way or another in the acknowledgments at the back of the book. Four individuals, however, deserve special mention. George Andreou, our preternaturally patient editor at Knopf, wrote much more of this book—the good bits—than either of us would ever let on. Kiryn Haslinger, our superbly efficient assistant at Cold Spring Harbor Lab, cajoled, bullied, edited, researched, nit-picked, mediated, wrote—all in approximately equal measure. The book simply would not have happened without her. Jan Witkowski, also of Cold Spring Harbor Lab, did a marvelous job of pulling together chapters 10, 11, and 12 in record time and provided indispensable guidance throughout the project. Maureen Berejka, JDW's assistant, rendered sterling service as usual in her capacity as the sole inhabitant of Planet Earth capable of interpreting JDW's handwriting.

James D. Watson
Cold Spring Harbor, New York

Andrew Berry
Cambridge, Massachusetts

INTRODUCTION

THE SECRET OF LIFE

As was normal for a Saturday morning, I got to work at Cambridge University's Cavendish Laboratory earlier than Francis Crick on February 28, 1953. I had good reason for being up early. I knew that we were close—though I had no idea just how close—to figuring out the structure of a then little-known molecule called deoxyribonucleic acid: DNA. This was not any old molecule: DNA, as Crick and I appreciated, holds the very key to the nature of living things. It stores the hereditary information that is passed on from one generation to the next, and it orchestrates the incredibly complex world of the cell. Figuring out its 3-D structure—the molecule's architecture—would, we hoped, provide a glimpse of what Crick referred to only half-jokingly as "the secret of life."

We already knew that DNA molecules consist of multiple copies of a single basic unit, the nucleotide, which comes in four forms: adenine (A), thymine (T), guanine (G), and cytosine (C). I had spent the previous afternoon making cardboard cutouts of these various components, and now, undisturbed on a quiet Saturday morning, I could shuffle around the pieces of the 3-D jigsaw puzzle. How did they all fit together? Soon I realized that a simple pairing scheme worked exquisitely well: A fitted neatly with T, and G with C. Was this it? Did the molecule consist of two chains linked together by A-T and G-C pairs? It was so simple, so elegant, that it almost had to be right. But I had made mistakes in the past, and before I could get too excited, my pairing scheme would have to survive the scrutiny of Crick's critical eye. It was an anxious wait.

But I need not have worried: Crick realized straightaway that my pairing idea implied a double-helix structure with the two molecular chains running in opposite directions. Everything known about DNA and its properties—the facts we had been wrestling with as we tried to solve the problem—made sense in light of those gentle complementary twists. Most important, the way the molecule was organized immediately suggested solutions to two of biology's oldest mysteries: how hereditary information is stored, and how it is replicated. Despite this, Crick's brag in the Eagle, the pub where we habitually ate lunch, that we had indeed discovered that "secret of life," struck me as somewhat immodest, especially in England, where understatement is a way of life.

Crick, however, was right. Our discovery put an end to a debate as old as the human species: Does life have some magical, mystical essence, or is it, like any chemical reaction carried out in a science class, the product of normal physical and chemical processes? Is there something divine at the heart of a cell that brings it to life? The double helix answered that question with a definitive No.

Charles Darwin's theory of evolution, which showed how all of life is interrelated, was a major advance in our understanding of the world in materialistic—physicochemical—terms. The breakthroughs of biologists Theodor Schwann and Louis Pasteur during the second half of the nineteenth century were also an important step forward. Rotting meat did not spontaneously yield maggots; rather, familiar biological agents and processes were responsible—in this case egg-laying flies. The idea of spontaneous generation had been discredited.

Despite these advances, various forms of vitalism—the belief that physicochemical processes cannot explain life and its processes—lingered on. Many biologists, reluctant to accept natural selection as the sole determinant of the fate of evolutionary lineages, invoked a poorly defined overseeing spiritual force to account for adaptation. Physicists, accustomed to dealing with a simple, pared-down world—a few particles, a few forces—found the messy complexity of biology bewildering. Maybe, they suggested, the processes at the heart of the cell, the ones governing the basics of life, go beyond the familiar laws of physics and chemistry.

That is why the double helix was so important. It brought the Enlightenment's revolution in materialistic thinking into the cell. The intellectual journey that had begun with Copernicus displacing humans from the center of the universe and continued with Darwin's insistence that humans are merely modified monkeys had finally focused in on the very essence of life. And there was nothing special about it. The double helix is an elegant structure, but its message is downright prosaic: life is simply a matter of chemistry.

Crick and I were quick to grasp the intellectual significance of our discovery, but there was no way we could have foreseen the explosive impact of the double helix on science and society. Contained in the molecule's graceful curves was the key to molecular biology, a new science whose progress over the subsequent fifty years has been astounding. Not only has it yielded a stunning array of insights into fundamental biological processes, but it is now having an ever more profound impact on medicine, on agriculture, and on the law. DNA is no longer a matter of interest only to white-coated scientists in obscure university laboratories; it affects us all.

By the mid-sixties, we had worked out the basic mechanics of the cell, and we knew how, via the "genetic code," the four-letter alphabet of DNA sequence is translated into the twenty-letter alphabet of the proteins. The next explosive spurt in the new science's growth came in the 1970s with the introduction of techniques for manipulating DNA and reading its sequence of base pairs. We were no longer condemned to watch nature from the sidelines but could actually tinker with the DNA of living organisms, and we could actually read life's basic script. Extraordinary new scientific vistas opened up: we would at last come to grips with genetic diseases from cystic fibrosis to cancer; we would revolutionize criminal justice through genetic fingerprinting methods; we would profoundly revise ideas about human origins—about who we are and where we came from—by using DNA-based approaches to prehistory; and we would improve agriculturally important species with an effectiveness we had previously only dreamed of.

But the climax of the first fifty years of the DNA revolution came on Monday, June 26, 2000, with the announcement by U.S. president Bill Clinton of the completion of the rough draft sequence of the human genome: "Today, we are learning the language in which God created life. With this profound new knowledge, humankind is on the verge of gaining immense, new power to heal." The genome project was a coming-of-age for molecular biology: it had become "big science," with big money and big results. Not only was it an extraordinary technological achievement—the amount of information mined from the human complement of twenty-three pairs of chromosomes is staggering—but it was also a landmark in terms of our idea of what it is to be human. It is our DNA that distinguishes us from all other species, and that makes us the creative, conscious, dominant, destructive creatures that we are. And here, in its entirety, was that set of DNA—the human instruction book.

DNA has come a long way from that Saturday morning in Cambridge. However, it is also clear that the science of molecular biology—what DNA can do for us—still has a long way to go. Cancer still has to be cured; effective gene therapies for genetic diseases still have to be developed; genetic engineering still has to realize its phenomenal potential for improving our food. But all these things will come. The first fifty years of the DNA revolution witnessed a great deal of remarkable scientific progress as well as the initial application of that progress to human problems. The future will see many more scientific advances, but increasingly the focus will be on DNA's ever greater impact on the way we live.

DNA

The key to Mendel's triumph: genetic variation in pea plants

CHAPTER ONE

. . .

BEGINNINGS OF GENETICS: FROM MENDEL TO HITLER

My mother, Bonnie Jean, believed in genes. She was proud of her father's Scottish origins, and saw in him the traditional Scottish virtues of honesty, hard work, and thriftiness. She, too, possessed these qualities and felt that they must have been passed down to her from him. His tragic early death meant that her only nongenetic legacy was a set of tiny little girl's kilts he had ordered for her from Glasgow. Perhaps therefore it is not surprising that she valued her father's biological legacy over his material one.

Growing up, I had endless arguments with Mother about the relative roles played by nature and nurture in shaping us. By choosing nurture over nature, I was effectively subscribing to the belief that I could make myself into whatever I wanted to be. I did not want to accept that my genes mattered that much, preferring to attribute my Watson grandmother's extreme fatness to her having overeaten. If her shape was the product of her genes, then I too might have a hefty future. However, even as a teenager, I would not have disputed the evident basics of inheritance, that like begets like. My arguments with my mother concerned complex characteristics like aspects of personality, not the simple attributes that, even as an obstinate adolescent, I could see were passed down over the generations, resulting in "family likeness." My nose is my mother's and now belongs to my son Duncan.

Sometimes characteristics come and go within a few generations, but sometimes they persist over many. One of the most famous examples of a long-lived trait is known as the "Hapsburg Lip." This distinctive elongation of the jaw and

At age eleven, with my sister Elizabeth and my father, James

droopiness to the lower lip—which made the Hapsburg rulers of Europe such a nightmare assignment for generations of court portrait painters—was passed down intact over at least twenty-three generations.

The Hapsburgs added to their genetic woes by intermarrying. Arranging marriages between different branches of the Hapsburg clan and often among close relatives may have made political sense as a way of building alliances and ensuring dynastic succession, but it was anything but astute in genetic terms. Inbreeding of this kind can result in genetic disease, as the Hapsburgs found out to their cost. Charles II, the last of the Hapsburg monarchs in Spain, not only boasted a prize-worthy example of the family lip—he could not even chew his own food—but was also a complete invalid, and incapable, despite two marriages, of producing children.

Genetic disease has long stalked humanity. In some cases, such as Charles II's, it has had a direct impact on history. Retrospective diagnosis has suggested that George III, the English king whose principal claim to fame is to have lost the American colonies in the Revolutionary War, suffered from an inherited disease, porphyria, which causes periodic bouts of madness. Some historians—mainly British ones—have argued that it was the distraction caused by George's illness that permitted the Americans' against-the-odds military success. While most hereditary diseases have no such geopolitical impact, they nevertheless have brutal and often tragic consequences for the afflicted families, sometimes for many generations. Understanding genetics is not just about understanding why we look like our parents. It is also about coming to grips with some of humankind's oldest enemies: the flaws in our genes that cause genetic disease.

Our ancestors must have wondered about the workings of heredity as soon as evolution endowed them with brains capable of formulating the right kind of question. And the readily observable principle that close relatives tend to be similar can carry you a long way if, like our ancestors, your concern with the application of genetics is limited to practical matters like improving domesticated animals (for, say, milk yield in cattle) and plants (for, say, the size of fruit). Generations of careful selection—breeding initially to domesticate appropriate species, and then breeding only from the most productive cows and from the trees with the largest fruit—resulted in animals and plants tailor-made for human purposes. Underlying this enormous unrecorded effort is that simple rule of thumb: that the most productive cows will produce highly productive offspring and from the seeds of trees with large fruit large-fruited trees will grow. Thus, despite the extraordinary advances of the past hundred years or so, the twentieth and twenty-first centuries by no means have a monopoly on genetic insight. Although it wasn't until 1909 that the British biologist William Bateson gave the science of inheritance a name, genetics, and although the DNA revolution has opened up new and extraordinary vistas of potential progress, in fact the single greatest application of genetics to human well-being was carried out eons ago by anonymous ancient farmers. Almost everything we eat—cereals, fruit, meat, dairy products—is the legacy of that earliest and most far-reaching application of genetic manipulations to human problems.

An understanding of the actual mechanics of genetics proved a tougher nut to crack. Gregor Mendel (1822–1884) published his famous paper on the subject in 1866 (and it was ignored by the scientific community for another thirty-four years). Why did it take so long? After all, heredity is a major aspect of the natural world, and, more important, it is readily, and universally, observable: a dog owner sees how a cross between a brown and black dog turns out, and all parents consciously or subconsciously track the appearance of their own characteristics in their children. One simple reason is that genetic mechanisms turn out to be complicated. Mendel's solution to the problem is not intuitively obvious: children are not, after all, simply a *blend* of their parents' characteristics. Perhaps most important was the failure by early biologists to distinguish between two fundamentally different processes, heredity and development. Today we understand that a fertilized egg contains the genetic information, contributed by both parents, that determines whether someone will be afflicted with, say, porphyria. That is heredity. The subsequent process, the *development* of a new individual from that humble starting point of a single cell, the fertilized egg, involves implementing that information. Broken down in terms of academic disciplines, genetics focuses on the information and developmental biology focuses on the use of that information. Lumping heredity and development together into a single phenomenon, early scientists never asked the questions that might have steered them toward the secret of heredity. Nevertheless, the effort had been under way in some form since the dawn of Western history.

The Greeks, including Hippocrates, pondered heredity. They devised a theory of "pangenesis," which claimed that sex involved the transfer of miniaturized body parts: "Hairs, nails, veins, arteries, tendons and their bones, albeit invisible as their particles are so small. While growing, they gradually separate from each other." This idea enjoyed a brief renaissance when Charles Darwin, desperate to support his theory of evolution by natural selection with a viable hypothesis of inheritance, put forward a modified version of pangenesis in the second half of the nineteenth century. In Darwin's scheme, each organ—eyes, kidneys, bones—contributed circulating "gemmules" that accumulated in the sex organs, and were ultimately exchanged in the course of sexual reproduction. Because these gemmules were produced throughout an organism's lifetime, Darwin argued any change that occurred in the individual after birth, like the stretch of a giraffe's neck imparted by craning for the highest foliage, could be passed on to the next generation. Ironically, then, to buttress his theory of natural selection Darwin came to champion aspects of Jean-Baptiste Lamarck's theory of inheritance of acquired characteristics—the very theory that his evolutionary ideas did so much to discredit. Darwin was invoking only Lamarck's theory of inheritance; he continued to believe that natural selection was the driving force behind evolution, but supposed that natural selection operated on the variation produced by pangenesis. Had Darwin known about Mendel's work (although Mendel published his results shortly after *The Origin of Species* appeared, Darwin was never aware of them), he might have been spared the embarrassment of this late-career endorsement of some of Lamarck's ideas.

Whereas pangenesis supposed that embryos were assembled from a set of minuscule components, another approach, "preformationism," avoided the assembly step altogether: either the egg or the sperm (exactly which was a contentious issue) contained a complete *preformed* individual called a homunculus. Development was therefore merely a matter of enlarging this into a fully formed being. In the days of preformationism, what we now recognize as genetic disease was variously interpreted: sometimes as a manifestation of the wrath of God or the mischief of demons and devils; sometimes as evidence of either an excess of or a deficit of the father's "seed"; sometimes as the result of "wicked thoughts" on the part of the mother during pregnancy. On the premise that fetal malformation can result when a pregnant mother's desires are thwarted, leaving her feeling stressed and frustrated, Napoleon passed a law permitting expectant mothers to shoplift. None of these notions, needless to say, did much to advance our understanding of genetic disease.

By the early nineteenth century, better microscopes had defeated preformationism. Look as hard as you like, you will never see a tiny homunculus curled up inside a sperm or egg cell. Pangenesis, though an earlier misconception, lasted rather longer—the argument would persist that the gemmules were simply too small to visualize—but was eventually laid to rest by August Weismann, who argued that inheritance depended on the continuity of germ plasm between generations and thus changes to the body over an individual's lifetime could *not* be transmitted to subsequent generations. His simple experiment involved cutting the tails off several

Genetics before Mendel: a homunculus, a preformed miniature person imagined to exist in the head of a sperm cell

generations of mice. According to Darwin's pangenesis, tailless mice would produce gemmules signifying "no tail" and so their offspring should develop a severely stunted hind appendage or none at all. When Weismann showed that the tail kept appearing after many generations of amputees, pangenesis bit the dust.

Gregor Mendel was the one who got it right. By any standards, however, he was an unlikely candidate for scientific superstardom. Born to a farming family in what is now the Czech Republic, he excelled at the village school and, at twenty-one, entered the Augustinian monastery at Brünn. After proving a disaster as a parish priest—his response to the ministry was a nervous breakdown—he tried his hand at teaching. By all accounts he was a good teacher, but in order to qualify to teach a full range of subjects, he had to take an exam. He failed it. Mendel's father superior, Abbot Napp, then dispatched him to the University of Vienna, where he was to bone up full-time for the retesting. Despite apparently doing well in physics at Vienna, Mendel again failed the exam, and so never rose above the rank of substitute teacher.

Around 1856, at Abbot Napp's suggestion, Mendel undertook some scientific experiments on heredity. He chose to study a number of characteristics of the pea plants he grew in his own patch of the monastery garden. In 1865 he presented his results to the local natural history society in two lectures, and, a year later, published them in the society's journal. The work was a tour de force: the experiments were brilliantly designed and painstakingly executed, and his analysis of the results was insightful and deft. It seems that his training in physics contributed to his breakthrough because, unlike other biologists of that time, he approached the problem quantitatively. Rather than simply noting that crossbreeding of red and white flowers resulted in some red and some white offspring, Mendel actually counted them, realizing that the ratios of red to white progeny might be significant—as indeed they are. Despite sending copies of his article to various prominent scientists, Mendel found himself completely ignored by the scientific community. His attempt to draw attention to his results merely backfired. He wrote to his one contact among the ranking scientists of the day, botanist Karl Nägeli in Munich, asking him to replicate the experiments, and he duly sent off 140 carefully labeled packets of seeds. He should not have bothered. Nägeli believed that the obscure monk should be of service to him, rather than the other way around, so he sent Mendel seeds of his own favorite plant, hawkweed, challenging the monk to re-create his results with a different species. Sad to say, for various reasons, hawkweed is not well-suited to breeding experiments such as those Mendel had performed on the peas. The entire exercise was a waste of his time.

Mendel's low-profile existence as monk-teacher-researcher ended abruptly in 1868 when, on Napp's death, he was elected abbot of the monastery. Although he continued his research—increasingly on bees and the weather—administrative duties were a burden, especially as the monastery became embroiled in a messy dispute over back taxes. Other factors, too, hampered him as a scientist. Portliness eventually curtailed his fieldwork: as he wrote, hill climbing had become "very difficult for me in a world where universal gravitation prevails." His doctors prescribed tobacco to keep his weight in check, and he obliged them by smoking twenty cigars a day, as many as Winston Churchill. It was not his lungs, however, that let him down: in 1884, at the age of sixty-one, Mendel succumbed to a combination of heart and kidney disease.

Not only were Mendel's results buried in an obscure journal, but they would have been unintelligible to most scientists of the era. He was far ahead of his time with his combination of careful experiment and sophisticated quantitative analysis. Little wonder, perhaps, that it was not until 1900 that the scientific community caught up with him. The rediscovery of Mendel's work, by three plant geneticists interested in similar problems, provoked a revolution in biology. At last the scientific world was ready for the monk's peas.

Mendel realized that there are specific factors—later to be called "genes"—that are passed from parent to offspring. He worked out that these factors come in pairs and that the offspring receives one from each parent.

Noticing that peas came in two distinct colors, green and yellow, he deduced that there were two versions of the pea-color gene. A pea has to have two copies of the G version if it is to become green, in which case we say that it is GG for the pea-color gene. It must therefore have received a G pea-color gene from both of its parents. However, yellow peas can result both from YY and YG combinations. Having only one copy of the Y version is sufficient to produce yellow peas. Y trumps G. Because in the YG case the Y signal dominates the G signal, we call Y "dominant." The subordinate G version of the pea-color gene is called "recessive."

Each parent pea plant has two copies of the pea-color gene, yet it contributes only one copy to each offspring; the other copy is furnished by the other parent. In plants, pollen grains contain sperm cells—the male contribution to the next generation—and each sperm cell contains just one copy of the pea-color gene. A parent pea plant with a YG combination will produce sperm that contain either a Y version or a G one. Mendel discovered that the process is random: 50 percent of the sperm produced by that plant will have a Y and 50 percent will have a G.

Suddenly many of the mysteries of heredity made sense. Characteristics, like the Hapsburg Lip, that are transmitted with a high probability (actually 50 percent) from generation to generation are dominant. Other characteristics that appear in family trees much more sporadically, often skipping generations, may be recessive. When a gene is recessive an individual has to have two copies of it for the corresponding trait to be expressed. Those with one copy of the gene are carriers: they don't themselves exhibit the characteristic, but they can pass the gene on. Albinism, in which the body fails to produce pigment so the skin and hair are strikingly white, is an example of a recessive characteristic that is transmitted in this way. Therefore, to be albino you have to have two copies of the gene, one from each parent. (This was the case with the Reverend Dr. William Archibald Spooner, who was also—perhaps only by coincidence—prone to a peculiar form of linguistic confusion whereby, for example, "a well-oiled bicycle" might become "a well-boiled icicle." Such reversals would come to be termed "spoonerisms" in his honor.) Your parents, meanwhile, may have shown no sign of the gene at all. If, as is often the case, each has only one copy, then they are both carriers. The trait has skipped at least one generation.

Mendel's results implied that *things*—material objects—were transmitted from generation to generation. But what was the nature of these things?

At about the time of Mendel's death in 1884, scientists using ever-improving

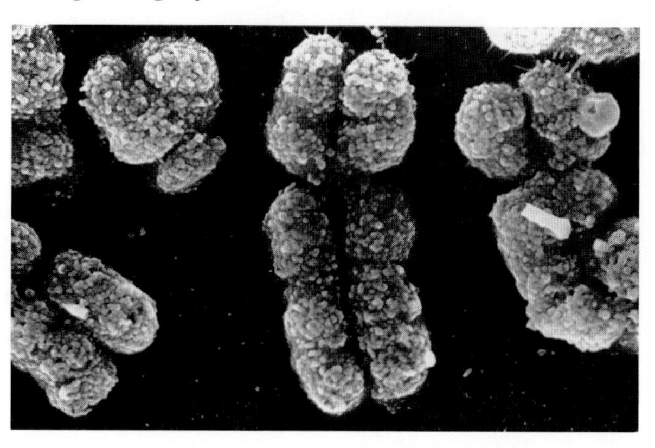

The human X chromosome, as seen with an electron microscope

optics to study the minute architecture of cells coined the term "chromosome" to describe the long stringy bodies in the cell nucleus. But it was not until 1902 that Mendel and chromosomes came together.

A medical student at Columbia University, Walter Sutton, realized that chromosomes had a lot in common with Mendel's mysterious factors. Studying grasshopper chromosomes, Sutton noticed that most of the time they are doubled up—just like Mendel's paired factors. But Sutton also identified one type of cell in which chromosomes were not paired: the sex cells. Grasshopper sperm have only a single set of chromosomes, not a double set. This was exactly what Mendel had described: his pea plant sperm cells also only carried a single copy of each of his factors. It was clear that Mendel's factors, now called genes, must be on the chromosomes.

In Germany Theodor Boveri independently came to the same conclusions as Sutton, and so the biological revolution their work had precipitated came to be called the Sutton-Boveri chromosome theory of inheritance. Suddenly genes were real. They were on chromosomes, and you could actually see chromosomes through the microscope.

Not everyone bought the Sutton-Boveri theory. One skeptic was Thomas Hunt Morgan, also at Columbia. Looking down the microscope at those stringy chromosomes, he could not see how they could account for all the

Notoriously camera shy T. H. Morgan was photographed surreptitiously while at work in the fly room.

changes that occur from one generation to the next. If all the genes were arranged along chromosomes, and all chromosomes were transmitted intact from one generation to the next, then surely many characteristics would be inherited together. But since empirical evidence showed this not to be the case, the chromosomal theory seemed insufficient to explain the variation observed in nature. Being an astute experimentalist, however, Morgan had an idea how he might resolve such discrepancies. He turned to the fruit fly, *Drosophila melanogaster,* the drab little beast that, ever since Morgan, has been so beloved by geneticists.

In fact, Morgan was not the first to use the fruit fly in breeding experiments—that distinction belonged to a lab at Harvard that first put the critter to work in 1901—but it was Morgan's work that put the fly on the scientific map. *Drosophila* is a good choice for genetic experiments. It is easy to find (as anyone who has left out a bunch of overripe bananas during the summer well knows); it is easy to raise (bananas will do as feed); and you can accommodate hundreds of flies in a single milk bottle (Morgan's students had no difficulty acquiring milk bottles, pinching them at dawn from doorsteps in their Manhattan neighborhood); and it breeds and breeds and breeds (a whole generation takes about ten days, and each female lays several hundred eggs). Starting in 1907 in a famously squalid, cockroach-infested, banana-stinking lab that came to be known affectionately as the "fly room," Morgan and his students ("Morgan's boys" as they were called) set to work on fruit flies.

Unlike Mendel, who could rely on the variant strains isolated over the years by farmers and gardeners—yellow peas as opposed to green ones, wrinkled skin as opposed to smooth—Morgan had no menu of established genetic differences in the fruit fly to draw upon. And you cannot do genetics until you have isolated some distinct characteristics to track through the generations. Morgan's first goal therefore was to find "mutants," the fruit fly equivalents of yellow or wrinkled peas. He was looking for genetic novelties, random variations that somehow simply appeared in the population.

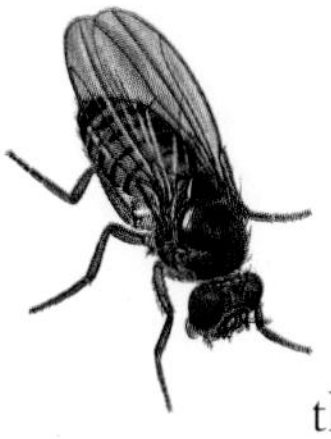

One of the first mutants Morgan observed turned out to be one of the most instructive. While normal fruit flies have red eyes, these had white ones. And he noticed that the white-eyed flies were typically male. It was known that the sex of a fruit fly—or, for that matter, the sex of a human—is determined chromosomally: females have two copies of the X chromosome, whereas males have one copy of the X and one copy of the much smaller Y. In light of this information, the white-eye result suddenly made sense: the eye-color gene is located on the X chromosome and the white-eye mutation, W, is recessive. Because males have only a single X chromosome, even recessive genes, in the absence of a dominant counterpart to suppress them, are automatically expressed. White-eyed females were relatively rare because they typically had only one copy of W so they expressed the dominant red eye color. By correlating a gene—the one for eye color—with a chromosome, the X, Morgan, despite his initial reservations, had effectively proved the Sutton-Boveri theory. He had also found an example of "sex-linkage," in which a particular characteristic is disproportionately represented in one sex.

Like Morgan's fruit flies, Queen Victoria provides a famous example of sex-linkage. On one of her X chromosomes, she had a mutated gene for hemophilia, the "bleeding disease" in whose victims proper blood clotting fails to occur. Because her other copy was normal, and the hemophilia gene is recessive, she herself did not have the disease. But she was a carrier. Her daughters did not have the disease either; evidently each possessed at least one copy of the normal version. But Victoria's sons were not all so lucky. Like all males (fruit fly males included), each had only one X chromosome; this was necessarily derived from Victoria (a Y chromosome could have come only from Prince Albert, Victoria's husband). Because Victoria had one mutated copy and one normal copy, each of her sons had a 50-50 chance of having the disease. Prince Leopold drew the short straw: he developed hemophilia, and died at thirty-one, bleeding to death after a minor fall. Two of Victoria's daughters, Princesses Alice and Beatrice, were carriers, having inherited the mutated gene from their mother. They each produced carrier daughters and sons with hemophilia. Alice's grandson Alexis, heir to the Russian throne, had hemophilia, and would doubtless have died young had the Bolsheviks not gotten to him first.

Morgan's fruit flies had other secrets to reveal. In the course of studying genes located on the same chromosome, Morgan and his students found that chromosomes actually break apart and re-form during the production of sperm and egg cells. This meant that Morgan's original objections to the Sutton-Boveri theory were unwarranted: the breaking and re-forming—"recombination," in modern genetic parlance—shuffles gene copies between members of a chromosome pair. This means that, say, the copy of chromosome 12 I got from my mother (the other, of course, comes from my father) is in fact a mix of my mother's two copies of chromosome 12, one of which came from her mother and one from her father. Her two 12s recombined—exchanged material—during the production of the egg cell that eventually turned into me. Thus my maternally derived chromosome 12 can be viewed as a mosaic of my grandparents' 12s. Of course, my mother's maternally derived 12 was itself a mosaic of her grandparents' 12s, and so on.

Recombination permitted Morgan and his students to map out the positions of particular genes along a given chromosome. Recombination involves breaking (and re-forming) chromosomes. Because genes are arranged like beads along a chromosome string, a break is statistically much more likely to occur between two genes that are far apart (with more potential break points intervening) on the chromosome than between two genes that are close together. If, therefore, we see a lot of reshuffling for any two genes on a single chromosome, we can conclude that they are a long way apart; the rarer the reshuffling, the closer the genes likely are. This basic and immensely powerful principle underlies all of genetic mapping. One of the primary tools of scientists involved in the Human Genome Project and of researchers at the forefront of the battle against genetic disease was thus developed all those years ago in the filthy, cluttered Columbia fly room. Each new headline in the science section of the newspaper these days along the lines of "Gene for Something Located" is a tribute to the pioneering work of Morgan and his boys.

The rediscovery of Mendel's work, and the breakthroughs that followed it, sparked a surge of interest in the social significance of genetics. While scientists had been grappling with the precise mechanisms of heredity through the eighteenth and nineteenth centuries, public concern had been mounting about the burden placed on society by what came to be called the "degenerate classes"—the inhabitants of poorhouses, workhouses, and insane asylums. What could be done with these people? It remained a matter of controversy whether they should be treated charitably—which, the less charitably inclined claimed, ensured such folk would never exert themselves and would therefore remain forever dependent on the largesse of the state or of private institutions—or whether they should be simply ignored, which, according to the charitably inclined, would result only in perpetuating the inability of the unfortunate to extricate themselves from their blighted circumstances.

The publication of Darwin's *Origin of Species* in 1859 brought these issues into sharp focus. Although Darwin carefully omitted to mention human evolution, fearing that to do so would only further inflame an already raging controversy, it required no great leap of imagination to apply his idea of natural selection to humans. Natural selection is the force that determines the fate of all genetic variations in nature—mutations like the one Morgan found in the fruit fly eye-color gene, but also perhaps differences in the abilities of human individuals to fend for themselves.

Natural populations have an enormous reproductive potential. Take fruit flies, with their generation time of just ten days, and females that produce some three hundred eggs apiece (half of which will be female): starting with a single fruit fly couple, after a month (i.e., three generations later), you will have 150 × 150 × 150 fruit flies on your hands—that's more than 3 million flies, all of them derived from just one pair in just one month. Darwin made the point by choosing a species from the other end of the reproductive spectrum:

> The elephant is reckoned to be the slowest breeder of all known animals, and I have taken some pains to estimate its probable minimum rate of natural increase: it will be under the mark to assume that it breeds when thirty years old, and goes on breeding till ninety years old, bringing forth three

> pairs of young in this interval; if this be so, at the end of the fifth century there would be alive fifteen million elephants, descended from the first pair.

All these calculations assume that all the baby fruit flies and all the baby elephants make it successfully to adulthood. In theory, therefore, there must be an infinitely large supply of food and water to sustain this kind of reproductive overdrive. In reality, of course, those resources are limited, and not all baby fruit flies or baby elephants make it. There is competition among individuals within a species for those resources. What determines who wins the struggle for access to the resources? Darwin pointed out genetic variation means that some individuals have advantages in what he called "the struggle for existence." To take the famous example of Darwin's finches from the Galápagos Islands, those individuals with genetic advantages—like the right size of beak for eating the most abundant seeds—are more likely to survive and reproduce. So the advantageous genetic variant—having a bill the right size—tends to be passed on to the next generation. The result is that natural selection enriches the next generation with the beneficial mutation so that eventually, over enough generations, every member of the species ends up with that characteristic.

The Victorians applied the same logic to humans. They looked around and were alarmed by what they saw. The decent, moral, hardworking middle classes were being massively outreproduced by the dirty, immoral, lazy lower classes. The Victorians assumed that the virtues of decency, morality, and hard work ran in families just as the vices of filth, wantonness, and indolence did. Such characteristics must then be hereditary; thus, to the Victorians, morality and immorality were merely two of Darwin's genetic variants. And if the great unwashed were outreproducing the respectable classes, then the "bad" genes would be increasing in the human population. The species was doomed! Humans would gradually become more and more depraved as the "immorality" gene became more and more common.

Francis Galton had good reason to pay special attention to Darwin's book, as the author was his cousin and friend. Darwin, some thirteen years older, had provided guidance during Galton's rather rocky college experience. But it was *The Origin of Species* that would inspire Galton to start a social and genetic crusade that would ultimately have disastrous consequences. In 1883, a year after his cousin's death, Galton gave the movement a name: eugenics.

Eugenics was only one of Galton's many interests; Galton enthusiasts refer to him as a polymath, detractors as a dilettante. In fact, he made significant contributions to geography, anthropology, psychology, genetics, meteorology, statistics, and, by setting fingerprint analysis on a sound scientific footing, to criminology. Born in 1822 into a prosperous family, his education—partly in medicine and partly in mathematics—was mostly a chronicle of defeated expectations. The death of his father when he was twenty-one simultaneously freed him from paternal restraint and yielded a handsome inheritance; the young man duly took advantage of both. After a full six years of being what might be described today as a trust-fund dropout, however, Galton settled down to become a productive member of the Victorian establishment. He made his name leading an expedition to a then little known region of southwest Africa in 1850–52. In his account of his explorations, we encounter the first instance of the one strand that connects his many varied interests: he counted and measured everything. Galton was only happy when he could reduce a phenomenon to a set of numbers.

SARTLEE, THE HOTTENTOT VENUS.
NOW Exhibiting in London.
Drawn from Life

A nineteenth-century exaggerated view of a Nama woman

At a missionary station he encountered a striking specimen of steatopygia—a condition of particularly protuberant buttocks, common among the indigenous Nama women of the region—and realized that this woman was naturally endowed with the figure that was then fashionable in Europe. The only difference was that it required enormous (and costly) ingenuity on the part of European dressmakers to create the desired "look" for their clients.

> I profess to be a scientific man, and was exceedingly anxious to obtain accurate measurements of her shape; but there was a difficulty in doing this. I did not know a word of Hottentot [the Dutch name for the Nama], and could never therefore have explained to the lady what the object of my

> footrule could be; and I really dared not ask my worthy missionary host to interpret for me. I therefore felt in a dilemma as I gazed at her form, that gift of bounteous nature to this favoured race, which no mantua-maker, with all her crinoline and stuffing, can do otherwise than humbly imitate. The object of my admiration stood under a tree, and was turning herself about to all points of the compass, as ladies who wish to be admired usually do. Of a sudden my eye fell upon my sextant; the bright thought struck me, and I took a series of observations upon her figure in every direction, up and down, crossways, diagonally, and so forth, and I registered them carefully upon an outline drawing for fear of any mistake; this being done, I boldly pulled out my measuring tape, and measured the distance from where I was to the place she stood, and having thus obtained both base and angles, I worked out the results by trigonometry and logarithms.

Galton's passion for quantification resulted in his developing many of the fundamental principles of modern statistics. It also yielded some clever observations. For example, he tested the efficacy of prayer. He figured that if prayer worked, those most prayed for should be at an advantage; to test the hypothesis he studied the longevity of British monarchs. Every Sunday, congregations in the Church of England following the *Book of Common Prayer* beseeched God to "Endue the king/queen plenteously with heavenly gifts; Grant him/her in health and wealth long to live." Surely, Galton reasoned, the cumulative effect of all those prayers should be beneficial. In fact, prayer seemed ineffectual: he found that on average the monarchs died somewhat younger than other members of the British aristocracy.

Because of the Darwin connection—their common grandfather, Erasmus Darwin, too was one of the intellectual giants of his day—Galton was especially sensitive to the way in which certain lineages seemed to spawn disproportionately large numbers of prominent and successful people. In 1869 he published what would become the underpinning of all his ideas on eugenics, a treatise called *Hereditary Genius: An Inquiry into Its Laws and Consequences.* In it he purported to show that talent, like simple genetic traits such as the Hapsburg Lip, does indeed run in families; he recounted, for example, how some families had produced generation after generation of judges. His analysis largely neglected to take into account the effect of the environment: the son of a prominent judge is, after all, rather more likely to become a judge—by virtue of his father's connections, if nothing else—than the son of a peasant farmer. Galton did not, however, completely overlook the effect of the environment, and it was he who first referred to the "nature/nurture" dichotomy, possibly in reference to Shakespeare's irredeemable villain, Caliban, "a devil, a born devil, on whose nature/Nurture can never stick."

The results of his analysis, however, left no doubt in Galton's mind.

> I have no patience with the hypothesis occasionally expressed, and often implied, especially in tales written to teach children to be good, that babies are born pretty much alike, and that the sole agencies in creating differences between boy and boy, and man and man, are steady application and moral effort. It is in the most unqualified manner that I object to pretensions of natural equality.

A corollary of his conviction that these traits are genetically determined, he argued, was that it would be possible to "improve" the human stock by preferentially breeding gifted individuals, and preventing the less gifted from reproducing.

> It is easy . . . to obtain by careful selection a permanent breed of dogs or horses gifted with peculiar powers of running, or of doing anything else, so it would be quite practicable to produce a highly-gifted race of men by judicious marriages during several consecutive generations.

Galton introduced the terms *eugenics* (literally "good in birth") to describe this application of the basic principle of agricultural breeding to humans. In time, eugenics came to refer to "self-directed human evolution": by making conscious choices about who should have children, eugenicists believed that they could head off the "eugenic crisis" precipitated in the Victorian imagination by the high rates of reproduction of inferior stock coupled with the typically small families of the superior middle classes.

Eugenics as it was perceived during the first part of the twentieth century: an opportunity for humans to control their own evolutionary destiny

Eugenics these days is a dirty word, associated with racists and Nazis—a dark, best-forgotten phase of the history of genetics. It is important to appreciate, however, that in the closing years of the nineteenth and early years of the twentieth centuries, eugenics was not tainted in this way, and was seen by many as offering genuine potential for improving not just society as a whole but the lot of individuals within society as well. Eugenics was embraced with particular enthusiasm by those who today would be termed the "liberal left." Fabian socialists—some the era's most progressive thinkers—flocked to the cause, including George Bernard Shaw, who wrote that "there is now no reasonable excuse for refusing to face the fact that nothing but a eugenic religion can save our civilisation." Eugenics seemed to offer a solution to one of society's most persistent woes: that segment of the population that is incapable of existing outside an institution.

Whereas Galton had preached what came to be known as "positive eugenics," encouraging genetically superior people to have children, the American eugenics movement preferred to focus on "negative eugenics," preventing genetically inferior people from doing so. The goals of each program were basically the same—the improvement of the human genetic stock—but these two approaches were very different.

The American focus on getting rid of bad genes, as opposed to increasing frequencies of good ones, stemmed from a few influential family studies of "degeneration" and "feeblemindedness"—two peculiar terms characteristic of the American obsession with genetic decline. In 1875 Richard Dugdale published his account of the Juke clan of upstate New York. Here, according to Dugdale, were several generations of seriously bad apples—murderers, alcoholics, and rapists. Apparently in the area near their home in New York State the very name "Juke" was a term of reproach.

Another highly influential study was published in 1912 by Henry Goddard, the psychologist who gave us the word "moron," on what he called "The Kallikak Family." This is the story of two family lines originating from a single male ancestor who had a child out of wedlock (with a "feebleminded" wench he met in a tavern while serving in the military during the American Revolutionary War), as well as siring a legitimate family. The illegitimate side of the Kallikak line, according to Goddard, was bad news indeed, "a race of defective degenerates," while the legitimate side comprised respectable, upstanding members of the community. To Goddard, this "natural experiment in heredity" was an exemplary tale of good genes versus bad. This view was reflected in the fictitious name he chose for the family. "Kallikak" is a hybrid of two Greek words, *kalos* (beautiful, of good repute) and *kakos* (bad).

"Rigorous" new methods for testing mental performance—the first IQ tests, which were introduced to the United States from Europe by the same Henry Goddard—seemed to confirm the general impression that the human species was gaining downward momentum on a genetic slippery slope. In those early days of IQ testing, it was thought that high intelligence and an alert mind inevitably implied a capacity to absorb large quantities of information. Thus how much you knew was considered a sort of index of your IQ. Following this line of reasoning, early IQ tests included lots of general knowledge questions. Here are a few from a standard test administered to U.S. Army recruits during World War I:

> Pick one of four:
>
> The Wyandotte is a kind of:
> 1) *horse* 2) *fowl* 3) *cattle* 4) *granite*
>
> The ampere is used in measuring:
> 1) *wind power* 2) *electricity* 3) *water power* 4) *rain fall*
>
> The number of a Zulu's legs is:
> 1) *two* 2) *four* 3) *six* 4) *eight*
>
> [Answers are 2, 2, 1]

Some half of the nation's army recruits flunked the test and were deemed "feebleminded." These results galvanized the eugenics movement in the United States: it seemed to concerned Americans that the gene pool really was becoming more and more awash in low-intelligence genes.

Scientists realized that eugenic policies required some understanding of the genetics underlying characteristics like feeblemindedness. With the rediscovery of Mendel's work, it seemed that this might actually be possible. The lead in this endeavor was taken on Long Island by one of my predecessors as director of Cold Spring Harbor Laboratory. His name was Charles Davenport.

In 1910, with funding from a railroad heiress, Davenport established the Eugenics Record Office at Cold Spring Harbor. Its mission was to collect basic information—pedigrees—on the genetics of traits ranging from epilepsy to criminality. It became the nerve center of the American eugenics movement. Cold Spring Harbor's mission was much the same then as it is now: today we strive to be at the forefront of genetic research, and Davenport had no less lofty aspirations—but in those days the forefront was eugenics. However, there is no

The staff of the Eugenics Record Office, pictured with members of the Cold Spring Harbor Laboratory. Davenport, seated in the very center, hired personnel on the basis of his belief that women were genetically suited to the task of gathering pedigree data.

doubt that the research program initiated by Davenport was deeply flawed from the outset and had horrendous, albeit unintended, consequences.

Eugenic thinking permeated everything Davenport did. He went out of his way, for instance, to hire women as field researchers because he believed them to have better observational and social skills than men. But, in keeping with the central goal of eugenics to reduce the number of bad genes, and increase the number of good ones, these women were hired for a maximum of three years. They were smart and educated, and therefore, by definition, the possessors of good genes. It would hardly be fitting for the Eugenics Record Office to hold them back too long from their rightful destiny of producing families and passing on their genetic treasure.

Davenport applied Mendelian analysis to pedigrees he constructed of human characteristics. Initially, he confined his attentions to a number of simple traits—like albinism (recessive) and Huntington disease (dominant)—whose mode of inheritance he identified correctly. After these early successes he plunged into a study of the genetics of human behavior. Everything was fair game: all he needed was a pedigree and some information about the family history (i.e., who in the line manifested the particular characteristic in question),

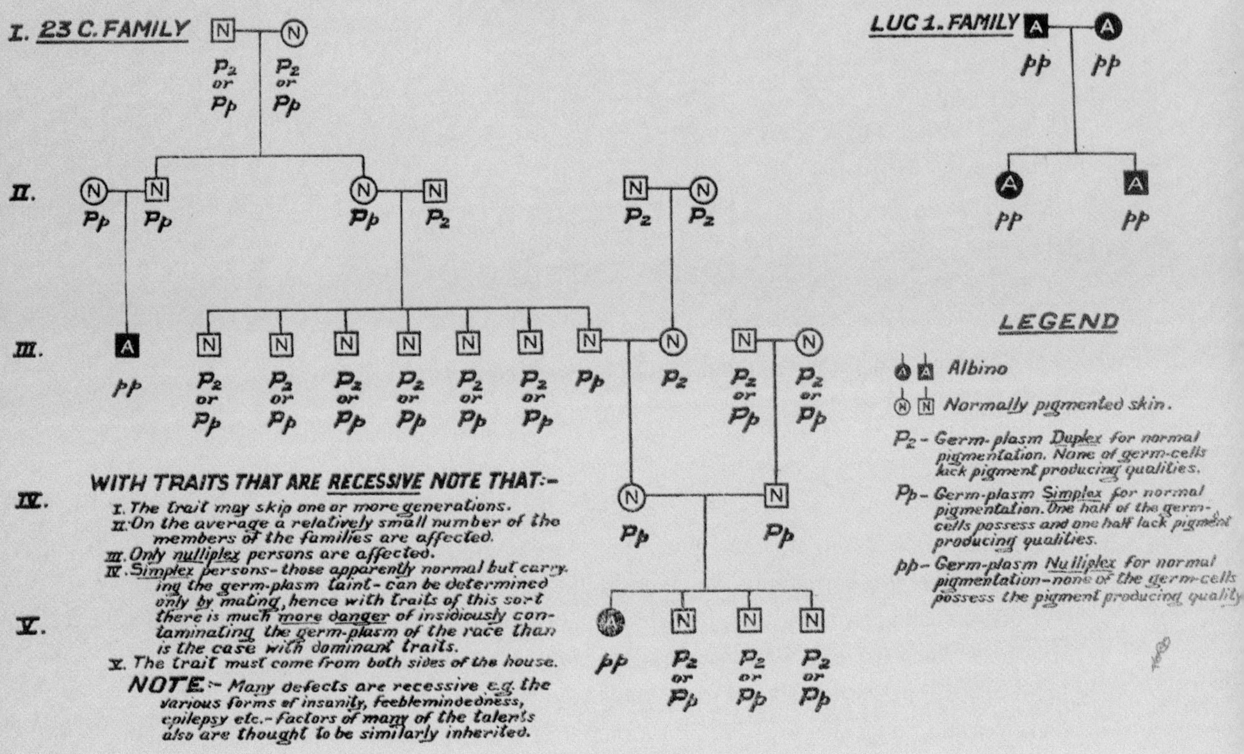

Sound genetics: Davenport's pedigree showing how albinism is inherited

and he would derive conclusions about the underlying genetics. The most cursory perusal of his 1911 book, *Heredity in Relation to Eugenics,* reveals just how wide-ranging Davenport's project was. He shows pedigrees of families with musical and literary ability, and of a "family with mechanical and inventive ability, particularly with respect to boat-building." (Apparently Davenport thought that he was tracking the transmission of the boat-building gene.) Davenport even claimed that he could identify distinct family types associated with different surnames. Thus people with the surname Twinings have these characteristics: "broad-shouldered, dark hair, prominent nose, nervous temperament, temper usually quick, not revengeful. Heavy eyebrows, humorous vein, and sense of ludicrous; lovers of music and horses."

The entire exercise was worthless. Today we know all the characteristics in question are readily affected by environmental factors. Davenport, like Galton, assumed unreasonably that nature unfailingly triumphed over nurture. In addition, whereas the traits he had studied earlier, albinism and Huntington disease, have a simple genetic basis—they are caused by a particular mutation in a particular gene—for most behavioral characteristics, the genetic basis, if any, is complex. They may be determined by a large number of different genes, each one contributing just a little to the final outcome. This situation makes the interpretation of pedigree data like Davenport's virtually impossible. Moreover, the genetic causes of poorly defined characteristics like "feeblemindedness" in one individual may be very different from those in another, so that any search for underlying genetic generalities is futile.

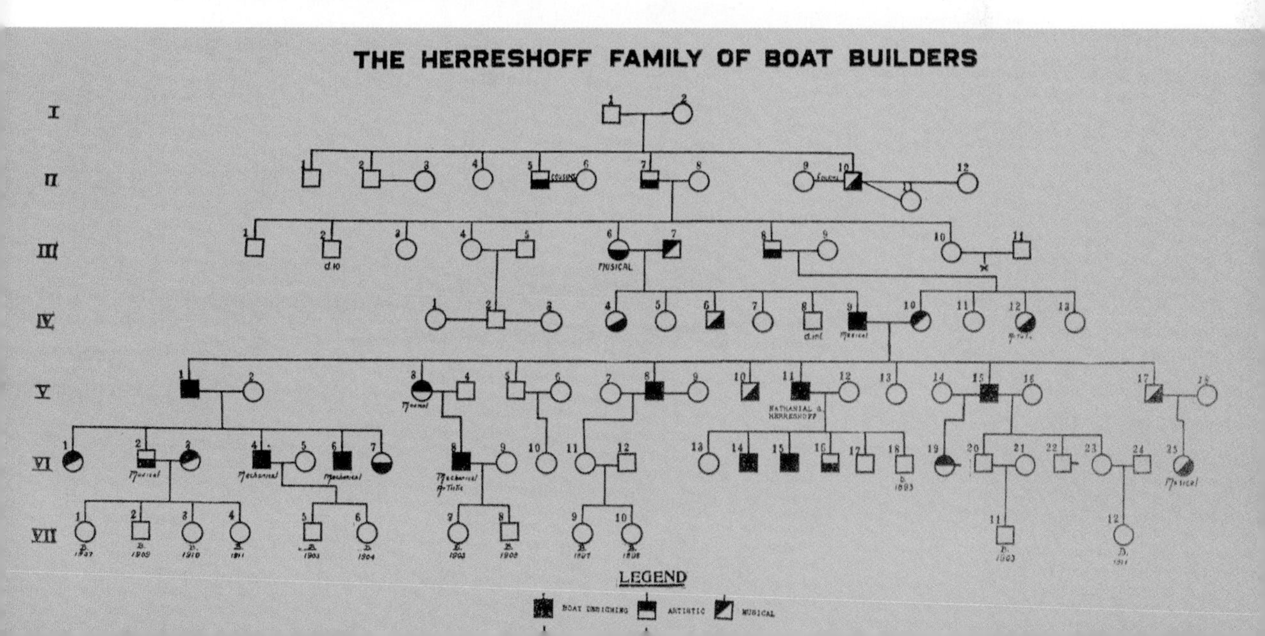

Unsound genetics: Davenport's pedigree showing how boat-building skills are inherited. He fails to factor in the effect of the environment; a boat-builder's son is likely to follow his father's trade because he has been raised in that environment.

Regardless of the success or failure of Davenport's scientific program, the eugenics movement had already developed a momentum of its own. Local chapters of the Eugenics Society organized competitions at state fairs, giving awards to families apparently free from the taint of bad genes. Fairs that had previously displayed only prize cattle and sheep now added "Better Babies" and "Fitter Families" contests to their programs. Effectively these were efforts to encourage positive eugenics—inducing the right kind of people to have children. Eugenics was even de rigueur in the nascent feminist movement. The feminist champions of birth control, Marie Stopes in Britain and, in the United States, Margaret Sanger, founder of Planned Parenthood, both viewed birth control as a form of eugenics. Sanger put it succinctly in 1919: "More children from the fit, less from the unfit—that is the chief issue of birth control."

Altogether more sinister was the growth of negative eugenics—preventing the wrong kind of people from having children. In this development, a watershed event occurred in 1899 when a young man called Clawson approached a

"Large family" winner, Fitter Families Contest, Texas State Fair (1925)

prison doctor in Indiana called Harry Sharp (appropriately named in light of his enthusiasm for the surgeon's knife). Clawson's problem—or so it was diagnosed by the medical establishment of the day—was compulsive masturbation. He reported that he had been hard at it ever since the age of twelve. Masturbation was seen as part of the general syndrome of degeneracy, and Sharp accepted the conventional wisdom (however bizarre it may seem to us today) that Clawson's mental shortcomings—he had made no progress in school—were caused by his compulsion. The solution? Sharp performed a vasectomy, then a recently invented procedure, and subsequently claimed that he had "cured" Clawson. As a result, Sharp developed his own compulsion: to perform vasectomies.

Sharp promoted his success in treating Clawson (for which, incidentally, we have only Sharp's own report as confirmation) as evidence of the procedure's efficacy for treating all those identified as being of Clawson's kind—all "degenerates." Sterilization had two things going for it. First, it might prevent degenerate behavior, as Sharp claimed it had in Clawson. This, if nothing else, would save society a lot of money because those who had required incarceration, whether in prisons or insane asylums, would be rendered "safe" for release. Second, it would prevent the likes of Clawson from passing their inferior (degenerate) genes on to subsequent generations. Sterilization, Sharp believed, offered the perfect solution to the eugenic crisis.

Sharp was an effective lobbyist, and in 1907 Indiana passed the first compulsory sterilization law, authorizing the sterilization of confirmed "criminals, idiots, rapists, and imbeciles." Indiana's was the first of many: eventually thirty American states had enacted similar statutes, and by 1941 some sixty thousand individuals in the United States had duly been sterilized, half of them in California alone. The laws, which effectively resulted in state governments deciding who could and who could not have children, were challenged in court, but in 1927 the Supreme Court upheld the Virginia statute in the landmark case of Carrie Buck. Oliver Wendell Holmes wrote the decision:

> It is better for all the world if, instead of waiting to execute degenerate offspring for crime, or to let them starve for their imbecility, society can prevent those who are manifestly unfit from continuing their kind . . . Three generations of imbeciles is enough.

Sterilization caught on outside the United States as well—and not only in Nazi Germany. Switzerland and the Scandinavian countries enacted similar legislation.

Racism is not implicit to eugenics—good genes, the ones eugenics seeks to promote, can in principle belong to people of any race. Starting with Galton, however, whose account of his African expedition had confirmed prejudices about "inferior races," the prominent practitioners of eugenics tended to be racists who used eugenics to provide a "scientific" justification for racist views. Henry Goddard, of Kallikak family fame, conducted IQ tests on immigrants at Ellis Island in 1913 and found as many as 80 percent of potential new Americans to be certifiably feebleminded. The IQ tests he carried out during World War I for the U.S. Army reached a similar conclusion: 45 percent of foreign-born draftees had a mental age of less than eight (only 21 percent of native-born draftees fell into this category). That the tests were biased—they were, after all, carried out in English—was not taken to be relevant: racists had the ammunition they required, and eugenics would be pressed into the service of the cause.

Although the term "white supremacist" had yet to be coined, America had plenty of them early in the twentieth century. White Anglo-Saxon Protestants, Theodore Roosevelt prominent among them, were concerned that immigration was corrupting the WASP paradise that America, in their view, was supposed to be. In 1916 Madison Grant, a wealthy New Yorker and friend of both Davenport and Roosevelt, published *The Passing of the Great Race,* in which he argued that the Nordic peoples are superior to all others, including other Europeans. To preserve the United States' fine Nordic genetic heritage, Grant campaigned for immigration restrictions on all non-Nordics. He championed racist eugenic policies, too:

> Under existing conditions the most practical and hopeful method of race improvement is through the elimination of the least desirable elements in the nation by depriving them of the power to contribute to future generations. It is well known to stock breeders that the color of a herd of cattle

> can be modified by continuous destruction of worthless shades and of course this is true of other characters. Black sheep, for instance, have been practically obliterated by cutting out generation after generation all animals that show this color phase.

Despite appearances, Grant's book was hardly a minor publication by a marginalized crackpot; it was an influential best-seller. Later translated into German, it appealed—not surprisingly—to the Nazis. Grant gleefully recalled having received a personal letter from Hitler, who wrote to say that the book was his Bible.

Although not as prominent as Grant, arguably the most influential of the era's exponents of "scientific" racism was Davenport's right-hand man, Harry Laughlin. Son of an Iowa preacher, Laughlin's expertise was in racehorse pedigrees and chicken breeding. He oversaw the operations of the Eugenics Record Office, but was at his most effective as a lobbyist. In the name of eugenics, he fanatically promoted forced sterilization measures and restrictions on the influx of genetically dubious foreigners (i.e., non–northern Europeans). Particularly important historically was his role as an expert witness at congressional hearings on immigration: Laughlin gave full rein to his prejudices, all of them of course dressed up as "science." When the data were problematic, he fudged them. When he unexpectedly found, for instance, that immigrant Jewish children did better than the native-born in public schools, Laughlin changed the categories he presented, lumping Jews in with whatever nation they had come from, thereby diluting away their superior performance. The passage in 1924 of the Johnson-Reed Immigration Act, which severely restricted immigration from southern Europe and elsewhere, was greeted as a triumph by the likes of Madison Grant; it was Harry Laughlin's finest hour. As vice president some years earlier, Calvin Coolidge had chosen to overlook both Native Americans and the nation's immigration history when he declared that "America must remain American." Now, as president, he signed his wish into law.

Like Grant, Laughlin had his fans among the Nazis, who modeled some of their own legislation on the American laws he had developed. In 1936 he enthusiastically accepted an honorary degree from Heidelberg University, which chose to honor him as "the farseeing representative of racial policy in

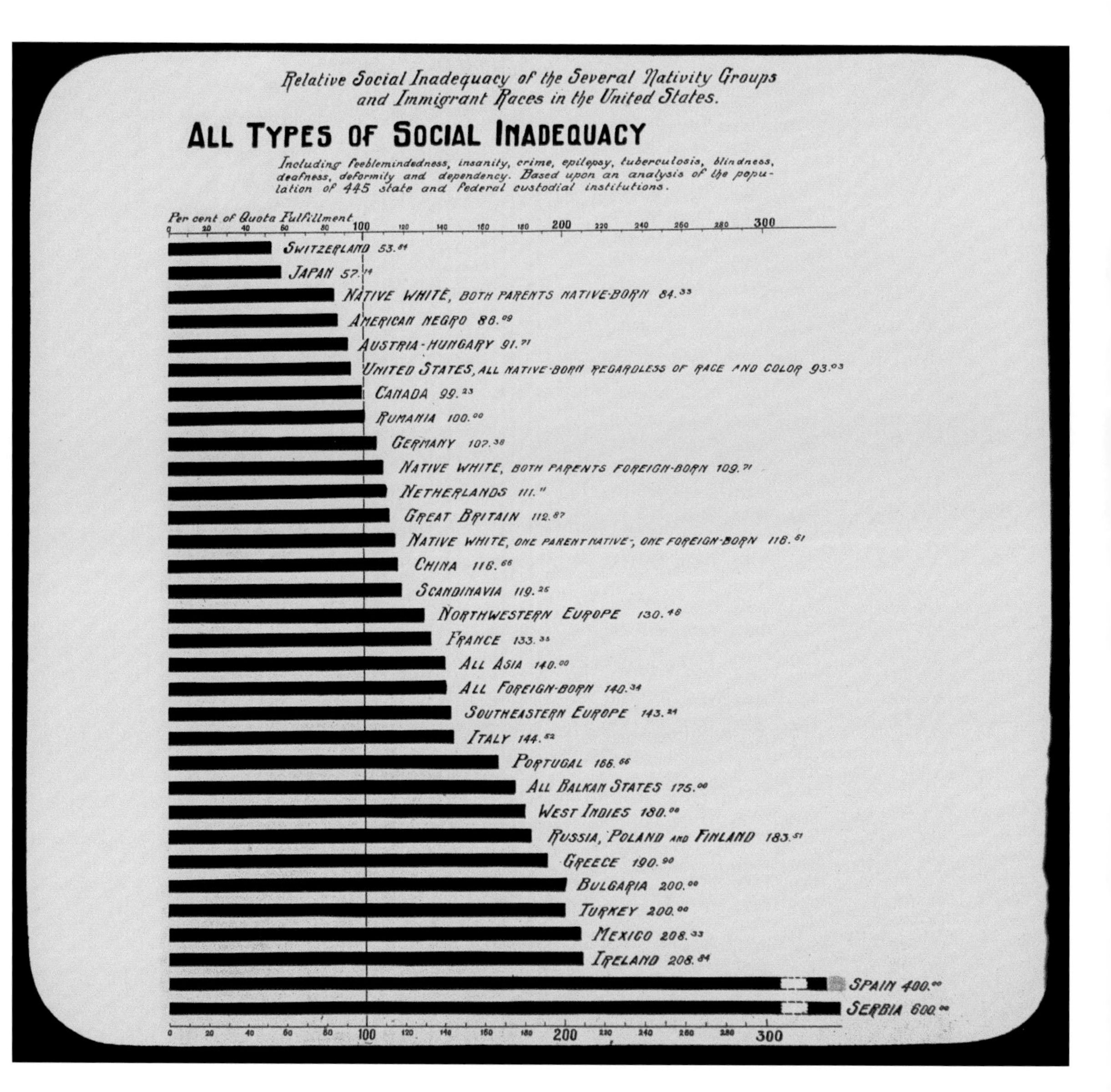

Scientific racism: social inadequacy in the United States analyzed by national group (1922). "Social inadequacy" is used here by Harry Laughlin as an umbrella term for a host of sins ranging from feeblemindedness to tuberculosis. Laughlin computed an institutional "quota" for each group on the basis of the proportion of that group in the U.S. population as a whole. Shown, as a percentage, is the number of institutionalized individuals from a particular group divided by the group's quota. Groups scoring over 100 percent are over-represented in institutions.

America." In time, however, a form of late-onset epilepsy ensured that Laughlin's later years were especially pathetic. All his professional life he had campaigned for the sterilization of epileptics on the grounds that they were genetically degenerate.

Hitler's *Mein Kampf* is saturated with pseudoscientific racist ranting derived from long-standing German claims of racial superiority and from some of the uglier aspects of the American eugenics movement. Hitler wrote that the state "must declare unfit for propagation all who are in any way visibly sick or who have inherited a disease and can therefore pass it on, and put this into actual practice," and elsewhere, "Those who are physically and mentally unhealthy and unworthy must not perpetuate their suffering in the body of their children." Shortly after coming to power in 1933, the Nazis had passed a comprehensive sterilization law—the "law for the prevention of progeny with hereditary defects"—that was explicitly based on the American model. (Laughlin proudly published a translation of the law.) Within three years, 225,000 people had been sterilized.

Positive eugenics, encouraging the "right" people to have children, also thrived in Nazi Germany, where "right" meant properly Aryan. Heinrich Himmler, head of the SS (the Nazi elite corps), saw his mission in eugenic terms: SS officers should ensure Germany's genetic future by having as many children as possible. In 1936, he established special maternity homes for SS wives to guarantee that they got the best possible care during pregnancy. The proclamations at the 1935 Nuremberg Rally included a "law for the protection of German blood and German honor," which prohibited marriage between Germans and Jews and even "extra-marital sexual intercourse between Jews and citizens of German or related blood." The Nazis were unfailingly thorough in closing up any reproductive loopholes.

Neither, tragically, were there any loopholes in the U.S. Johnson-Reed Immigration Act that Harry Laughlin had worked so hard to engineer. For many Jews fleeing Nazi persecution, the United States was the logical first choice of destination, but the country's restrictive—and racist—immigration policies resulted in many being turned away. Not only had Laughlin's sterilization law provided Hitler with the model for his ghastly program, but his impact on immigration legislation meant that the United States would in effect abandon German Jewry to its fate at the hands of the Nazis.

In 1939, with the war under way, the Nazis introduced euthanasia. Sterilization proved too much trouble. And why waste the food? The inmates of asylums were categorized as "useless eaters." Questionnaires were distributed among the mental hospitals where panels of experts were instructed to mark them with a cross in the cases of patients whose lives they deemed "not worth living." Seventy-five thousand came back so marked, and the technology of mass murder—the gas chamber—was duly developed. Subsequently, the Nazis expanded the definition of "not worth living" to include whole ethnic groups, among them the Gypsies and, in particular, the Jews. What came to be called the Holocaust was the culmination of Nazi eugenics.

Eugenics ultimately proved a tragedy for humankind. It also proved a disaster for the emerging science of genetics, which could not escape the taint. In fact, despite the prominence of eugenicists like Davenport, many scientists had criticized the movement and dissociated themselves from it. Alfred Russel Wallace, the co-discoverer with Darwin of natural selection, condemned eugenics in 1912 as "simply the meddlesome interference of an arrogant, scientific priestcraft." Thomas Hunt Morgan, of fruit fly fame, resigned on "scientific grounds" from the board of scientific directors of the Eugenics Record Office. Raymond Pearl, at Johns Hopkins, wrote in 1928 that "orthodox eugenicists are going contrary to the best established facts of genetical science."

Eugenics had lost its credibility in the scientific community long before the Nazis appropriated it for their own horrific purposes. The science underpinning it was bogus, and the social programs constructed upon it utterly reprehensible. Nevertheless, by midcentury the valid science of genetics, human genetics in particular, had a major public relations problem on its hands. When in 1948 I first came to Cold Spring Harbor, former home of the by-then-defunct Eugenics Record Office, nobody would even mention the "E word"; nobody was willing to talk about our science's past even though past issues of the German *Journal of Racial Hygiene* still lingered on the shelves of the library.

Realizing that such goals were not scientifically feasible, geneticists had long since forsaken the grand search for patterns of inheritance of human behavioral characteristics—whether Davenport's feeblemindedness or Galton's genius—and were now focusing instead on the gene and how it functioned in the cell. With the development during the 1930s and 1940s of new and more effective technologies for studying biological molecules in ever greater detail, the time had finally arrived for an assault on the greatest biological mystery of all: what is the chemical nature of the gene?

By ERWIN SCHRÖDINGER

WHAT IS LIFE ?

The Physicist's approach to the Subject—With an Epilogue on Determinism and Free Will

CAMBRIDGE UNIVERSITY PRESS

SCHRÖDINGER: What is Life?

CHAPTER TWO

. . .

THE DOUBLE HELIX: THIS IS LIFE

I got hooked on the gene during my third year at the University of Chicago. Until then, I had planned to be a naturalist and looked forward to a career far removed from the urban bustle of Chicago's South Side, where I grew up. My change of heart was inspired not by an unforgettable teacher but a little book that appeared in 1944, *What Is Life?*, by the Austrian-born father of wave mechanics, Erwin Schrödinger. It grew out of several lectures he had given the year before at the Institute for Advanced Study in Dublin. That a great physicist had taken the time to write about biology caught my fancy. In those days, like most people, I considered chemistry and physics to be the "real" sciences, and theoretical physicists were science's top dogs.

Schrödinger argued that life could be thought of in terms of storing and passing on biological information. Chromosomes were thus simply information bearers. Because so much information had to be packed into every cell, it must be compressed into what Schrödinger called a "hereditary code-script" embedded in the molecular fabric of chromosomes. To understand life, then, we would have to identify these molecules, and crack their code. He even speculated that understanding life—which would involve finding the gene—might take us beyond the laws of physics as we then understood them. Schrödinger's book was tremendously influential. Many of those who would become major players in Act 1 of molecular biology's great drama, including Francis Crick (a former physicist himself), had, like me, read *What Is Life?* and been impressed.

In my own case, Schrödinger struck a chord because I too was intrigued by

The physicist Erwin Schrödinger, whose book What Is Life? *turned me on to the gene*

the essence of life. A small minority of scientists still thought life depended upon a vital force emanating from an all-powerful god. But like most of my teachers, I disdained the very idea of vitalism. If such a "vital" force were calling the shots in nature's game, there was little hope life would ever be understood through the methods of science. On the other hand, the notion that life might be perpetuated by means of an instruction book inscribed in a secret code appealed to me. What sort of molecular code could be so elaborate as to convey all the multitudinous wonder of the living world? And what sort of molecular trick could ensure that the code is exactly copied every time a chromosome duplicates?

At the time of Schrödinger's Dublin lectures, most biologists supposed that proteins would eventually be identified as the primary bearers of genetic instruction. Proteins are molecular chains built up from twenty different building blocks, the amino acids. Because permutations in the order of amino acids along the chain are virtually infinite, proteins could, in principle, readily encode the information underpinning life's extraordinary diversity. DNA then was not considered a serious candidate for the bearer of code-scripts, even though it was exclusively located on chromosomes and had been known about for some seventy-five years. In 1869, Friedrich Miescher, a Swiss biochemist working in Germany, had isolated from pus-soaked bandages supplied by a local hospital a substance he called "nuclein." Because pus consists largely of white blood cells, which, unlike red blood cells, have nuclei and therefore DNA-containing chromosomes, Miescher had stumbled on a good source of DNA. When he later discovered that "nuclein" was to be found in chromosomes alone, Miescher understood that his discovery was indeed a big one. In 1893, he wrote: "Inheritance insures a continuity in form from generation to generation that lies even deeper than the chemical molecule. It lies in the structuring atomic groups. In this sense, I am a supporter of the chemical heredity theory."

Nevertheless, for decades afterward, chemistry would remain unequal to the task of analyzing the immense size and complexity of the DNA molecule. Only in the 1930s was DNA shown to be a long molecule containing four different chemical bases: adenine (A), guanine (G), thymine (T), and cytosine (C). But at the time of Schrödinger's lectures, it was still unclear just how the subunits (called deoxynucleotides) of the molecule were chemically linked. Nor was it known whether DNA molecules might vary in their sequences of the four different bases. If DNA were indeed Schrödinger's code-script, then the molecule would have to be capable of existing in an immense number of different forms. But back then it was still considered a possibility that one simple sequence like AGTC might be repeated over and over along the entire length of DNA chains.

DNA did not move into the genetic limelight until 1944, when Oswald Avery's lab at the Rockefeller Institute in New York City reported that the composition of the surface coats of pneumonia bacteria could be changed. This was not the result he and his junior colleagues, Colin MacLeod and Maclyn McCarty, expected.

For more than a decade Avery's group had been following up on another most unexpected observation made in 1928 by Fred Griffith, a scientist in the British Ministry of Health. Griffith was interested in pneumonia and studied its bacterial agent, *Pneumococcus.* It was known that there were two strains, designated "smooth" (S) and "rough" (R) according to their appearance under the microscope. These strains differed not only visually but also in their virulence. Inject S bacteria into a mouse, and within a few days the mouse dies; inject R bacteria and the mouse remains healthy. It turns out that S bacterial cells have a coating that prevents the mouse's immune system from recognizing the invader. The R cells have no such coating and are therefore readily attacked by the mouse's immune defenses.

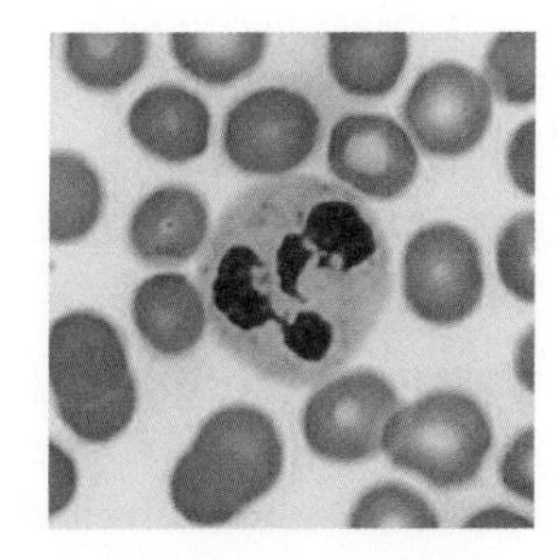

A view through the microscope of blood cells treated with a chemical that stains DNA. In order to maximize their oxygen-transporting capacity, red blood cells have no nucleus and therefore no DNA. But white blood cells, which patrol the bloodstream in search of intruders, have a nucleus containing chromosomes.

Through his involvement with public health, Griffith knew that multiple strains had sometimes been isolated from a single patient, and so he was curious about how different strains might interact in his unfortunate mice. With one combination, he made a remarkable discovery: when he injected heat-killed S bacteria (harmless) *and* normal R bacteria (also harmless), the mouse died. How could two harmless forms of bacteria conspire to become lethal? The clue came when he isolated the *Pneumococcus* bacteria retrieved from the dead mice and discovered living S bacteria. It appeared the living innocuous R bacteria had acquired something from the dead S variant; whatever it was, that something had allowed the R in the presence of the heat-killed S bacteria to transform itself into a living killer S strain. Griffith confirmed that this change was for real by culturing the S bacteria from the dead mouse over several generations: the bacteria bred true for the S type, just as any regular S strain would. A *genetic* change had indeed occurred to the R bacteria injected into the mouse.

Though this transformation phenomenon seemed to defy all understanding, Griffith's observations at first created little stir in the scientific world. This was partly because Griffith was intensely private and so averse to large gatherings that he seldom attended scientific conferences. Once, he had to be virtually forced to give a lecture. Bundled into a taxi and escorted to the hall by colleagues, he discoursed in a mumbled monotone, emphasizing an obscure corner of his microbiological work but making no mention of bacterial transformation. Luckily, however, not everyone overlooked Griffith's breakthrough.

Oswald Avery was also interested in the sugarlike coats of the *Pneumococcus*. He set out to duplicate Griffith's experiment in order to isolate and characterize whatever it was that had caused those R cells to change to the S type. In 1944 Avery, MacLeod, and McCarty published their results: an exquisite set of experiments showing unequivocally that DNA was the transforming principle. Culturing the bacteria in the test tube rather than in mice made it much easier to search for the chemical identity of the transforming factor in the heat-killed S cells. Methodically destroying one by one the biochemical components of the heat-treated S cells, Avery and his group looked to see whether transformation was prevented. First they degraded the sugarlike coat of the S bacteria. Transformation still occurred: the coat was not the transforming principle. Next they used a mixture of two protein-destroying enzymes, trypsin and chymotrypsin, to degrade virtually all the proteins in the S cells. To their surprise, transformation was again unaffected. Next they tried an enzyme (RNase) that breaks down RNA (ribonucleic acid), a second class of nucleic acids similar to DNA and possibly involved in protein synthesis. Again transformation occurred. Finally, they came to DNA, exposing the S bacterial extracts to the DNA-destroying enzyme, DNase. This time they hit a home run. All S-inducing activity ceased completely. The transforming factor was DNA.

In part because of its bombshell implications, the resulting February 1944 paper by Avery, MacLeod, and McCarty met with a mixed response. Many geneticists accepted their conclusions. After all, DNA was found on every chromosome; why shouldn't it be the genetic material? By contrast, however, most biochemists expressed doubt that DNA was a complex enough molecule to act as the repository of such a vast quantity of biological information. They continued to believe that proteins, the other component of chromosomes, would prove to be the hereditary substance. In principle, as the biochemists rightly noted, it would be much easier to encode a vast body of complex information using the twenty-letter amino-acid alphabet of proteins than the four-letter nucleotide alphabet of DNA. Particularly vitriolic in his rejection of DNA as the genetic substance was Avery's own colleague at the Rockefeller Institute, the protein chemist Alfred Mirsky. By then, however, Avery was no longer scientifically active. The Rockefeller Institute had mandatorily retired him at age sixty-five.

Avery missed out on more than the opportunity to defend his work against the attacks of his colleagues: He was never awarded the Nobel Prize, which was certainly his due, for identifying DNA as the transforming principle. Because the Nobel committee makes its records public fifty years following each award, we now know that Avery's candidacy was blocked by the Swedish physical chemist Einar Hammarsten. Though Hammarsten's reputation was based largely on his having produced DNA samples of unprecedented high quality, he still believed genes to be an undiscovered class of proteins. In fact, even after the double helix was found, Hammarsten continued to insist that Avery should not receive the prize until after the mechanism of DNA transformation had been completely worked out. Avery died in 1955; had he lived only a few more years, he would almost certainly have gotten the prize.

When I arrived at Indiana University in the fall of 1947 with plans to pursue the gene for my Ph.D. thesis, Avery's paper came up over and over in conversations. By then, no one doubted the reproducibility of his results, and more recent work coming out of the Rockefeller Institute made it all the less likely that proteins would prove to be the genetic actors in bacterial transformation. DNA had at last become an important objective for chemists setting their sights on the next breakthrough. In Cambridge, England, the canny Scottish chemist Alexander Todd rose to the challenge of identifying the chemical bonds that linked together nucleotides in DNA. By early 1951, his lab had proved that these links were always the same, such that the backbone of the DNA molecule was very regular. During the same period, the Austrian-born refugee Erwin Chargaff, at the College of Physicians and Surgeons of Columbia University, used the new technique of paper chromatography to measure the relative amounts of the four DNA bases in DNA samples extracted from a variety of vertebrates and bacteria. While some species had DNA in which adenine and thymine predominated, others had DNA with more guanine and cytosine. The possibility thus presented itself that no two DNA molecules had the same composition.

At Indiana I joined a small group of visionary scientists, mostly physicists and chemists, studying the reproductive process of the viruses that attack bacteria (bacteriophages—"phages" for short). The Phage Group was born when my Ph.D. supervisor, the Italian-trained medic Salvador Luria and his close friend, the German-born theoretical physicist Max Delbrück, teamed up with the American physical chemist Alfred Hershey. During World War II both Luria and Delbrück were considered enemy aliens, and thus ineligible to serve in the war effort of American science, even though Luria, a Jew, had been forced to leave France for New York City and Delbrück had fled Germany as an objector to Nazism. Thus excluded, they continued to work in their respective university labs—Luria at Indiana and Delbrück at Vanderbilt—and collaborated on phage experiments during successive summers at Cold Spring Harbor. In 1943, they joined forces with the brilliant but taciturn Hershey, then doing phage research of his own at Washington University in St. Louis.

The Phage Group's program was based on its belief that phages, like all viruses, were in effect naked genes. This concept had first been proposed in 1922 by the imaginative American geneticist Herman J. Muller, who three years later demonstrated that X rays cause mutations. His belated Nobel Prize came in 1946, just after he joined the faculty of Indiana University. It was his presence, in fact, that led me to Indiana. Having started his career under T. H. Morgan, Muller knew better than anyone else how genetics had evolved during the first half of the twentieth century, and I was enthralled by his lectures during my first term. His work on fruit flies (*Drosophila*), however, seemed to me to belong more to the past than to the future, and I only briefly considered doing thesis research under his supervision. I opted instead for Luria's phages, an even speedier experimental subject than *Drosophila*: genetic crosses of phages done one day could be analyzed the next.

For my Ph.D. thesis research, Luria had me follow in his footsteps by studying how X rays killed phage particles. Initially I had hoped to show that viral death was caused by damage to phage DNA. Reluctantly, however, I eventually had to concede that my experimental approach could never give unambiguous answers at the chemical level. I could draw only biological conclusions. Even though phages were indeed effectively naked genes, I realized that the deep answers the Phage Group was seeking could be arrived at only through advanced chemistry. DNA somehow had to transcend its status as an acronym; it had to be understood as a molecular structure in all its chemical detail.

Upon finishing my thesis, I saw no alternative but to move to a lab where I could study DNA chemistry. Unfortunately, however, knowing almost no pure chemistry, I would have been out of my depth in any lab attempting difficult experiments in organic or physical chemistry. I therefore took a postdoctoral fellowship in the Copenhagen lab of the biochemist Herman Kalckar in the fall of 1950. He was studying the synthesis of the small molecules that make up DNA, but I figured out quickly that his biochemical approach would never lead to an understanding of the essence of the gene. Every day spent in his lab would be one more day's delay in learning how DNA carried genetic information.

My Copenhagen year nonetheless ended productively. To escape the cold Danish spring, I went to the Zoological Station at Naples during April and May. During my last week there, I attended a small conference on X-ray diffraction methods for determining the 3-D structure of molecules. X-ray diffraction is a way of studying the atomic structure of any molecule that can be crystallized. The crystal is bombarded with X rays, which bounce off its atoms and are scattered. The scatter pattern gives information about the structure of the molecule but, taken alone, is not enough to solve the structure. The additional information needed is the "phase assignment," which deals with the wave properties of the molecule. Solving the phase problem was not easy, and at that time only the most audacious scientists were willing to take it on. Most of the successes of the diffraction method had been achieved with relatively simple molecules.

My expectations for the conference were low. I believed that a three-dimensional understanding of protein structure, or for that matter of DNA, was more than a decade away. Disappointing earlier X-ray photos suggested that DNA was particularly unlikely to yield up its secrets via the X-ray approach. These results were not surprising since the exact sequences of DNA were expected to differ from one individual molecule to another. The resulting irregularity of surface configurations would understandably prevent the long thin DNA chains from lying neatly side by side in the regular repeating patterns required for X-ray analysis to be successful.

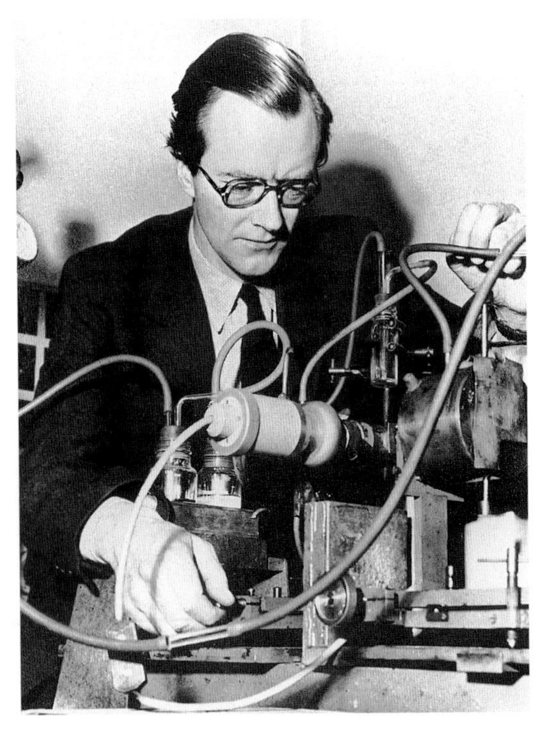

Maurice Wilkins in his lab at King's College, London

It was therefore a surprise and a delight to hear the last-minute talk on DNA by a thirty-four-year-old Englishman named Maurice Wilkins from the Biophysics Lab of King's College, London. Wilkins was a physicist who during the war had worked on the Manhattan Project. For him, as for many of the other scientists involved, the actual deployment of the bomb on Hiroshima and Nagasaki, supposedly the culmination of all their work, was profoundly disillusioning. He considered forsaking science altogether to become a painter in Paris, but biology intervened. He too had read Schrödinger's book, and was now tackling DNA with X-ray diffraction.

He displayed a photograph of an X-ray diffraction pattern he had recently obtained, and its many precise reflections indicated a highly regular crystalline packing. DNA, one had to conclude, must have a regular structure, the elucidation of which might well reveal the nature of the gene. Instantly I saw myself moving to London to help Wilkins find the structure. My attempts to converse with him after his talk, however, went nowhere. All I got for my efforts was a declaration of his conviction that much hard work lay ahead.

While I was hitting consecutive dead ends, back in America the world's preeminent chemist, Caltech's Linus Pauling, announced a major triumph: he had found the exact arrangement in which chains of amino acids (called *polypeptides*) fold up in proteins, and called his structure the α-helix (alpha helix). That it was Pauling who made this breakthrough was no surprise: he was a scientific superstar. His book *The Nature of the Chemical Bond* essentially laid the foundation of modern chemistry, and, for chemists of the day, it was the Bible. Pauling had been a precocious child. When he was nine, his father, a druggist in Oregon, wrote to the *Oregonian* newspaper requesting suggestions of reading matter for his bookish son, adding that he had already read the Bible and Darwin's *Origin of Species.* But the early death of Pauling's father, which brought the family to financial ruin, makes it remarkable that the promising young man managed to get an education at all.

As soon as I returned to Copenhagen I read about Pauling's α-helix. To my surprise, his model was not based on a deductive leap from experimental X-ray diffraction data. Instead, it was Pauling's long experience as a structural chemist that had emboldened him to infer which type of helical fold would be most compatible with the underlying chemical features of the polypeptide chain. Pauling made scale models of the different parts of the protein molecule, working out plausible schemes in three dimensions. He had reduced the problem to a kind of three-dimensional jigsaw puzzle in a way that was simple yet brilliant.

Whether the α-helix was correct—in addition to being pretty—was now the question. Only a week later, I got the answer. Sir Lawrence Bragg, the English inventor of X-ray crystallography and 1915 Nobel laureate in Physics, came to

Lawrence Bragg (left) with Linus Pauling, who is carrying a model of the α-helix

Copenhagen and excitedly reported that his junior colleague, the Austrian-born chemist Max Perutz, had ingeniously used synthetic polypeptides to confirm the correctness of Pauling's α-helix. It was a bittersweet triumph for Bragg's Cavendish Laboratory. The year before, they had completely missed the boat in their paper outlining possible helical folds for polypeptide chains.

By then Salvador Luria had tentatively arranged for me to take up a research position at the Cavendish. Located at Cambridge University, this was the most famous laboratory in all of science. Here Ernest Rutherford first described the structure of the atom. Now it was Bragg's own domain, and I was to work as apprentice to the English chemist John Kendrew, who was interested in determining the 3-D structure of the protein myoglobin. Luria advised me to visit the Cavendish as soon as possible. With Kendrew in the States, Max Perutz would check me out. Together, Kendrew and Perutz had earlier established the Medical Research Council (MRC) Unit for the Study of the Structure of Biological Systems.

A month later in Cambridge, Perutz assured me that I could quickly master the necessary X-ray diffraction theory and should have no difficulty fitting in with the others in their tiny MRC Unit. To my relief, he was not put off by my biology background. Nor was Lawrence Bragg, who briefly came down from his office to look me over.

I was twenty-three when I arrived back at the MRC Unit in Cambridge in early October. I found myself sharing space in the biochemistry room with a thirty-five-year-old ex-physicist, Francis Crick, who had spent the war working on magnetic mines for the Admiralty. When the war ended, Crick had planned to stay on in military research, but, on reading Schrödinger's *What Is Life?,* he had moved toward biology. Now he was at the Cavendish to pursue the 3-D structure of proteins for his Ph.D.

Francis Crick with the Cavendish X-ray tube

Crick was always fascinated by the intricacies of important problems. His endless questions as a child compelled his weary parents to buy him a children's encyclopedia, hoping that it would satisfy his curiosity. But it only made him insecure: he confided to his mother his fear that everything would have been discovered by the time he grew up, leaving him nothing to do. His mother reassured him (correctly, as it happened) that there would still be a thing or two for him to figure out.

A great talker, Crick was invariably the center of attention in any gathering. His booming laugh was forever echoing down the hallways of the Cavendish. As the MRC Unit's resident theoretician, he used to come up with a novel insight at least once a month, and he would explain his latest idea at great length to anyone willing to listen. The morning we met he lit up when he learned that my objective in coming to Cambridge was to learn enough crystallography to have a go at the DNA structure. Soon I was asking Crick's opinion about using Pauling's model-building approach to go directly for the structure. Would we need many more years of diffraction experimentation before modeling would be practicable? To bring us up to speed on the status of DNA structural studies, Crick invited Maurice Wilkins, a friend since the end of the war, up from London for Sunday lunch. Then we could learn what progress Wilkins had made since his talk in Naples.

Wilkins expressed his belief that DNA's structure was a helix, formed by several chains of linked nucleotides twisted around each other. All that remained to be settled was the number of chains. At the time, Wilkins favored three on the basis of his density measurements of DNA fibers. He was keen to start model-building, but he had run into a roadblock in the form of a new addition to the King's College Biophysics Unit, Rosalind Franklin.

A thirty-one-year-old Cambridge-trained physical chemist, Franklin was an obsessively professional scientist; for her twenty-ninth birthday all she requested was her own subscription to her field's technical journal, *Acta Crystallographica*. Logical and precise, she was impatient with those who acted otherwise. And she was given to strong opinions, once describing her Ph.D. thesis adviser, Ronald Norrish, a future Nobel Laureate, as "stupid, bigoted, deceitful, ill-mannered and tyrannical." Outside the laboratory, she was a determined and gutsy mountaineer, and, coming from the upper echelons of London society, she belonged to a more rarefied social world than most scientists. At the end of a hard day at the bench, she would occasionally change out of her lab coat into an elegant evening gown and disappear into the night.

Just back from a four-year X-ray crystallographic investigation of graphite in Paris, Franklin had been assigned to the DNA project while Wilkins was away from King's. Unfortunately, the pair soon proved incompatible. Franklin, direct and data-focused, and Wilkins, retiring and speculative, were destined never to collaborate. Shortly before Wilkins accepted our lunch invitation, the two had had a big blowup in which Franklin had insisted that no model-building could

Rosalind Franklin on one of the mountain hiking vacations she loved

commence before she collected much more extensive diffraction data. Now they effectively didn't communicate, and Wilkins would have no chance to learn of her progress until Franklin presented her lab seminar scheduled for the beginning of November. If we wanted to listen, Crick and I were welcome to go as Wilkins's guests.

Crick was unable to make the seminar, so I attended alone and briefed him later on what I believed to be its key take-home messages on crystalline DNA.

In particular, I described from memory Franklin's measurements of the crystallographic repeats and the water content. This prompted Crick to begin sketching helical grids on a sheet of paper, explaining that the new helical X-ray theory he had devised with Bill Cochran and Vladimir Vand would permit even me, a former bird-watcher, to predict correctly the diffraction patterns expected from the molecular models we would soon be building at the Cavendish.

As soon as we got back to Cambridge, I arranged for the Cavendish machine shop to construct the phosphorous atom models needed for short sections of the sugar phosphate backbone found in DNA. Once these became available, we tested different ways the backbones might twist around each other in the center of the DNA molecule. Their regular repeating atomic structure should allow the atoms to come together in a consistent, repeated conformation. Following Wilkins's hunch, we focused on three-chain models. When one of these appeared to be almost plausible, Crick made a phone call to Wilkins to announce we had a model we thought might be DNA.

The next day both Wilkins and Franklin came up to see what we had done. The threat of unanticipated competition briefly united them in common purpose. Franklin wasted no time in faulting our basic concept. My memory was that she had reported almost no water present in crystalline DNA. In fact, the opposite was true. Being a crystallographic novice, I had confused the terms "unit cell" and "asymmetric unit." Crystalline DNA was in fact water-rich. Consequently, Franklin pointed out, the backbone had to be on the outside and not, as we had it, in the center, if only to accommodate all the water molecules she had observed in her crystals.

That unfortunate November day cast a very long shadow. Franklin's opposition to model-building was reinforced. Doing experiments, not playing with Tinkertoy representations of atoms, was the way she intended to proceed. Even worse, Sir Lawrence Bragg passed down the word that Crick and I should desist from all further attempts at building a DNA model. It was further decreed that DNA research should be left to the King's lab, with Cambridge continuing to focus solely on proteins. There was no sense in two MRC-funded labs competing against each other. With no more bright ideas up our sleeves, Crick and I were reluctantly forced to back off, at least for the time being.

It was not a good moment to be condemned to the DNA sidelines. Linus

Pauling had written Wilkins to request a copy of the crystalline DNA diffraction pattern. Though Wilkins had declined, saying he wanted more time to interpret it himself, Pauling was hardly obliged to depend upon data from King's. If he wished, he could easily start serious X-ray diffraction studies at Caltech.

The following spring, I duly turned away from DNA and set about extending prewar studies on the pencil-shaped tobacco mosaic virus using the Cavendish's powerful new X-ray beam. This light experimental workload gave me plenty of time to wander through various Cambridge libraries. In the zoology building, I read Erwin Chargaff's paper describing his finding that the DNA bases adenine and thymine occurred in roughly equal amounts, as did the bases guanine and cytosine. Hearing of these one-to-one ratios Crick wondered whether, during DNA duplication, adenine residues might be attracted to thymine and vice versa, and whether a corresponding attraction might exist between guanine and cytosine. If so, base sequences on the "parental" chains (e.g., ATGC) would have to be complementary to those on "daughter" strands (yielding in this case TACG).

These remained idle thoughts until Erwin Chargaff came through Cambridge in the summer of 1952 on his way to the International Biochemical Congress in Paris. Chargaff expressed annoyance that neither Crick nor I saw the need to know the chemical structures of the four bases. He was even more upset when we told him that we could simply look up the structures in textbooks as the need arose. I was left hoping that Chargaff's data would prove irrelevant. Crick, however, was energized to do several experiments looking for molecular "sandwiches" that might form when adenine and thymine (or alternatively, guanine and cytosine) were mixed together in solution. But his experiments went nowhere.

Like Chargaff, Linus Pauling also attended the International Biochemical Congress, where the big news was the latest result from the Phage Group. Alfred Hershey and Martha Chase at Cold Spring Harbor had just confirmed Avery's transforming principle: DNA was the hereditary material! Hershey and Chase proved that only the DNA of the phage virus enters bacterial cells; its protein coat remains on the outside. It was more obvious than ever that DNA must be understood at the molecular level if we were to uncover the essence of the gene. With Hershey and Chase's result the talk of the town, I was sure that Pauling would now bring his formidable intellect and chemical wisdom to bear on the problem of DNA.

Early in 1953, Pauling did indeed publish a paper outlining the structure of DNA. Reading it anxiously I saw that he was proposing a three-chain model with sugar phosphate backbones forming a dense central core. Superficially it was similar to our botched model of fifteen months earlier. But instead of using positively charged atoms (e.g., Mg^{2+}) to stabilize the negatively charged backbones, Pauling made the unorthodox suggestion that the phosphates were held together by hydrogen bonds. But it seemed to me, the biologist, that such hydrogen bonds required extremely acidic conditions never found in cells. With a mad dash to Alexander Todd's nearby organic chemistry lab my belief was confirmed: The impossible had happened. The world's best-known, if not best, chemist had gotten his chemistry wrong. In effect, Pauling had knocked the A off of DNA. Our quarry was deoxyribonucleic acid, but the structure he was proposing was not even acidic.

Hurriedly I took the manuscript to London to inform Wilkins and Franklin they were still in the game. Convinced that DNA was not a helix, Franklin had no wish even to read the article and deal with the distraction of Pauling's helical ideas, even when I offered Crick's arguments for helices. Wilkins, however, was very interested indeed in the news I brought; he was now more certain than ever that DNA was helical. To prove the point, he showed me a photograph obtained more than six months earlier by Franklin's graduate student Raymond Gosling, who had X-rayed the so-called B form of DNA. Until that moment, I didn't know a B form even existed. Franklin had put this picture aside, preferring to concentrate on the A form, which she thought would more likely yield useful data. The X-ray pattern of this B form was a distinct cross. Since Crick and others had already deduced that such a pattern of reflections would be created by a helix, this evidence made it clear that DNA had to be a helix! In fact, despite Franklin's reservations, this was no surprise. Geometry itself suggested that a helix was the most logical arrangement for a long string of repeating units such as the nucleotides of DNA. But we still did not know what that helix looked like, nor how many chains it contained.

The time had come to resume building helical models of DNA. Pauling was bound to realize soon enough that his brainchild was wrong. I urged Wilkins to

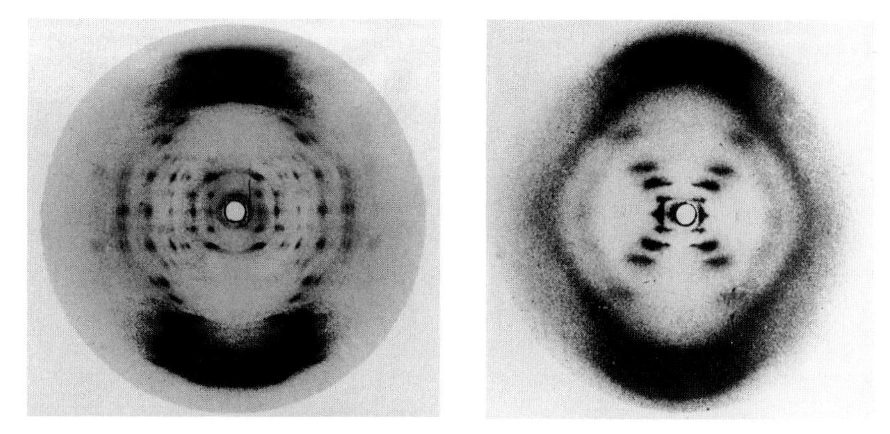

X-ray photos of the A and B forms of DNA from, respectively, Maurice Wilkins and Rosalind Franklin. The differences in molecular structure are caused by differences in the amount of water associated with each DNA molecule.

waste no time. But he wanted to wait until Franklin had completed her scheduled departure for another lab later that spring. She had decided to move on to avoid the unpleasantness at King's. Before leaving, she had been ordered to stop further work with DNA and had already passed on many of her diffraction images to Wilkins.

When I returned to Cambridge and broke the news of the DNA B form, Bragg no longer saw any reason for Crick and me to avoid DNA. He very much wanted the DNA structure to be found on his side of the Atlantic. So we went back to model-building, looking for a way the known basic components of DNA—the backbone of the molecule and the four different bases, adenine, thymine, guanine, and cytosine—could fit together to make a helix. I commissioned the shop at the Cavendish to make us a set of tin bases, but they couldn't produce them fast enough for me: I ended up cutting out rough approximations from stiff cardboard.

By this time I realized the DNA density-measurement evidence actually slightly favored a two-chain, rather than three-chain, model. So I decided to search out plausible double helices. As a biologist, I preferred the idea of a genetic molecule made of two, rather than three, components. After all, chromosomes, like cells, increase in number by duplicating, not triplicating.

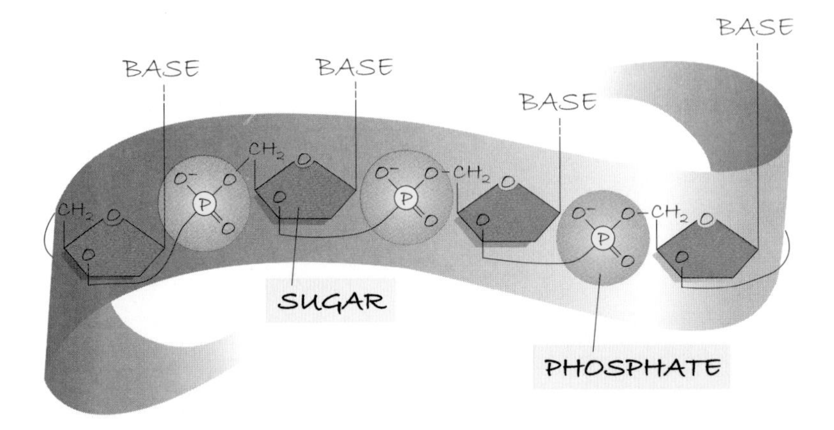

The chemical backbone of DNA

I knew that our previous model with the backbone on the inside and the bases hanging out was wrong. Chemical evidence from the University of Nottingham, which I had too long ignored, indicated that the bases must be hydrogen-bonded to each other. They could only form bonds like this in the regular manner implied by the X-ray diffraction data if they were in the center of the molecule. But how could they come together in pairs? For two weeks I got nowhere, misled by an error in my nucleic acid chemistry textbook. Happily, on February 27, Jerry Donahue, a theoretical chemist visiting the Cavendish from Caltech, pointed out that the textbook was wrong. So I changed the locations of the hydrogen atoms on my cardboard cutouts of the molecules.

The next morning, February 28, 1953, the key features of the DNA model all fell into place. The two chains were held together by strong hydrogen bonds between adenine-thymine and guanine-cytosine base pairs. The inferences Crick had drawn the year before based on Chargaff's research had indeed been correct. Adenine does bond to thymine and guanine does bond to cytosine, but not through flat surfaces to form molecular sandwiches. When Crick arrived, he took it all in rapidly, and gave my base-pairing scheme his blessing. He realized right away that it would result in the two strands of the double helix running in opposite directions.

It was quite a moment. We felt sure that this was it. Anything that simple, that elegant just had to be right. What got us most excited was the complementarity of the base sequences along the two chains. If you knew the sequence—the order of bases—along one chain, you automatically knew the sequence along the other. It was immediately apparent that this must be how the genetic messages of genes are copied so exactly when chromosomes duplicate prior to cell division. The molecule would "unzip" to form two separate strands. Each separate strand then could serve as the template for the synthesis of a new strand, one double helix becoming two.

In *What Is Life?* Schrödinger had suggested that the language of life might be like Morse code, a series of dots and dashes. He wasn't far off. The language of DNA is a linear series of As, Ts, Gs, and Cs. And just as transcribing a page out

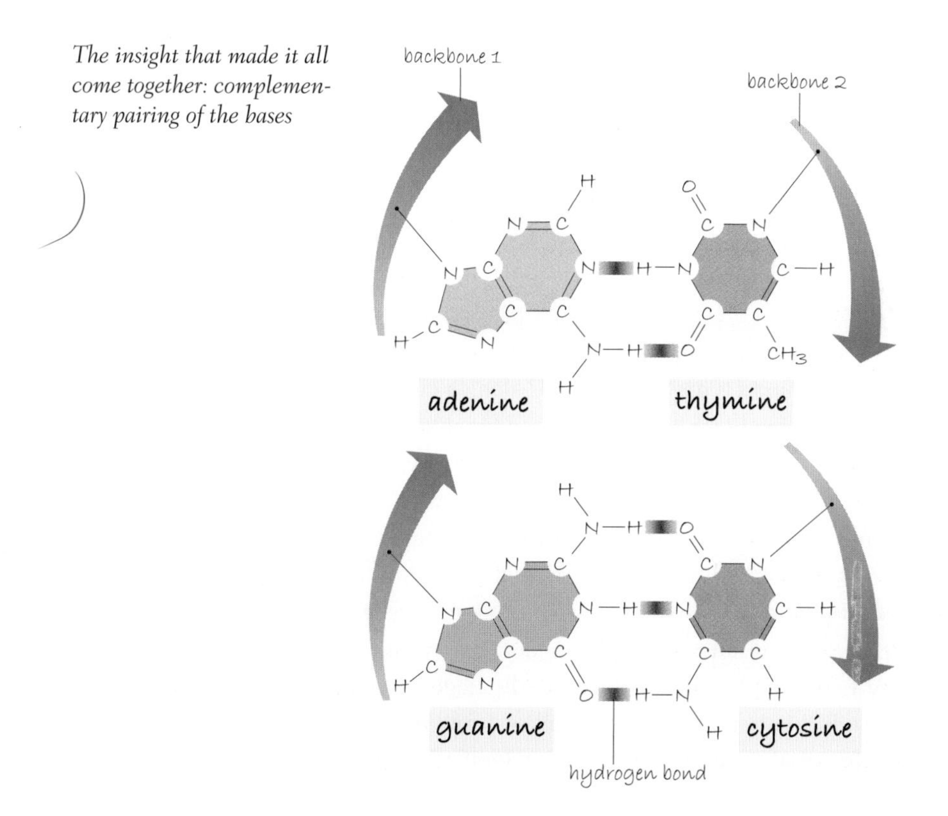

The insight that made it all come together: complementary pairing of the bases

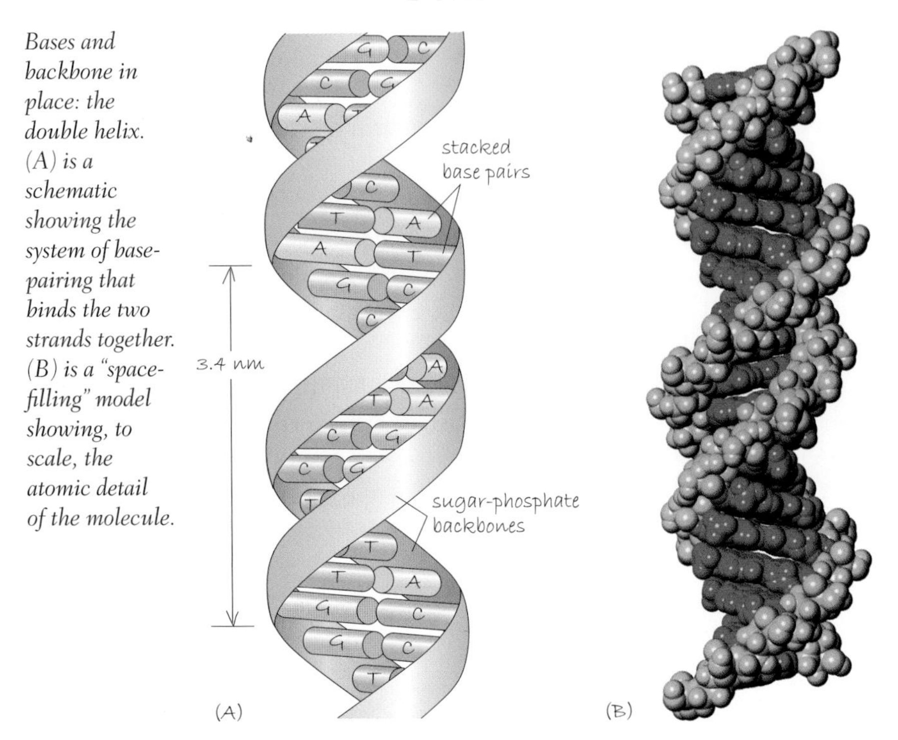

Bases and backbone in place: the double helix. (A) is a schematic showing the system of base-pairing that binds the two strands together. (B) is a "space-filling" model showing, to scale, the atomic detail of the molecule.

of a book can result in the odd typo, the rare mistake creeps in when all these As, Ts, Gs, and Cs are being copied along a chromosome. These errors are the mutations geneticists had talked about for almost fifty years. Change an "i" to an "a" and "Jim" becomes "Jam" in English; change a T to a C and "ATG" becomes "ACG" in DNA.

The double helix made sense chemically and it made sense biologically. Now there was no need to be concerned about Schrödinger's suggestion that new laws of physics might be necessary for an understanding of how the hereditary code-script is duplicated: genes in fact were no different from the rest of chemistry. Later that day, during lunch at the Eagle, the pub virtually adjacent to the Cavendish Lab, Crick, ever the talker, could not help but tell everyone we had just found the "secret of life." I myself, though no less electrified by the thought, would have waited until we had a pretty three-dimensional model to show off.

Among the first to see our demonstration model was the chemist Alexander Todd. That the nature of the gene was so simple both surprised and pleased him. Later, however, he must have asked himself why his own lab, having established the general chemical structure of DNA chains, had not moved on to asking how the chains folded up in three dimensions. Instead the essence of the molecule was left to be discovered by a two-man team, a biologist and a physicist, neither of whom possessed a detailed command even of undergraduate chemistry. But paradoxically, this was, at least in part, the key to our success: Crick and I arrived at the double helix first precisely because most chemists at that time thought DNA too big a molecule to understand by chemical analysis.

At the same time, the only two chemists with the vision to seek DNA's 3-D structure made major tactical mistakes: Rosalind Franklin's was her resistance to model-building; Linus Pauling's was a matter of simply neglecting to read the existing literature on DNA, particularly the data on its base composition published by Chargaff. Ironically, Pauling and Chargaff sailed across the Atlantic on the same ship following the Paris Biochemical Congress in 1952, but failed to hit it off. Pauling was long accustomed to being right. And he believed there was no chemical problem he could not work out from first principles by himself. Usually this confidence was not misplaced. During the Cold War, as a prominent critic of the American nuclear weapons development program, he was questioned by the FBI after giving a talk. How did he know how much plutonium there is in an atomic bomb? Pauling's response was "Nobody told me. I figured it out."

Over the next several months Crick and (to a lesser extent) I relished showing off our model to an endless stream of curious scientists. However, the Cambridge biochemists did not invite us to give a formal talk in the biochemistry building. They started to refer to it as the "WC," punning our initials with those used in Britain for the toilet or water closet. That we had found the double helix without doing experiments irked them.

The manuscript that we submitted to *Nature* in early April was published just over three weeks later, on April 25, 1953. Accompanying it were two longer papers by Franklin and Wilkins, both supporting the general correctness of our model. In June, I gave the first presentation of our model at the Cold Spring

MOLECULAR STRUCTURE OF NUCLEIC ACIDS

A Structure for Deoxyribose Nucleic Acid

WE wish to suggest a structure for the salt of deoxyribose nucleic acid (D.N.A.). This structure has novel features which are of considerable biological interest.

A structure for nucleic acid has already been proposed by Pauling and Corey[1]. They kindly made their manuscript available to us in advance of publication. Their model consists of three intertwined chains, with the phosphates near the fibre axis, and the bases on the outside. In our opinion, this structure is unsatisfactory for two reasons: (1) We believe that the material which gives the X-ray diagrams is the salt, not the free acid. Without the acidic hydrogen atoms it is not clear what forces would hold the structure together, especially as the negatively charged phosphates near the axis will repel each other. (2) Some of the van der Waals distances appear to be too small.

Another three-chain structure has also been suggested by Fraser (in the press). In his model the phosphates are on the outside and the bases on the inside, linked together by hydrogen bonds. This structure as described is rather ill-defined, and for this reason we shall not comment on it.

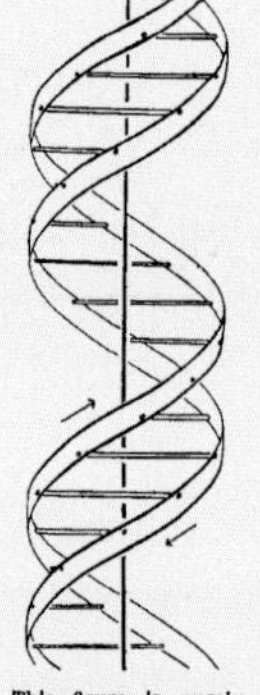

This figure is purely diagrammatic. The two ribbons symbolize the two phosphate—sugar chains, and the horizontal rods the pairs of bases holding the chains together. The vertical line marks the fibre axis

We wish to put forward a radically different structure for the salt of deoxyribose nucleic acid. This structure has two helical chains each coiled round the same axis (see diagram). We have made the usual chemical assumptions, namely, that each chain consists of phosphate diester groups joining β-D-deoxyribofuranose residues with 3′,5′ linkages. The two chains (but not their bases) are related by a dyad perpendicular to the fibre axis. Both chains follow right-handed helices, but owing to the dyad the sequences of the atoms in the two chains run in opposite directions. Each chain loosely resembles Furberg's[2] model No. 1; that is, the bases are on the inside of the helix and the phosphates on the outside. The configuration of the sugar and the atoms near it is close to Furberg's 'standard configuration', the sugar being roughly perpendicular to the attached base. There is a residue on each chain every 3·4 A. in the z-direction. We have assumed an angle of 36° between adjacent residues in the same chain, so that the structure repeats after 10 residues on each chain, that is, after 34 A. The distance of a phosphorus atom from the fibre axis is 10 A. As the phosphates are on the outside, cations have easy access to them.

The structure is an open one, and its water content is rather high. At lower water contents we would expect the bases to tilt so that the structure could become more compact.

The novel feature of the structure is the manner in which the two chains are held together by the purine and pyrimidine bases. The planes of the bases are perpendicular to the fibre axis. They are joined together in pairs, a single base from one chain being hydrogen-bonded to a single base from the other chain, so that the two lie side by side with identical z-co-ordinates. One of the pair must be a purine and the other a pyrimidine for bonding to occur. The hydrogen bonds are made as follows: purine position 1 to pyrimidine position 1; purine position 6 to pyrimidine position 6.

If it is assumed that the bases only occur in the structure in the most plausible tautomeric forms (that is, with the keto rather than the enol configurations) it is found that only specific pairs of bases can bond together. These pairs are: adenine (purine) with thymine (pyrimidine), and guanine (purine) with cytosine (pyrimidine).

In other words, if an adenine forms one member of a pair, on either chain, then on these assumptions the other member must be thymine; similarly for guanine and cytosine. The sequence of bases on a single chain does not appear to be restricted in any way. However, if only specific pairs of bases can be formed, it follows that if the sequence of bases on one chain is given, then the sequence on the other chain is automatically determined.

It has been found experimentally[3,4] that the ratio of the amounts of adenine to thymine, and the ratio of guanine to cytosine, are always very close to unity for deoxyribose nucleic acid.

It is probably impossible to build this structure with a ribose sugar in place of the deoxyribose, as the extra oxygen atom would make too close a van der Waals contact.

The previously published X-ray data[5,6] on deoxyribose nucleic acid are insufficient for a rigorous test of our structure. So far as we can tell, it is roughly compatible with the experimental data, but it must be regarded as unproved until it has been checked against more exact results. Some of these are given in the following communications. We were not aware of the details of the results presented there when we devised our structure, which rests mainly though not entirely on published experimental data and stereochemical arguments.

It has not escaped our notice that the specific pairing we have postulated immediately suggests a possible copying mechanism for the genetic material.

Full details of the structure, including the conditions assumed in building it, together with a set of co-ordinates for the atoms, will be published elsewhere.

We are much indebted to Dr. Jerry Donohue for constant advice and criticism, especially on interatomic distances. We have also been stimulated by a knowledge of the general nature of the unpublished experimental results and ideas of Dr. M. H. F. Wilkins, Dr. R. E. Franklin and their co-workers at King's College, London. One of us (J. D. W.) has been aided by a fellowship from the National Foundation for Infantile Paralysis.

J. D. WATSON
F. H. C. CRICK

Medical Research Council Unit for the Study of the Molecular Structure of Biological Systems,
Cavendish Laboratory, Cambridge.
April 2.

[1] Pauling, L., and Corey, R. B., *Nature*, **171**, 346 (1953); *Proc. U.S. Nat. Acad. Sci.*, **39**, 84 (1953).
[2] Furberg, S., *Acta Chem. Scand.*, **6**, 634 (1952).
[3] Chargaff, E., for references see Zamenhof, S., Brawerman, G., and Chargaff, E., *Biochim. et Biophys. Acta*, **9**, 402 (1952).
[4] Wyatt, G. R., *J. Gen. Physiol.*, **36**, 201 (1952).
[5] Astbury, W. T., Symp. Soc. Exp. Biol. 1, Nucleic Acid, 66 (Camb. Univ. Press, 1947).
[6] Wilkins, M. H. F., and Randall, J. T., *Biochim. et Biophys. Acta*, **10**, 192 (1953).

Short and sweet: our Nature *paper announcing the discovery. The same issue also carried longer articles by Rosalind Franklin and Maurice Wilkins.*

Unveiling the double helix: my lecture at Cold Spring Harbor Laboratory, June 1953

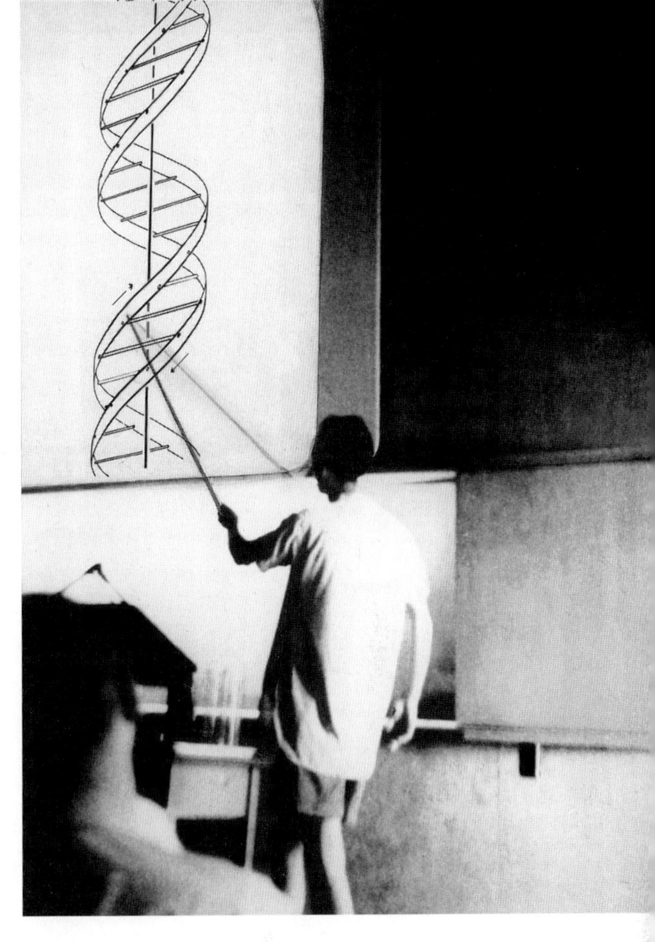

Harbor symposium on viruses. Max Delbrück saw to it that I was offered, at the last minute, an invitation to speak. To this intellectually high-powered meeting I brought a three-dimensional model built in the Cavendish, the adenine-thymine base pairs in red and the guanine-cytosine base pairs in green.

In the audience was Seymour Benzer, yet another ex-physicist who had heeded the clarion call of Schrödinger's book. He immediately understood what our breakthrough meant for his studies of mutations in viruses. He realized that he could now do for a short stretch of bacteriophage DNA what Morgan's boys had done forty years earlier for fruit fly chromosomes: he would map mutations—determine their order—along a gene, just as the fruit fly pioneers had mapped genes along a chromosome. Like Morgan, Benzer would have to depend on recombination to generate new genetic combinations, but, whereas Morgan had the advantage of a ready mechanism of recombination—the production of sex cells in a fruit fly—Benzer had to induce recombination by simultaneously infecting a single bacterial host cell with two different strains of bacteriophage, which differed by one or more mutations in the region of interest. Within the bacterial cell, recombination—the exchange of segments of molecules—would occasionally occur between the different viral DNA molecules, producing new permutations of mutations—so-called "recombinants." Within a single astonishingly productive year in his Purdue

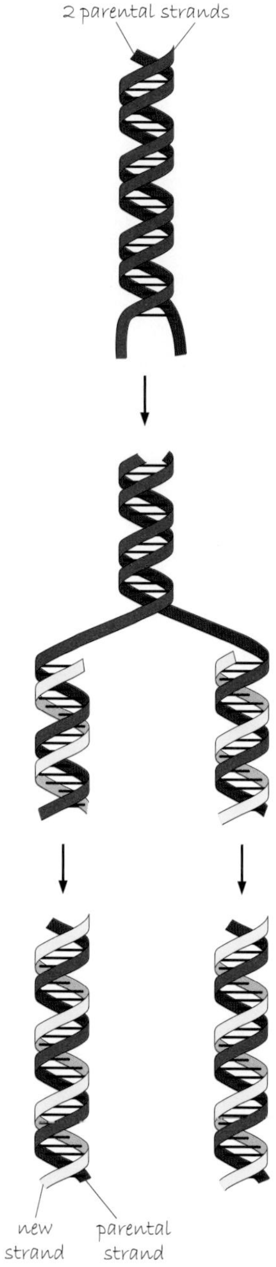

University lab, Benzer produced a map of a single bacteriophage gene, *rII,* showing how a series of mutations—all errors in the genetic script—were laid out linearly along the virus DNA. The language was simple and linear, just like a line of text on the written page.

The response of the Hungarian physicist Leo Szilard to my Cold Spring Harbor talk on the double helix was less academic. His question was, "Can you patent it?" At one time Szilard's main source of income had been a patent that he held with Einstein, and he had later tried unsuccessfully to patent with Enrico Fermi the nuclear reactor they built at the University of Chicago in 1942. But then as now patents were given only for useful inventions and at the time no one could conceive of a practical use for DNA. Perhaps then, Szilard suggested, we should copyright it.

There remained, however, a single missing piece in the double helical jigsaw puzzle: our unzipping idea for DNA replication had yet to be experimentally verified. Max Delbrück, for example, was unconvinced. Though he liked the double helix as a model, he worried that unzipping it might generate horrible knots. Five years later, a former student of Pauling's, Matt Meselson, and the equally bright young phage worker Frank Stahl put to rest such fears when they published the results of a single elegant experiment.

They had met in the summer of 1954 at the Marine Biological Laboratory at Woods Hole, Massachusetts, where I was then lecturing, and agreed—over a good many gin martinis—that they should get together to do some science. The

DNA replication: the double helix is unzipped and each strand copied.

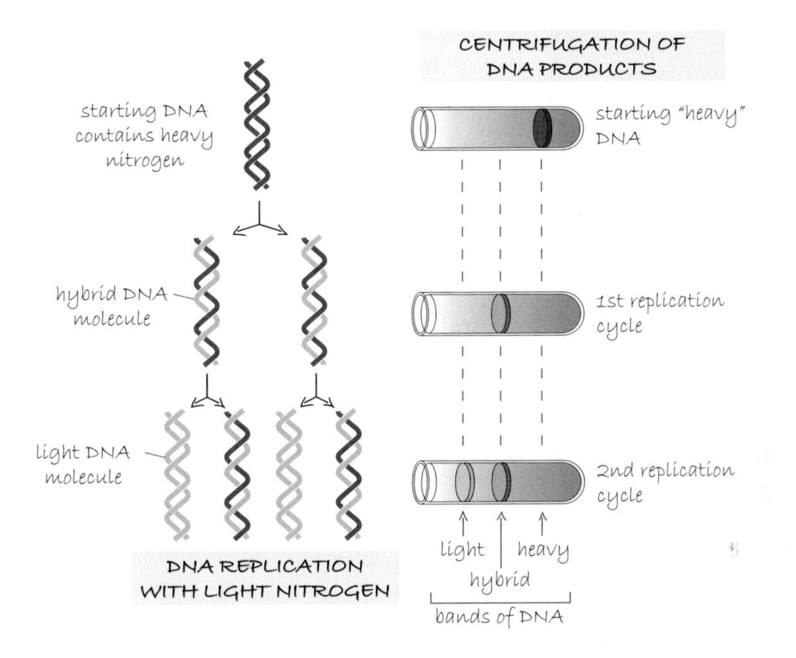

The Meselson-Stahl experiment

result of their collaboration has been described as "the most beautiful experiment in biology."

They used a centrifugation technique that allowed them to sort molecules according to slight differences in weight; following a centrifugal spin, heavier molecules end up nearer the bottom of the test tube than lighter ones. Because nitrogen atoms (N) are a component of DNA, and because they exist in two distinct forms, one light and one heavy, Meselson and Stahl were able to tag segments of DNA and thereby track the process of its replication in bacteria. Initially all the bacteria were raised in a medium containing heavy N, which was thus incorporated in both strands of the DNA. From this culture they took a sample, transferring it to a medium containing only light N, ensuring that the next round of DNA replication would have to make use of light N. If, as Crick and I had predicted, DNA replication involves unzipping the double helix and

Matt Meselson beside an ultra-centrifuge, the hardware at the heart of "the most beautiful experiment in biology"

copying each strand, the resultant two "daughter" DNA molecules in the experiment would be hybrids, each consisting of one heavy N strand (the template strand derived from the "parent" molecule) and one light N strand (the one newly fabricated from the new medium). Meselson and Stahl's centrifugation procedure bore out these expectations precisely. They found three discrete bands in their centrifuge tubes, with the heavy-then-light sample halfway between the heavy-heavy and light-light samples. DNA replication works just as our model supposed it would.

The biochemical nuts and bolts of DNA replication were being analyzed at around the same time in Arthur Kornberg's laboratory at Washington University in St. Louis. By developing a new, "cell-free" system for DNA synthesis, Kornberg discovered an enzyme (DNA polymerase) that links the DNA components and makes the chemical bonds of the DNA backbone. Kornberg's enzymatic synthesis of DNA was such an unanticipated and important event that he was awarded the 1959 Nobel Prize in Physiology or Medicine, less than two years after the key experiments. After his prize was announced, Kornberg was photographed holding a copy of the double helix model I had taken to Cold Spring Harbor in 1953.

It was not until 1962 that Francis Crick, Maurice Wilkins, and I were to receive our own Nobel Prize in Physiology or Medicine. Four years earlier, Rosalind Franklin had died of ovarian cancer at the tragically young age of thirty-seven. Before then Crick had become a close colleague and a real friend of Franklin's. Following the two operations that would fail to stem the advance of her cancer, Franklin convalesced with Crick and his wife, Odile, in Cambridge.

It was and remains a long-standing rule of the Nobel Committee never to split a single prize more than three ways. Had Franklin lived, the problem would have arisen whether to bestow the award upon her or Maurice Wilkins. The Swedes might have resolved the dilemma by awarding them both the Nobel Prize in Chemistry that year. Instead, it went to Max Perutz and John Kendrew, who had elucidated the three-dimensional structures of hemoglobin and myoglobin respectively.

Arthur Kornberg at the time of winning his Nobel Prize

The discovery of the double helix sounded the death knell for vitalism. Serious scientists, even those religiously inclined, realized that a complete understanding of life would not require the revelation of new laws of nature. Life was just a matter of physics and chemistry, albeit exquisitely organized physics and chemistry. The immediate task ahead would be to figure out how the DNA-encoded script of life went about its work. How does the molecular machinery of cells read the messages of DNA molecules? As the next chapter will reveal, the unexpected complexity of the reading mechanism led to profound insights into how life first came about.

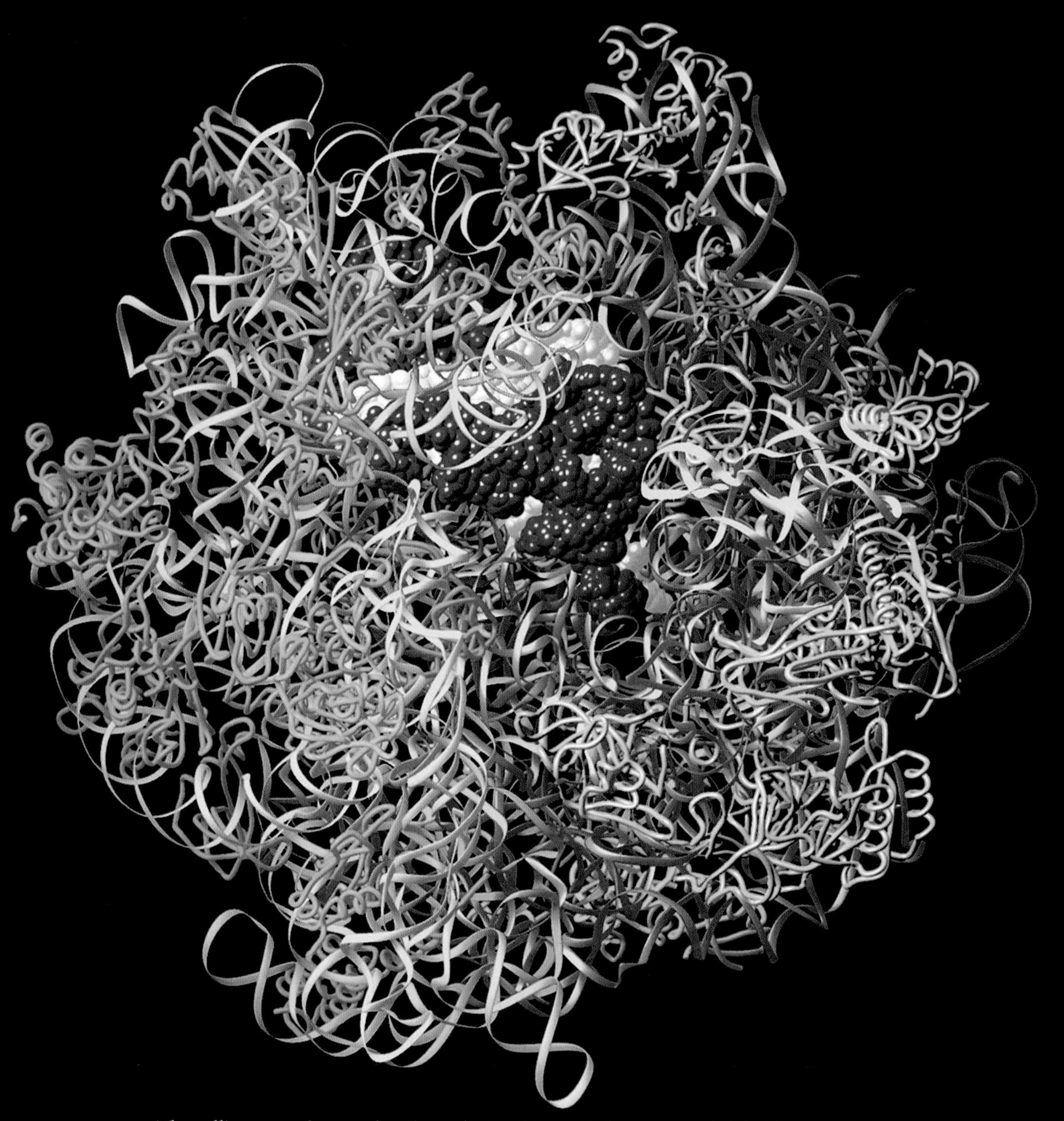

The cell's protein factory, the ribosome, in all its 3-D glory as revealed by X-ray analysis. (For simplicity, this computer-generated image does not show individual atoms.) There are millions of ribosomes in every cell. It is here that the information encoded in DNA is used to produce proteins, the actors in life's molecular drama. The ribosome consists of two subunits (orange and yellow), each composed of RNA, plus some sixty proteins (blue and green) plastered over the outside. Here the ribosome is caught in the act of producing a protein. Specialized small RNA molecules (purple, white, and red) transport amino acids to the ribosome for incorporation into the growing protein chain.

CHAPTER THREE

. . .

READING THE CODE: BRINGING DNA TO LIFE

Long before Oswald Avery's experiments put DNA in the spotlight as the "transforming principle," geneticists were trying to understand just how the hereditary material—whatever it might be—was able to influence the characteristics of a particular organism. How did Mendel's "factors" affect the form of peas, making them either wrinkled or round?

The first clue came around the turn of the century, just after the rediscovery of Mendel's work. Archibald Garrod, an English physician whose slow progress through medical school and singular lack of a bedside manner had ensured him a career in research rather than patient care at St. Bartholomew's Hospital in London, was interested in a group of rare diseases of which a common marked symptom was strangely colored urine. One of these diseases, alkaptonuria, has been dubbed "black diaper syndrome" because those afflicted with it pass urine that turns black on exposure to air. Despite this alarming symptom, the disease is usually not lethal, though it can lead in later life to an arthritis-like condition as the black-urine pigments accumulate in the joints and spine. Contemporary science attributed the blackening to a substance produced by bacteria living in the gut, but Garrod argued that the appearance of black urine in newborns, whose guts lack bacterial colonies, implied that the substance was produced by the body itself. He inferred that it was the product of a flaw in the body's chemical machinery, an "error in metabolism" in his words, suggesting there might be a critical glitch in some biochemical pathway.

Garrod further observed that alkaptonuria, though very rare in the population as a whole, occurred more frequently among children of marriages between blood relatives. In 1902, he was able to explain the phenomenon in terms of Mendel's newly rediscovered laws. Here was the pattern of inheritance to be expected of a rare recessive gene: two first cousins, say, have both received a copy of the "alkaptonuria" gene from the same grandparent, creating a one-in-four chance that their union will produce a child homozygous for the gene (i.e., a child with two copies of the recessive gene) who will therefore develop alkaptonuria. Combining his biochemical and genetic analyses, Garrod concluded that alkaptonuria is an "inborn error in metabolism." Though nobody really appreciated it at the time, Garrod was thus the first to make the causal connection between genes and their physiological effect. Genes in some way governed metabolic processes, and an error in a gene—a mutation—could result in a defective metabolic pathway.

The next significant step would not occur until 1941, when George Beadle and Ed Tatum published their study of induced mutations in a tropical bread mold. Beadle had grown up outside Wahoo, Nebraska, and would have taken over the family farm had a high-school science teacher not encouraged him to consider an alternative career. Through the thirties, first at Caltech in association with T. H. Morgan of fruit fly fame and then at the Institut de Biologie Physico-Chimique in Paris, Beadle had applied himself to discovering how genes work their magic in affecting, for example, eye color in fruit flies. Upon his arrival at Stanford University in 1937, he recruited Tatum, who joined the effort against the advice of his academic advisers. Ed Tatum had been both an undergraduate and graduate student at the University of Wisconsin, doing studies of bacteria that lived in milk (of which there was no shortage in the Cheese State). Though the job with Beadle might be intellectually challenging, Tatum's Wisconsin professors counseled in favor of the financial security to be found in a career with the dairy industry. Fortunately for science, Tatum chose Beadle over butter.

Beadle and Tatum came to realize that fruit flies were too complex for the kind of research at hand: finding the effect of a single mutation in an animal as complicated as *Drosophila* would be like looking for a needle in a haystack. They chose instead to work with an altogether simpler species, *Neurospora crassa,* the orange-red mold that grows on bread in tropical countries. The plan was simple: subject the mold to X rays to cause mutations—just as Muller had done with fruit flies—and then try to determine the impact of the resulting mutations on the fungi. They would track the effects of the mutations in this way: Normal (i.e., unmutated) *Neurospora,* it was known, could survive on a so-called minimal culture medium; on this basic "diet" they could evidently synthesize biochemically all the larger molecules they required to live, constructing them from the simpler ones in the nutrient medium. Beadle and Tatum theorized that a mutation that knocked out any of those synthetic pathways would result in the irradiated mold strain being unable to grow on minimal medium; that same strain should, however, still manage to thrive on a "complete" medium, one containing all the molecules necessary for life, like amino acids and vitamins. In other words, the mutation preventing the synthesis of a key nutrient would be rendered harmless if the nutrient were available directly from the culture medium.

Beadle and Tatum irradiated some five thousand specimens, then set about testing each one to see whether it could survive on minimal medium. The first survived fine; so did the second, and the third . . . It was not until they tested strain number 299 that they found one that could no longer exist on minimal medium, though as predicted it could survive on the complete version. Number 299 would be but the first of many mutant strains that they would analyze. The next step was to see what exact capacity the mutants had lost. Maybe 299 could not synthesize essential amino acids. Beadle and Tatum tried adding amino acids to the minimal medium, but still 299 failed to grow. What about vitamins? They added a slew of them to the minimal medium, and this time 299 thrived. Now it was time to narrow the field, adding each vitamin individually and then gauging the growth response of 299. Niacin didn't work, nor riboflavin, but when they added vitamin B_6, 299 was able to survive on minimal medium. 299's X-ray-induced mutation had somehow disrupted the synthetic pathway involved in the production of B_6. But how? Knowing that biochemical syntheses of this kind are governed by protein enzymes that promote the individual incremental chemical reactions along the pathway, Beadle and Tatum suggested that each mutation they discovered had knocked out a particular enzyme. And since mutations occur in genes, genes must produce enzymes. When it appeared in 1941, their study inspired a slogan that summarized what had become the understanding of how genes work: "One gene, one enzyme."

But since all enzymes were then thought to be proteins, the question soon arose whether genes also encoded the many cellular proteins that were *not* enzymes. The first suggestion that genes might provide the information for all proteins came from Linus Pauling's lab at Caltech. He and his student Harvey Itano studied hemoglobin, the protein in red blood cells that transports oxygen from the lung to metabolically active tissues, like muscle, where it is needed. In particular, they focused on the hemoglobin of people with sickle-cell disease, also known as sickle-cell anemia, a genetic disorder common in Africans, and therefore among African Americans as well. The red blood cells of sickle-cell victims tend to become deformed, assuming a distinctive "sickle" shape under the microscope, and the resulting blockages in capillaries can be horribly painful, even lethal. Later research would uncover an evolutionary rationale for the disease's prevalence among Africans: because part of the malaria parasite's life cycle is spent in red blood cells, people with sickle-cell hemoglobin suffer less severely from malaria. Human evolution seems to have struck a Faustian bargain on behalf of some inhabitants of tropical regions: the sickle-cell affliction confers some protection against the ravages of malaria.

Itano and Pauling compared the hemoglobin proteins of sickle-cell patients with those of non-sickle-cell individuals and found that the two molecules differed in their electrical charge. Around that time, the late forties, geneticists determined that sickle-cell disease is transmitted as a classical Mendelian recessive character. Sickle-cell disease, they therefore inferred, must be caused by a mutation in the hemoglobin gene, a mutation that affects the chemical composition of the resultant hemoglobin protein. And so it was that Pauling was able to refine Garrod's notion of "inborn errors of metabolism" by recognizing some to be what he called "molecular diseases." Sickle-cell was just that, a molecular disease.

In 1956, the sickle-cell hemoglobin story was taken a step further by Vernon Ingram, working in the Cavendish Laboratory where Francis Crick and I had found the double helix. Using recently developed methods of identifying the specific amino acids in the chain that makes up a protein, Ingram was able to specify precisely the molecular difference that Itano and Pauling had noted as affecting the overall charge of the molecule. It amounted to a single amino acid: Ingram determined that glutamic acid, found at position 6 in the normal protein chain, is replaced, in sickle-cell hemoglobin, by valine. Here, conclusively, was

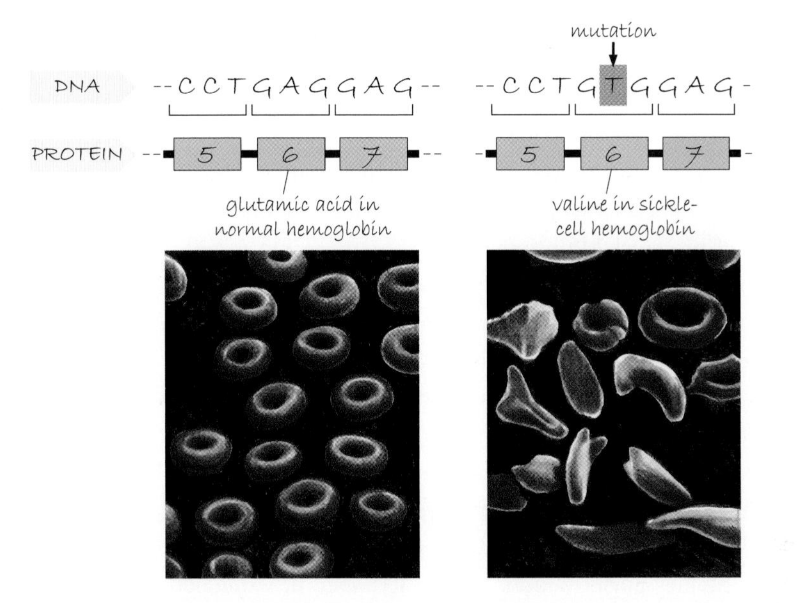

The impact of mutation. A single base change in the DNA sequence of the human beta hemoglobin gene results in the incorporation of the amino acid valine rather than glutamic acid into the protein. This single difference causes sickle-cell disease, in which the red blood cells become distorted into a characteristic sickle shape.

evidence that genetic mutations—differences in the sequence of As, Ts, Gs, and Cs in the DNA code of a gene—could be "mapped" directly to differences in the amino acid sequences of proteins. Proteins are life's active molecules: they form the enzymes that catalyze biochemical reactions, and they also provide the body's major structural components, like keratin, of which skin, hair, and nails are composed. And so the way DNA exerts its controlling magic over cells, over development, over life as a whole, is through proteins.

But how is the information encoded in DNA—a molecular string of nucleotides, As, Ts, Gs, and Cs—converted into a protein—a string of amino acids?

Shortly after Francis Crick and I published our account of the double helix, we began to hear from the well-known Russian-born theoretical physicist George Gamow. His letters—invariably handwritten and embellished with cartoons and other squiggles, some quite relevant, others less so—were always signed simply "Geo" (pronounced "Jo," as we would later discover). He'd become interested in DNA and, even before Ingram had conclusively demonstrated the connection between the DNA base sequence and the amino acid sequence of proteins, in the relationship between DNA and protein. Sensing that biology was at last becoming an exact science, Gamow foresaw a time when every organism could be described genetically by a very long number represented exclusively by the numerals 1, 2, 3, and 4, each one standing for one of the bases, A, T, G, and C. At first, we took him for a buffoon; we ignored his first letter. A few months later, however, when Crick met him in New York City, the magnitude of his gifts became clear and we promptly welcomed him aboard the DNA bandwagon as one of its earliest recruits.

Gamow had come to the United States in 1934 to escape the engulfing tyranny of Stalin's Soviet Union. In a 1948 paper, he explained the abundance of different chemical elements present throughout the universe in relation to thermonuclear processes that had taken place in the early phases of the Big Bang. The research, having been carried out by Gamow and his graduate student Ralph Alpher, would have been published with the byline of "Alpher and Gamow" had Gamow not decided to include as well the name of his friend Hans Bethe, an eminently talented physicist to be sure, but one who had contributed nothing to the study. It delighted the inveterate prankster Gamow that the paper appeared attributed to "Alpher, Bethe, and Gamow," no less than that its publication date was, fortuitously, April 1. To this day, cosmologists still refer to it as the αβγ (Alpha-Beta-Gamma) paper.

By the time I first met Gamow in 1954, he had already devised a formal scheme in which he proposed that overlapping triplets of DNA bases served to specify certain amino acids. Underlying his theory was a belief that there existed on the surface of each base pair a cavity that was complementary in shape to part of the surface of one of the amino acids. I told Gamow I was skeptical: DNA could not be the direct template along which amino acids arranged themselves before being connected into polypeptide chains, as lengths of linked amino acids are called. Being a physicist, Gamow had not, I supposed, read the scientific papers refuting the notion that protein synthesis occurs where DNA is located—in the nucleus. In fact, it had been observed that the removal of the nucleus from a cell has no immediate effect on the rate at which proteins are made. Today we know that amino acids are actually assembled into proteins in ribosomes, small cellular particles containing a second form of nucleic acid called RNA.

RNA's exact role in life's biochemical puzzle was unclear at that time. In some viruses, like tobacco mosaic virus, it seemed to play a role similar to DNA in other species, encoding the proteins specific to that organism. And in cells, RNA had to be involved somehow in protein synthesis, since cells that made lots of proteins were always RNA-rich. Even before we found the double helix, I thought it likely that the genetic information in chromosomal DNA was used to make RNA chains of complementary sequences. These RNA chains might in turn serve as the templates that specified the order of amino acids in their respective proteins. If so, RNA was thus an intermediate between DNA and protein. Francis Crick would later refer to this DNA → RNA → protein flow of information as the "central dogma." The view soon gained support with the discovery in 1959 of the enzyme RNA polymerase. In virtually all cells, it catalyzes the production of single-stranded RNA chains from double-stranded DNA templates.

It appeared the essential clues to the process by which proteins are made would come from further studies of RNA, not DNA. To advance the cause of "cracking the code"—deciphering that elusive relationship between DNA sequence and the amino acid sequence of proteins—Gamow and I formed the RNA Tie Club. Its members would be limited to twenty, one for each of the twenty different amino acids. Gamow designed a club necktie and commissioned the production of the amino-acid-specific tiepins. These were badges of office, each bearing the standardized three-letter abbreviation of an amino acid, the one the member wearing the pin was responsible for studying. I had PRO for proline and Gamow had ALA for alanine. In an era when tiepins with letters usually advertised one's initials, Gamow took pleasure in confusing people with his ALA pin. His joke backfired when a sharp-eyed hotel clerk refused to honor his check, noting that the name printed on the check bore no relation to the initials on the gentleman's jewelry.

The fact that most of the scientists interested in the coding problem at that time could be squeezed into the club's membership of twenty showed how small the DNA-RNA world was. Gamow easily found room for a nonbiologist

The RNA Tie Club: Geo Gamow's characteristic scrawl in a letter; the man himself; a 1955 club meeting, with ties in evidence (Francis Crick, Alex Rich, Leslie Orgel, and me)

-2-

and mail to G. Gamow, M. L. B., Woods Hole, Mass. For a negative vote: do not do anything.

If the amount of collected dollars is sufficient to buy an RNA tie and a special base pin for the proposed candidates (for, of course, honorary members do not pay) they will be considered electe.

The latest date for receiving the votes is September 1st, 1955.

Sincerely yours, Ala,

The Synthesizer

... next week, and untill ... will be: San Diego. Calif. ... probably droping ... ften, will you ... the adress of ... sher. I may have ... more ties.

...ours Geo.

Rnatie Club

"Do or die, or don't try"

OFFICERS
GEO GAMOW - SYNTHESISER
GEORGE WASHINGTON UNIVERSITY
JIM WATSON - OPTIMIST
HARVARD UNIVERSITY
FRANCIS CRICK - PESSIMIST
CAMBRIDGE UNIVERSITY
MARTINAS YCAS - ARCHIVIST
QUARTERMASTER R. & D. LABS.
ALEX RICH - LORD PRIVY SEAL
NAT. INST. MENTAL HEALTH

July 4, 1955

Dear Pro,

This is the first official club circular. First, the assignments of tie pins (which, as you know, were randomized):

1) ALA - G. Gamow	8) GLY - R. Feynman	15) PRO - J. Watson
2) ARG - A. Rich	9) HIS - M. Calvin	16) SER - H. Gordon
3) ASP - P. Doty	10) ISO - N. Simons	17) THR - L. Orgel
4) ASN - R. Ledley	11) LEU - E. Teller	18) TRY - M. Delbrück
5) CYS - M. Ycas	12) LYS - E. Chargaff	19) TYR - F. Crick
6) GLU - R. Williams	13) MET - N. Metropolis	20) VAL - S. Brenner
7) GLN - A. Dounce	14) PHE - G. Stent	

From this list, 13 members have obtained their tie pins while the remaining 7 are still stubbornly holding out.

For RNA ties, please write to Jim Watson at ~~Harvard~~ Cambridge Universit...

The first matter of business is the election of honorary base...

...out ... lgnm... mad... ice) ... vot... onl... u ar...

friend, the physicist Edward Teller (LEU—leucine), while I inducted Richard Feynman (GLY—glycine), the extraordinarily imaginative Caltech physicist who, when momentarily frustrated in his exploration of inner atomic forces, often visited me in the biology building where I was then working.

One element of Gamow's 1954 scheme had the virtue of being testable: because it involved overlapping DNA triplets, it predicted that many pairs of amino acids would in fact never be found adjacent in proteins. So Gamow eagerly awaited the sequencing of additional proteins. To his disappointment, more and more amino acids began to be found next to each other, and his scheme became increasingly untenable. The coup de grâce for all Gamow-type codes came in 1956 when Sydney Brenner (VAL—valine) analyzed every amino acid sequence then available.

Brenner had been raised in a small town outside Johannesburg, South Africa, in two rooms at the back of his father's cobbler's shop. Though the elder Brenner, a Lithuanian immigrant, was illiterate, his precocious son discovered a love of reading at the age of four and, led by this passion, would be turned on to biology by a textbook called *The Science of Life*. Though he was one day to admit having stolen the book from the public library, neither larceny nor poverty could slow Brenner's progress: he entered the University of Witwatersrand's undergraduate medical program at fourteen, and was working on his Ph.D. at Oxford when he came to Cambridge a month after our discovery of the double helix. He recalls his reaction to our model: "That's when I saw that this was it. And in a flash you just knew that this was very fundamental."

Gamow was not the only one whose theories were biting the dust: I had my own share of disappointments. Having gone to Caltech in the immediate aftermath of the double helix, I wanted to find the structure of RNA. To my despair, Alexander Rich (ARG—arginine) and I soon discovered that X-ray diffraction of RNA yielded uninterpretable patterns: the molecule's structure was evidently not as beautifully regular as that of DNA. Equally depressing, in a note sent out early in 1955 to all Tie Club members, Francis Crick (TYR—tyrosine) predicted that the structure of RNA would not, as I supposed, hold the secret of the DNA → protein transformation. Rather, he suggested that amino acids were likely ferried to the actual site of protein synthesis by what he called "adaptor molecules," of which there existed one specific to every amino acid.

He speculated that these adaptors themselves might be very small RNA molecules. For two years I resisted his reasoning. Then a most unexpected biochemical finding proved that his novel idea was right on the mark.

It came from work at the Massachusetts General Hospital in Boston, where Paul Zamecnik had for several years been developing cell-free systems for studying protein synthesis. Cells are highly compartmentalized bodies, and Zamecnik correctly saw the need to study what was going on inside them without the complications posed by their various membranes. Using material derived from rat liver tissue, he and his collaborators were able to re-create in a test tube a simplified version of the cell interior in which they could track radioactively tagged amino acids as they were assembled into proteins. In this way Zamecnik was able to identify the ribosome as the site of protein synthesis, a fact that George Gamow did not accept initially.

Soon, with his colleague Mahlon Hoagland, Zamecnik made the even more unexpected discovery that amino acids, prior to being incorporated into polypeptide chains, were bound to small RNA molecules. This result puzzled them until they heard from me of Crick's adaptor theory. They then quickly confirmed Crick's suggestion that a specific RNA adaptor (called transfer RNA) existed for each amino acid. And each of these transfer RNA molecules also had on its surface a specific sequence of bases that permitted it to bind to a corresponding segment of the RNA template, thereby lining up the amino acids for protein synthesis.

Until the discovery of transfer RNA, all cellular RNA was thought to have a template role. Now we realized RNA could come in several different forms, though the two major RNA chains that comprised the ribosomes predominated. Puzzling at the time was the observation that these two RNA chains were of constant sizes. If these chains were the actual templates for protein synthesis, we would have expected them to vary in length in relation to the different sizes of their protein products. Equally disturbing, these chains proved very stable metabolically: once synthesized they did not break down. Yet experiments at the Institut Pasteur in Paris suggested that many templates for bacterial protein synthesis were short-lived. Even stranger, the sequences of the bases in the two ribosomal RNA chains showed no correlation to sequences of bases along the respective chromosomal DNA molecules.

Resolution of these paradoxes came in 1960 with discovery of a third form of RNA, messenger RNA. This was to prove the true template for protein synthesis. Experiments done in my lab at Harvard and at both Caltech and Cambridge by Matt Meselson, François Jacob, and Sydney Brenner showed that ribosomes were, in effect, molecular factories. Messenger RNA passed between the two ribosomal subunits like ticker tape being fed into an old-fashioned computer. Transfer RNAs, each with its amino acid, attached to the messenger RNA in the ribosome so that the amino acids were appropriately ordered before being chemically linked to form polypeptide chains.

Still unclear was the genetic code, the rules for translating a nucleic acid sequence into an ordered polypeptide sequence. In a 1956 RNA Tie Club manuscript, Sydney Brenner laid out the theoretical issues. In essence they boiled down to this: how could the code specify which one of 20 amino acids was to be incorporated into a protein chain at a particular point when there are only four DNA letters, A, T, G, C? Obviously a single nucleotide, with only four possible identities, was insufficient, and even two—which would allow for 16 (4 × 4) possible permutations—wouldn't work. It would take at minimum three nucleotides, a triplet, to code for a single amino acid. But this also supposed a puzzling redundant capacity. With a triplet, there could exist 64 permutations (4 × 4 × 4); since the code needed only 20, was it the case that most amino acids could be encoded by more than one triplet? If that were so, in principle, a "quadruplet" code (4 × 4 × 4 × 4) yielding 256 permutations was also perfectly feasible, though it implied even greater redundancy.

In 1961 at Cambridge University, Brenner and Crick did the definitive experiment that demonstrated that the code was triplet-based. By a clever use of chemical mutagens they were able to delete or insert DNA base pairs. They found that inserting or deleting a single base pair results in a harmful "frameshift" because the entire code beyond the site of the mutation is scrambled. Imagine a three-letter word code as follows: JIM ATE THE FAT CAT. Now imagine that the first "T" is deleted. If we are to preserve the three-letter word structure of the sentence, we have JIM AET HEF ATC AT—gibberish beyond the site of the deletion. The same thing happens when two base pairs are deleted or inserted: removing the first "T" and "E," we get JIM ATH EFA TCA T—more gibberish. Now what happens if we delete (or insert) *three* letters? Removing the first "A," "T," and "E," we get JIM THE FAT CAT; although we have lost one "word"—ATE—we have nevertheless retained the sense of the rest of the sentence. And even if our deletion straddles "words"—say we delete the first "T" and "E," and the second "T"—we still lose only those two words, and are again able to recover the intended sentence beyond them: JIM AHE FAT CAT. So it is with DNA sequence: a single insertion/deletion massively disrupts the protein because of the frameshift effect, which changes every single amino acid beyond the insertion/deletion point; so does a double insertion/deletion. But a triple insertion/deletion along a DNA molecule will not necessarily have a catastrophic effect; they will add/eliminate one amino acid but this does not necessarily disrupt all biological activity.

Crick came into the lab late one night with his colleague Leslie Barnett to check on the final result of the triple-deletion experiment, and realized at once the significance of the result, telling Barnett, "We're the only two who know it's a triplet code!" With me, Crick had been the first to glimpse the double helical secret of life; now he was the first to know for sure that the secret is written in three-letter words.

So the code came in threes, and the links from DNA to protein were RNA-mediated. But we still had to crack the code. What pair of amino acids was specified by a stretch of DNA with, say, sequence ATA TAT or GGT CAT? The first glimpse of the solution came in a talk given by Marshall Nirenberg at the International Congress of Biochemistry in Moscow in 1961.

After hearing about the discovery of messenger RNA, Nirenberg, working at the U.S. National Institutes of Health, wondered whether RNA synthesized in vitro would work as well as the naturally occurring messenger form when it came to protein synthesis in cell-free systems. To find out, he used RNA tailored according to procedures developed at New York University six years earlier by the French biochemist Marianne Grunberg-Manago. She had discovered an RNA-specific enzyme that could produce strings like AAAAAA or GGGGGG. And because one key chemical difference between RNA and DNA is RNA's substitution of uracil, "U," for thymine, "T," this enzyme would also produce strings of U, UUUUU . . . —poly-U, in the biochemical jargon. It was poly-U that Nirenberg and his German collaborator, Heinrich Matthaei, added to their cell-free system on May 22, 1961. The result was striking: the ribosomes started to pump out a simple protein, one consisting of a string of a single amino acid, phenylalanine. They had discovered that poly-U encodes polyphenylalanine. Therefore, one of the three-letter words by which the genetic code specified phenylalanine had to be UUU.

The International Congress that summer of 1961 brought together all the major players in molecular biology. Nirenberg, then a young scientist nobody had heard of, was slated to speak for just ten minutes, and hardly anyone,

THE GENETIC CODE

AMINO ACID	RNA CODON
alanine	GCA GCC GCG GCU
arginine	AGA AGG CGA CGC CGG CGU
asparagine	AAC AAU
aspartic acid	GAC GAU
cysteine	UGC UGU
glutamic acid	GAA GAG
glutamine	CAA CAG
glycine	GGA GGC GGG GGU
histidine	CAC CAU
isoleucine	AUA AUC AUU
leucine	UUA UUG CUA CUC CUG CUU
lysine	AAA AAG
methionine	AUG
phenylalanine	UUC UUU
proline	CCA CCC CCG CCU
serine	AGC AGU UCA UCC UCG UCU
threonine	ACA ACC ACG ACU
tryptophan	UGG
tyrosine	UAC UAU
valine	GUA GUC GUG GUU
STOP CODONS	UAA UAG UGA

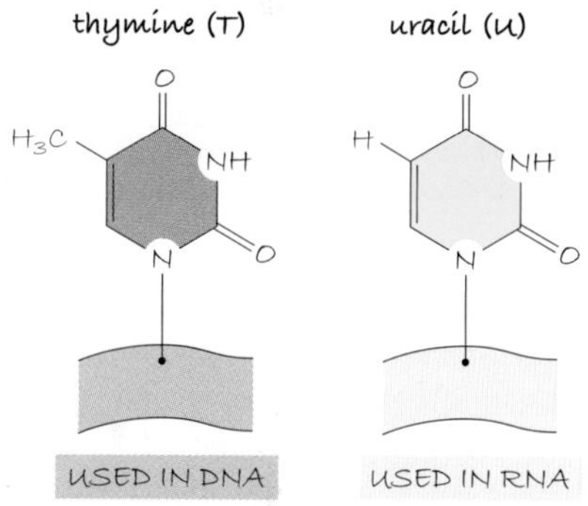

thymine (T)

H_3C O NH O N

USED IN DNA

uracil (U)

H O NH O N

USED IN RNA

The genetic code, showing the triplet sequences for messenger RNA. An important difference between DNA and RNA is that DNA uses thymine and RNA uracil. Both bases are complementary to adenine. Stop codons do what their name suggests: they mark the end of the coding part of a gene.

Francis Crick (center) with Gobind Khorana and Marianne Grunberg-Manago. Khorana unraveled much of the genetic code after Nirenberg's initial breakthrough, which was based on Grunberg-Manago's pioneering research.

including myself, attended his talk. But when news of his bombshell began to spread, Crick promptly inserted him into a later session of the conference so that Nirenberg could make his announcement to a now-expectant capacity audience. It was an extraordinary moment. A quiet, self-effacing young no-name speaking before a who's who crowd of molecular biology had shown the way toward finding the complete genetic code.

Practically speaking, Nirenberg and Matthaei had solved but one sixty-fourth of the problem—all we now knew was what UUU codes for phenylalanine. There remained sixty-three other three-letter triplets (codons) to figure out, and the following years would see a frenzy of research as we labored to discover what amino acids these other codons represented. The tricky part was synthesizing the various permutations of RNA: poly-U was relatively straightforward to produce, but what about AGG? A lot of ingenious chemistry went into solving these problems, much of it done at the University of Wisconsin by Gobind Khorana. By 1966, what each of the sixty-four codons specifies (in other words, the genetic code itself) had been established; Khorana and Nirenberg received the Nobel Prize for Physiology or Medicine in 1968.

Let's now put the whole story together and look at how a particular protein, hemoglobin, is produced.

Red blood cells are specialized as oxygen transporters: they use hemoglobin to transport oxygen from the lungs to the tissues where it is needed. Red blood cells are produced in the bone marrow by stem cells—at a rate of about two and a half million per second.

When the need arises to produce hemoglobin, the relevant segment of the

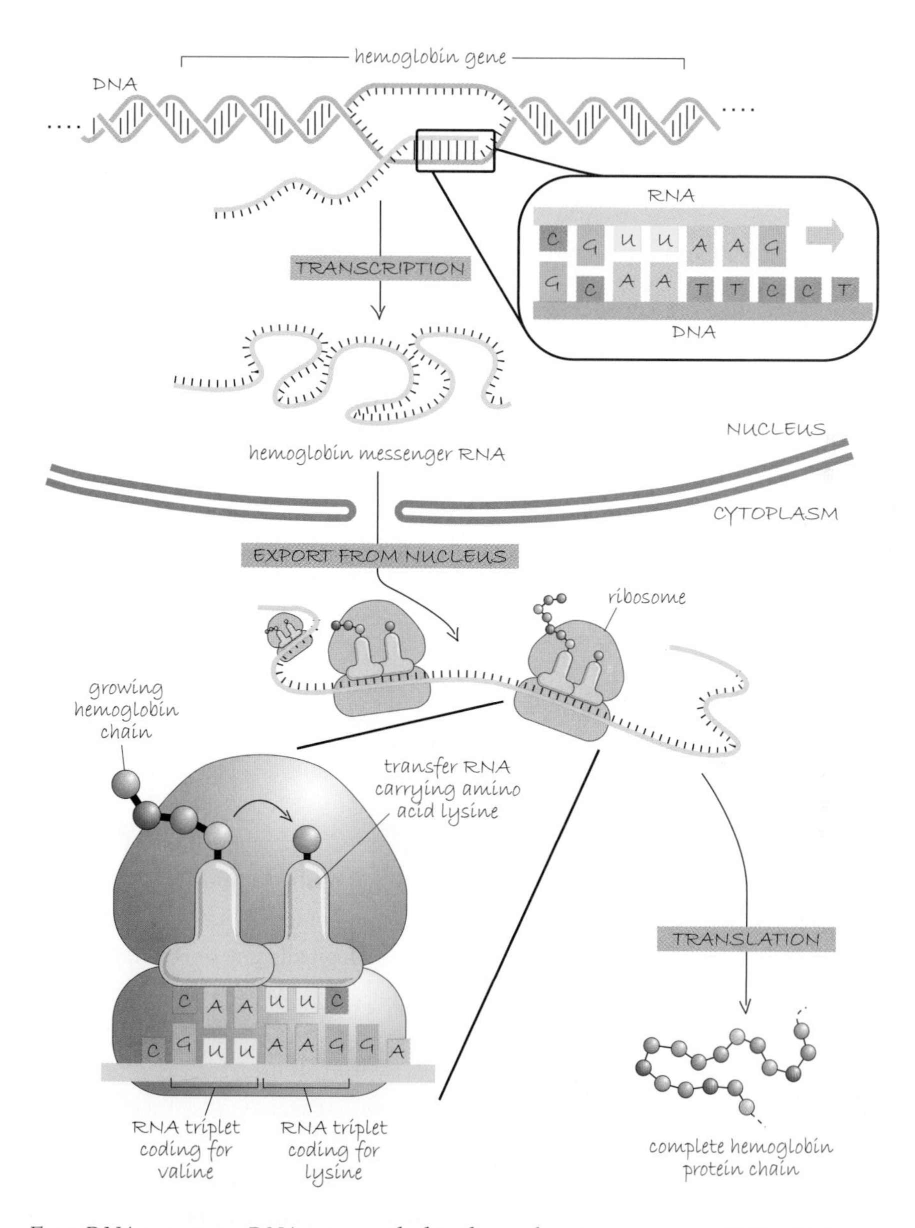

From DNA to protein. DNA is transcribed in the nucleus into messenger RNA, which is then exported to the cytoplasm for translation into protein. Translation occurs in ribosomes: transfer RNAs complementary to each base pair triplet codon in the messenger RNA deliver amino acids, which are bonded together to form a protein chain.

bone-marrow DNA—the hemoglobin gene—unzips just as DNA unzips when it is replicating. This time, instead of copying both strands, only one is copied or, to use the technical term, *transcribed;* and rather than a new strand of DNA, the product created with the help of the enzyme RNA polymerase is a new single strand of messenger RNA, which corresponds to the hemoglobin gene. The DNA from which the RNA has been derived now zips itself up again.

The messenger RNA is transported out of the nucleus and delivered to a ribosome, itself composed of RNA and proteins, where the information in the sequence of the messenger RNA will be used to generate a new protein molecule. This process is known as *translation.* Amino acids are delivered to the scene attached to transfer RNA. At one end of the transfer RNA is a particular triplet (in the case given in the diagram, CAA) that recognizes its opposite corresponding triplet in the messenger RNA, GUU. At its other end the transfer RNA is towing its matching amino acid, in this case valine. At the next triplet along the messenger RNA, because the DNA sequence is TTC (which specifies lysine), we have a lysine transfer RNA. All that remains now is to glue the two amino acids together biochemically. Do that 100 times, and you have a protein chain 100 amino acids long; the order of the amino acids has been specified by the order of As, Ts, Gs, and Cs in the DNA from which the messenger RNA was created. The two kinds of hemoglobin chains are 141 and 146 amino acids in length.

Proteins, however, are more than just linear chains of amino acids. Once the chain has been made, proteins fold into complex configurations, sometimes by themselves, sometimes assisted by "helper" molecules. It is only once they assume this configuration that they become biologically active. In the case of hemoglobin, it takes four chains, two of one kind and two of a slightly different kind, before the molecule is in business. And loaded into the center of each twisted chain is the key to oxygen transport, an iron atom.

It has been possible to use today's molecular biological tricks to go back and reconsider some of the classic examples of early genetics. For Mendel, the mechanism that caused some peas to be wrinkled and others round was mysterious; as far as he was concerned, these were merely characteristics that obeyed the laws of inheritance he had worked out. Now, however, we understand the difference in molecular detail.

In 1990, scientists in England found that wrinkled peas lack a certain enzyme involved in the processing of starch, the carbohydrate that is stored in seeds. It turns out that the gene for that enzyme in wrinkled-pea plants is non-functional owing to a mutation (in this case an intrusion of irrelevant DNA into the middle of the gene). Because wrinkled peas contain, as a result of this mutation, less starch and more sugar, they tend to lose more water as they are maturing. The outside seed coat of the pea, however, fails to shrink as the water escapes (and the volume of the pea decreases), and the result is the characteristic wrinkling—the contents being too little to fill out the coat.

Archibald Garrod's alkaptonuria has also entered the molecular era. In 1995, Spanish scientists working with fungi found a mutated gene that resulted in the accumulation of the same substance that Garrod had noted in the urine of alkaptonurics. The gene in question ordinarily produces an enzyme that turns out to be a basic feature of many living systems, and is present in humans. By comparing the sequence of the fungal gene to human sequences, it was possible to find the human gene, which encodes an enzyme called homogentisate dioxygenase. The next step was to compare the gene in normal individuals with the one in alkaptonurics. Lo and behold, the alkaptonurics' gene was nonfunctional, courtesy of single base pair mutations. Garrod's "inborn error in metabolism" is caused by a single difference in DNA sequence.

At the 1966 Cold Spring Harbor Symposium on the genetic code, there was a sense that we had done it all. The code was cracked, and we knew in outline how DNA exerted control of living processes through the proteins it specifies. Some of the old hands decided that it was time to move beyond the study of the gene per se. Francis Crick decided to move into neurobiology; never one to shy away from big problems, he was particularly interested in figuring out how the human brain works. Sydney Brenner turned to developmental biology, choosing to concentrate on a simple nematode worm in the belief that precisely so simple a creature would most readily permit scientists to unravel the connections between genes and development. Today, the worm, as it is known in the trade, is indeed the source of many of our insights into how organisms are put together. The worm's contribution was recognized by the Nobel Committee in 2002 when Brenner and two longstanding worm stalwarts, John Sulston at Cambridge and Bob Horvitz at MIT, were awarded the Nobel Prize in Physiology or Medicine.

Most of the early pioneers in the DNA game, however, chose to remain focused on the basic mechanisms of gene function. Why are some proteins much more abundant than others? Many genes are switched on only in specific cells or only at particular times in the life of a cell; how is that switching achieved? A muscle cell is hugely different from a liver cell, both in its function and in its appearance under the microscope. Changes in gene expression create this cellular diversity and differentiation: in essence, muscle cells and liver cells produce different sets of proteins. The simplest way to produce different proteins is to regulate which genes are transcribed in each cell. Thus some so-called housekeeping proteins—the ones essential for the functioning of the cell, such as those involved in the replication of DNA—are produced by all cells. Beyond that, particular genes are switched on at particular moments in particular cells to produce appropriate proteins. It is also possible to think of development—the process of growth from a single fertilized egg into a staggeringly complex adult human—as an enormous exercise in gene-switching: as tissues arise through development, so whole suites of genes must be switched on and off.

The first important advances in our understanding of how genes are switched on and off came from experiments in the 1960s by François Jacob and Jacques Monod at the Institut Pasteur in Paris. Monod had started slowly in science because, poor fellow, he was talented in so many fields that he had difficulty focusing. During the thirties, he spent time at Caltech's biology department under T. H. Morgan, father of fruit fly genetics, but not even daily exposure to Morgan's no-longer-so-boyish "boys" could turn Monod into a fruit fly convert. He preferred conducting Bach concerts at the university—which later offered him a job teaching undergraduate music appreciation—and in the lavish homes of local millionaires. Not until 1940 did he complete his Ph.D. at the Sorbonne in Paris, by which time he was already heavily involved in the French Resistance. In one of the few instances of biology's complicity in espionage, Monod

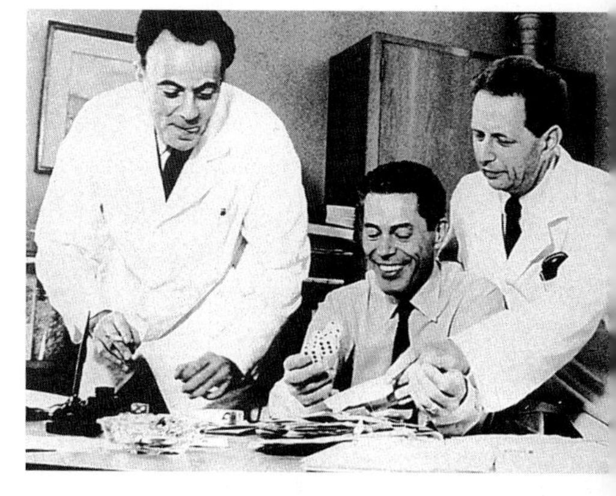

François Jacob, Jacques Monod, and André Lwoff

was able to conceal vital secret papers in the hollow leg bones of a giraffe skeleton on display outside his lab. As the war progressed, so did his importance to the Resistance (and with it his vulnerability to the Nazis). By D-day he was playing a major role in facilitating the Allied advance and harrying the German retreat.

Jacob too was involved in the war effort, having escaped to Britain and joined General de Gaulle's Free French Army. He served in North Africa and participated in the D-day landings. Shortly thereafter, he was nearly killed by a bomb; twenty pieces of shrapnel were removed, but he retains to this day another eighty. Because his arm was damaged, his injuries ended his ambition to be a surgeon, and, inspired like so many of our generation by Schrödinger's *What Is Life?,* he drifted toward biology. His attempts to join Monod's research group were, however, repeatedly rebuffed. But after seven or eight tries, by Jacob's own count, Monod's boss, the microbiologist André Lwoff, caved in in June 1950:

> Without giving me a chance to explain anew my wishes, my ignorance, my eagerness, [Lwoff] announced, "You know, we have discovered the induction of the prophage!" [i.e., how to activate bacteriophage DNA that has been incorporated into the host bacterium's DNA].
>
> I said, "Oh!" putting into it all the admiration I could and thinking to myself, "What the devil is a prophage?"
>
> Then he asked, "Would it interest you to work on phage?" I stammered out that that was exactly what I had hoped. "Good; come along on the first of September."

Jacob apparently went straight from the interview to a bookshop to find a dictionary that might tell him what he had just committed himself to.

Despite its inauspicious beginnings, the Jacob-Monod collaboration produced science of the very highest caliber. They tackled the gene-switching problem in *E. coli,* the familiar intestinal bacterium, focusing on its ability to make use of lactose, a kind of sugar. In order to digest lactose, the bacterium produces an enzyme called beta-galactosidase, which breaks the nutrient into two subunits, simpler sugars called galactose and glucose. When lactose is absent in the bacterial medium, the cell produces no beta-galactosidase; when, however, lactose is introduced, the cell starts to produce the enzyme. Concluding that it is the presence of lactose that induces the production of beta-galactosidase, Jacob and Monod set about discovering how that induction occurs.

In a series of elegant experiments, they found evidence of a "repressor" molecule that, in the absence of lactose, prevents the transcription of the beta-galactosidase gene. When, however, lactose is present, it binds to the repressor, thereby keeping it from blocking the transcription; thus the presence of lactose enables the transcription of the gene. In fact, Jacob and Monod found that lactose metabolism is coordinately controlled: it is not simply a matter of one gene being switched on or off at a given time. Other genes participate in digesting lactose, and the single repressor system serves to regulate all of them. While *E. coli* is a relatively simple system in which to investigate gene-switching, subsequent work on more complicated organisms, including humans, has revealed that the same basic principles apply across the board.

Jacob and Monod obtained their results by studying mutant strains of *E. coli.* They had no direct evidence of a repressor molecule: its existence was merely a logical inference from their solution to the genetic puzzle. Their ideas were not validated in the molecular realm until the late sixties, when Walter (Wally) Gilbert and Benno Müller-Hill at Harvard set out to isolate and analyze the repressor molecule itself. Jacob and Monod had only predicted its existence; Gilbert and Müller-Hill actually found it. Because the repressor is normally present only in tiny amounts, just a few molecules per cell, gathering a sample large enough to analyze proved technically challenging. But they got it in the end. At the same time, Mark Ptashne, working down the hall in another lab, managed to isolate and characterize another repressor molecule, this one in a bacteriophage gene-switching system. Repressor molecules turn out to be proteins that can bind to DNA. In the absence of lactose, then, that is exactly what the beta-galactosidase repressor does: by binding to a site on the *E. coli* DNA close to the point at which transcription of the beta-galactosidase gene starts, the repressor prevents the enzyme that produces messenger RNA from the gene from doing its job. When, however, lactose is introduced, that sugar binds to the repressor, preventing it from occupying the site on the DNA molecule close to the beta-galactosidase gene; transcription is then free to proceed.

The characterization of the repressor molecule completed a loop in our understanding of the molecular processes underpinning life. We knew that DNA produces protein via RNA; now we also knew that protein could interact directly with DNA, in the form of DNA-binding proteins, to regulate a gene's activity.

The discovery of the central role of RNA in the cell raised an interesting (and long-unanswered) question: why does the information in DNA need to go through an RNA intermediate before it can be translated into a polypeptide sequence? Shortly after the genetic code was worked out, Francis Crick proposed a solution to this paradox, suggesting that RNA predated DNA. He imagined RNA to have been the first genetic molecule, at a time when life was RNA-based: there would have been an "RNA world" prior the familiar "DNA world" of today (and of the past few billion years). Crick imagined that the different chemistry of RNA (based on its possession of the sugar ribose in its backbone, rather than the deoxyribose of DNA) might endow it with enzymatic properties that would permit it to catalyze its own self-replication.

Crick argued that DNA had to be a later development, probably in response to the relative instability of RNA molecules, which degrade and mutate much more easily than DNA molecules. If you want a good stable, long-term storage molecule for genetic data, then DNA is a much better bet than RNA.

Crick's ideas about an RNA world preceding the DNA one went largely unnoticed until 1983. That's when Tom Cech at the University of Colorado and Sidney Altman at Yale independently showed that RNA molecules do indeed have catalytic properties, a discovery that earned them the Nobel Prize in Chemistry in 1989. Even more compelling evidence of a pre-DNA RNA world came a decade later, when Harry Noller at the University of California, Santa Cruz, showed that the formation of peptide bonds, which link amino acids

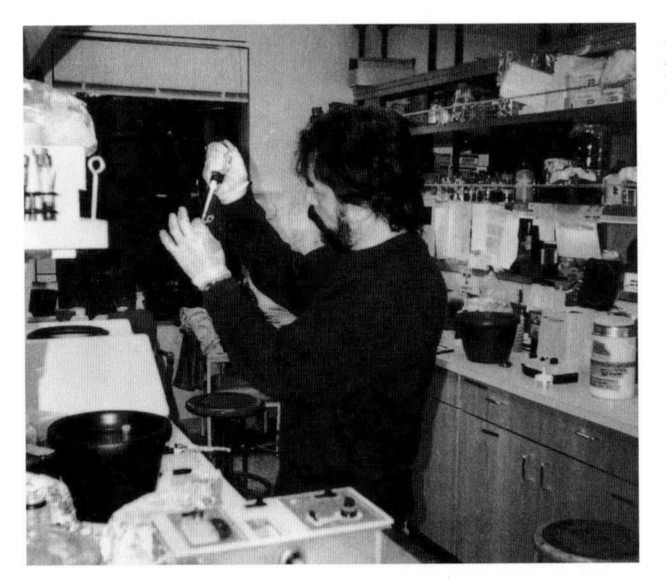

Harry Noller grappling with the ribosome

together in proteins, is not catalyzed by any of the sixty different proteins found associated with the ribosome, the site of protein synthesis. Instead, peptide bond formation is catalyzed by RNA. He arrived at this conclusion by stripping away all the proteins from the ribosome and finding that it was still capable of forming peptide bonds. Exquisitely detailed analysis of the 3-D structure of the ribosome by Noller and others shows why: the proteins are scattered over the surface, far from the scene of action at the heart of the ribosome.

These discoveries inadvertently resolved the chicken-and-egg problem of the origin of life. The prevailing assumption that the original life-form consisted of a DNA molecule posed an inescapable contradiction: DNA cannot assemble itself; it requires proteins to do so. Which came first? Proteins, which have no known means of duplicating information, or DNA, which can duplicate information but only in the presence of proteins? The problem was insoluble: you cannot, we thought, have DNA without proteins, and you cannot have proteins without DNA.

RNA, however, being a DNA equivalent (it can store and replicate genetic information) as well as a protein equivalent (it can catalyze critical chemical

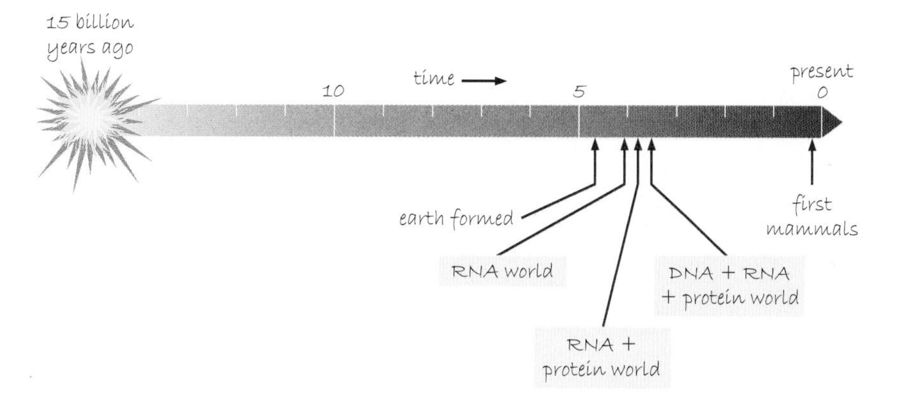

The evolution of life post–Big Bang. Exactly when life originated will likely never be known for sure but the first life-forms were probably entirely RNA based.

reactions) offers an answer. In fact, in the "RNA world" the chicken-and-egg problem simply disappears. RNA is both the chicken and the egg.

RNA is an evolutionary heirloom. Once natural selection has solved a problem, it tends to stick with that solution, in effect following the maxim "If it ain't broke, don't fix it." In other words, in the absence of selective pressure to change, cellular systems do not innovate and so bear many imprints of the evolutionary past. A process may be carried out in a certain way simply because it first evolved that way, not because that is absolutely the best and most efficient way.

Molecular biology had come a long way in its first twenty years after the discovery of the double helix. We understood the basic machinery of life, and we even had a grasp on how genes are regulated. But all we had been doing so far was observing; we were molecular naturalists for whom the rain forest was the cell—all we could do was describe what was there. The time had come to become proactive. Enough observation: we were beckoned by the prospect of intervention, of manipulating living things. The advent of recombinant DNA technologies, and with them the ability to tailor DNA molecules, would make all this possible.

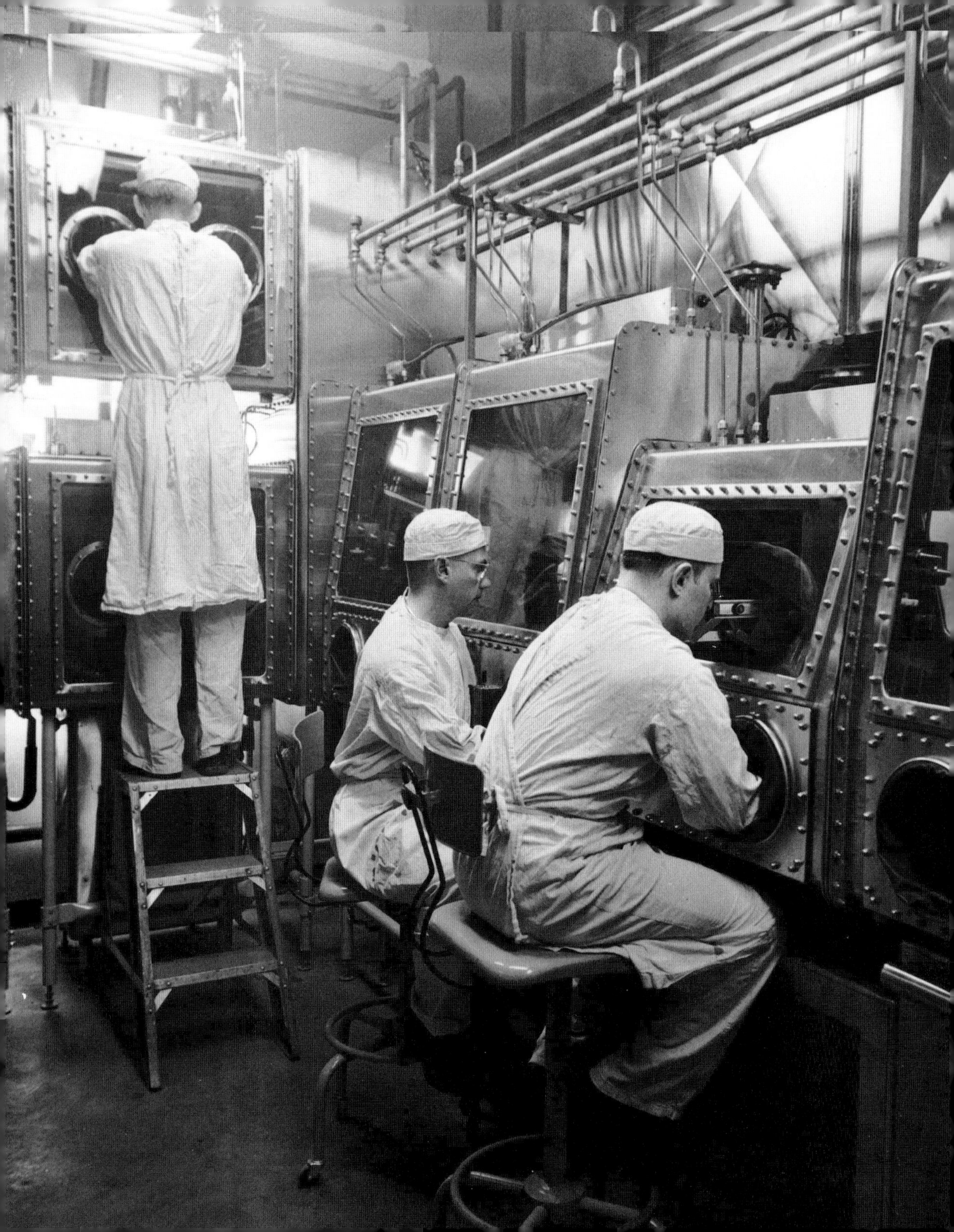

CHAPTER FOUR

PLAYING GOD: CUSTOMIZED DNA MOLECULES

DNA molecules are immensely long. Only one continuous DNA double helix is present in any given chromosome. Popular commentators like to evoke the vastness of these molecules through comparisons to the number of entries in the New York City phone book or the length of the River Danube. Such comparisons don't help me—I have no sense of how many phone numbers there are in New York City, and mention of the Danube more readily suggests a Strauss waltz than any sense of linear distance.

Except for the sex chromosomes, X and Y, the human chromosomes are numbered according to size. Chromosome 1 is the largest and chromosomes 21 and 22 are the smallest. In chromosome 1 there resides 8 percent of each cell's total DNA, about a quarter of a billion base pairs. Chromosomes 21 and 22 contain some 40 and 45 million base pairs respectively. Even the smallest DNA molecules, those from small viruses, have no fewer than several thousand base pairs.

The great size of DNA molecules posed a big problem in the early days of molecular biology. To come to grips with a particular gene—a particular stretch of DNA—we would have to devise some way of isolating it from all the rest of the DNA that sprawled around it in either direction. But it was not only a matter of isolating the gene; we also needed some way of "amplifying" it: obtaining a large enough sample of it to work with. In essence we needed a molecular editing system: a pair of molecular scissors that could cut the DNA text into manageable sections; a kind of molecular glue pot that would allow us to

A P4 laboratory, the ultrasafe facility required for biomedical research on lethal bugs like the Ebola virus or for developing biological weapons. During the late 1970s, scientists using genetic engineering methods to do research on human DNA were also required to use a P4 laboratory.

manipulate those pieces; and finally a molecular duplicating machine to amplify the pieces that we had cut out and isolated. We wanted to do the equivalent of what a word processor can now achieve: to cut, paste, and copy DNA.

Developing the basic tools to perform these procedures seemed a tall order even after we cracked the genetic code. A number of discoveries made in the late sixties and early seventies, however, serendipitously came together in 1973 to give us so-called "recombinant DNA" technology—the capacity to edit DNA. This was no ordinary advance in lab techniques. Scientists were suddenly able to tailor DNA molecules, creating ones that had never before been seen in nature. We could "play God" with the molecular underpinning of all of life. This was an unsettling idea to many people. Jeremy Rifkin, an alarmist for whom every new genetic technology has about it the whiff of Dr. Frankenstein's monster, had it right when he remarked that recombinant DNA "rivaled the importance of the discovery of fire itself."

Arthur Kornberg was the first to "make life" in a test tube. In the 1950s, as we have seen, he discovered DNA polymerase, the enzyme that replicates DNA through the formation of a complementary copy from an unzipped "parent" strand. Later he would work with a form of viral DNA; he was ultimately able to induce the replication of all of the virus's 5,300 base pairs of DNA. But the product was not "alive"; though identical in DNA sequence to its parent, it was biologically inert. Something was missing. The missing ingredient would remain a mystery until 1967, when Martin Gellert at the National Institutes of Health and Bob Lehman at Stanford simultaneously identified it. This enzyme was named "ligase." Ligase made it possible to "glue" the ends of DNA molecules together.

Kornberg could replicate the viral DNA using DNA polymerase and, by adding ligase, join the two ends together so that the entire molecule formed a continuous loop, just as it did in the original virus. Now the "artificial" viral DNA behaved exactly as the natural one did: the virus normally multiplies in *E. coli,* and Kornberg's test-tube DNA molecule did just that. Using just a couple of enzymes, some basic chemical ingredients, and viral DNA from which to make the copy, Kornberg had made a biologically active molecule. The media reported that he had created life in a test tube, inspiring President Lyndon Johnson to hail the breakthrough as an "awesome achievement."

The contributions of Werner Arber in the 1960s to the development of recombinant DNA technology were less expected. Arber, a Swiss biochemist, was interested not in grand questions about the molecular basis of life but in a puzzling aspect of the natural history of viruses. He studied the process whereby some viral DNAs are broken down after insertion into bacterial host cells. Some, but not all (otherwise viruses could not reproduce), host cells recognized certain viral DNAs as foreign, and selectively attacked them. But how—and why? All DNA throughout the natural world is the same basic molecule, whether found in bacteria, viruses, plants, or animals. What kept the bacteria from attacking their own DNA even as they went after the virus's?

The first answer came from Arber's discovery of a new group of DNA-degrading enzymes, restriction enzymes. Their presence in bacterial cells *restricts* viral growth by cutting foreign DNA. This DNA-cutting is a sequence-specific reaction: a given enzyme will cut DNA only when it recognizes a particular sequence. *Eco*R1, one of the first restriction enzymes to be discovered, recognizes and cuts the specific sequence of bases GAATTC.

But why is it that bacteria do not end up cutting up their own DNA in every place where the sequence GAATTC appears? Here Arber made a second big discovery. While making the restriction enzyme that targets specific sequences, the bacterium also produces a second enzyme that chemically modifies those very same sequences in its own DNA wherever they may occur.* Modified GAATTC sequences present in the bacterial DNA will pass unrecognized by *Eco*R1, even as the enzyme goes its marauding way, snipping the sequence wherever it occurs in the viral DNA.

The next ingredient of the recombinant DNA revolution emerged from studies of antibiotic resistance in bacteria. During the sixties, it was discovered that many bacteria developed resistance to an antibiotic not in the standard way (through a mutation in the bacterial genome) but by the import of an otherwise extraneous piece of DNA, called a "plasmid." Plasmids are small loops of DNA that live within bacteria and are replicated and passed on, along with the rest of

*The enzyme achieves this chemical modification by adding methyl groups, CH_3, to the bases.

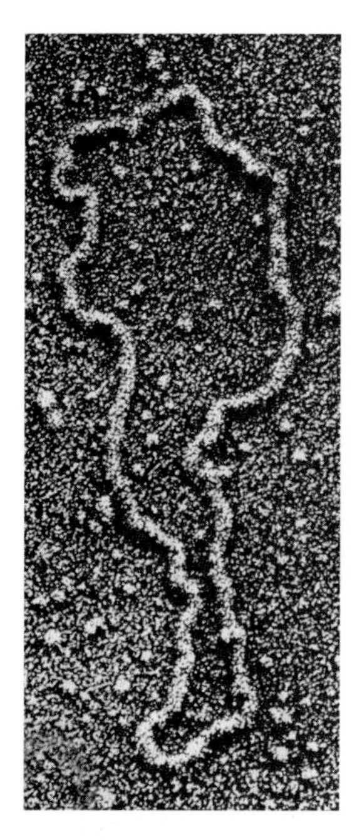

A plasmid as viewed by the electron microscope

the bacterial genome, during cell division. Under certain circumstances plasmids may also be passed from bacterium to bacterium, allowing the recipient instantly to acquire a whole cassette of genetic information it did not receive "at birth." That information often encompasses the genes conferring antibiotic resistance. Natural selection imposed by antibiotics favors those bacterial cells that have the resistance factor (the plasmid) on board.

Stanley Cohen, at Stanford University, was a plasmid pioneer. Thanks to the encouragement of his high-school biology teacher, Cohen opted for a medical career. Upon graduation from medical school, his plans to practice internal medicine were shelved when the prospect of being drafted as an army doctor inspired him to accept a research position at the National Institutes of Health. He soon found that he preferred research over practicing medicine. His big breakthrough came in 1971, when he devised a method to induce *E. coli* bacterial cells to import plasmids from outside the cell. Cohen was, in effect, "transforming" the *E. coli* as Fred Griffith, forty years before, had converted strains of nonlethal pneumonia bacteria into lethal ones through the uptake of DNA. In Cohen's case, however, it was the plasmid, with its antibiotic resistance genes, that was taken up by a strain that had previously been susceptible to the antibiotic. The strain would remain resistant to the antibiotic over subsequent generations, with copies of the plasmid DNA passed along intact during every cell division.

By the early seventies, all the ingredients to make recombinant DNA were in place. First we could cut DNA molecules using restriction enzymes and isolate the sequences (genes) we were interested in; then, using ligase, we could "glue" that sequence into a plasmid (which would thus serve as a kind of floppy disk containing our desired sequence); finally, we could copy our piece of DNA by inserting that same plasmid floppy into a bacterial cell. Ordinary bacterial cell division would take care of replicating the plasmid with our piece of DNA just as it would the cell's own inherited genetic materials. Thus, starting with a single plasmid transplanted into a single bacterial cell, bacterial reproduction could produce enormous quantities of our selected DNA sequence. As we let that cell reproduce and reproduce, ultimately to grow into a vast bacterial colony consisting of billions of bacteria, we would be simultaneously creating billions of copies of our piece of DNA. The colony was thus our DNA factory.

The three components—cutting, pasting, and copying—came together in November 1972, in Honolulu. The occasion was a conference on plasmids. Herb Boyer, a newly tenured young professor at the University of California, San Francisco, was there, and, not surprisingly, so was Stanley Cohen, first among plasmid pioneers. Boyer, like Cohen, was an East Coast boy. A former high-school varsity lineman from western Pennsylvania, Boyer was perhaps fortunate that his football coach was also his science teacher. Like Cohen, he would be part of a new generation of scientists who were reared on the double helix. His enthusiasm for DNA even inspired him to name his Siamese cats Watson and Crick. No one, certainly not the coach, was surprised when after college he took up graduate work in bacterial genetics.

Herb Boyer and Stanley Cohen, the world's first genetic engineers

Though Boyer and Cohen both now worked in the San Francisco Bay Area, they had not met before the Hawaii conference. Boyer was already an expert in restriction enzymes in an era when hardly anyone had even heard of them: it was he and his colleagues who had recently figured out the sequence of the cut site of the *Eco*R1 enzyme. Boyer and Cohen soon realized that between them they had the skills to push molecular biology to a whole new level, the world of cut, paste, and copy. In a deli near Waikiki, they set about late one evening dreaming up the birth of recombinant DNA technology, jotting their ideas down on napkins. That visionary mapping of the future has been described as "from corned beef to cloning."

Within a few months, Boyer's lab in San Francisco and Cohen's forty miles to the south in Palo Alto were collaborating. Naturally Boyer's carried out the restriction enzyme work and Cohen's the plasmid procedures. Fortuitously a technician in Cohen's lab, Annie Chang, lived in San Francisco and was able to ferry the precious cargo of experiments in progress between the two sites. The first experiment intended to make a hybrid, "a recombinant," of two different plasmids, each of which was known to confer resistance to a particular antibiotic. On one plasmid there was a gene, a stretch of DNA, for resistance to tetracycline, and on the other a gene for resistance to kanamycin. (Initially, as we might expect, bacteria carrying the first type of plasmid were killed by kanamycin while those with the second were killed by tetracycline.) The goal was to make a single "super-plasmid" that would confer resistance to both.

First, the two types of unaltered plasmid were snipped with restriction enzymes. Next the plasmids were mixed in the same test tube and ligase added to prompt the snipped ends to glue themselves together. For some molecules in the mix, the ligase would merely cause a snipped plasmid to make itself whole again—the two ends of the same plasmid would have been glued together. Sometimes, however, the ligase would cause a snipped plasmid to incorporate pieces of DNA from the other type of plasmid, thus yielding the desired hybrid. With this accomplished, the next step was to transplant all the plasmids into bacteria by using Cohen's plasmid-importing tricks. Colonies thus generated were then cultured on plates coated with both tetracycline and kanamycin. Plasmids that had simply re-formed would still confer resistance to only one of the antibiotics; bacteria carrying such plasmids would therefore not survive on the double-antibiotic medium. The only bacteria to survive were those with recombinant plasmids—those that had reassembled themselves from the two kinds of DNA present, the one coding for tetracycline resistance *and* the one coding for resistance to kanamycin.

The next challenge lay in creating a hybrid plasmid using DNA from a completely different sort of organism—a human being, for example. An early successful experiment involved putting a gene from the African clawed toad into an *E. coli* plasmid and transplanting that into bacteria. Every time cells in the bacterial colony divided, they duplicated the inserted segment of toad DNA. We had, in the rather confusing terminology of molecular biology, "cloned" the toad DNA.* Mammal DNA, too, proved eminently clonable. This is not terribly sur-

*"Cloning" is the term applied to producing multiple identical pieces of a piece of DNA inserted into a bacterial cell. The term is confusingly also applied to the cloning of whole animals, most notably Dolly the sheep. In the first type we are copying just a piece of DNA; in the other, we are copying an entire genome.

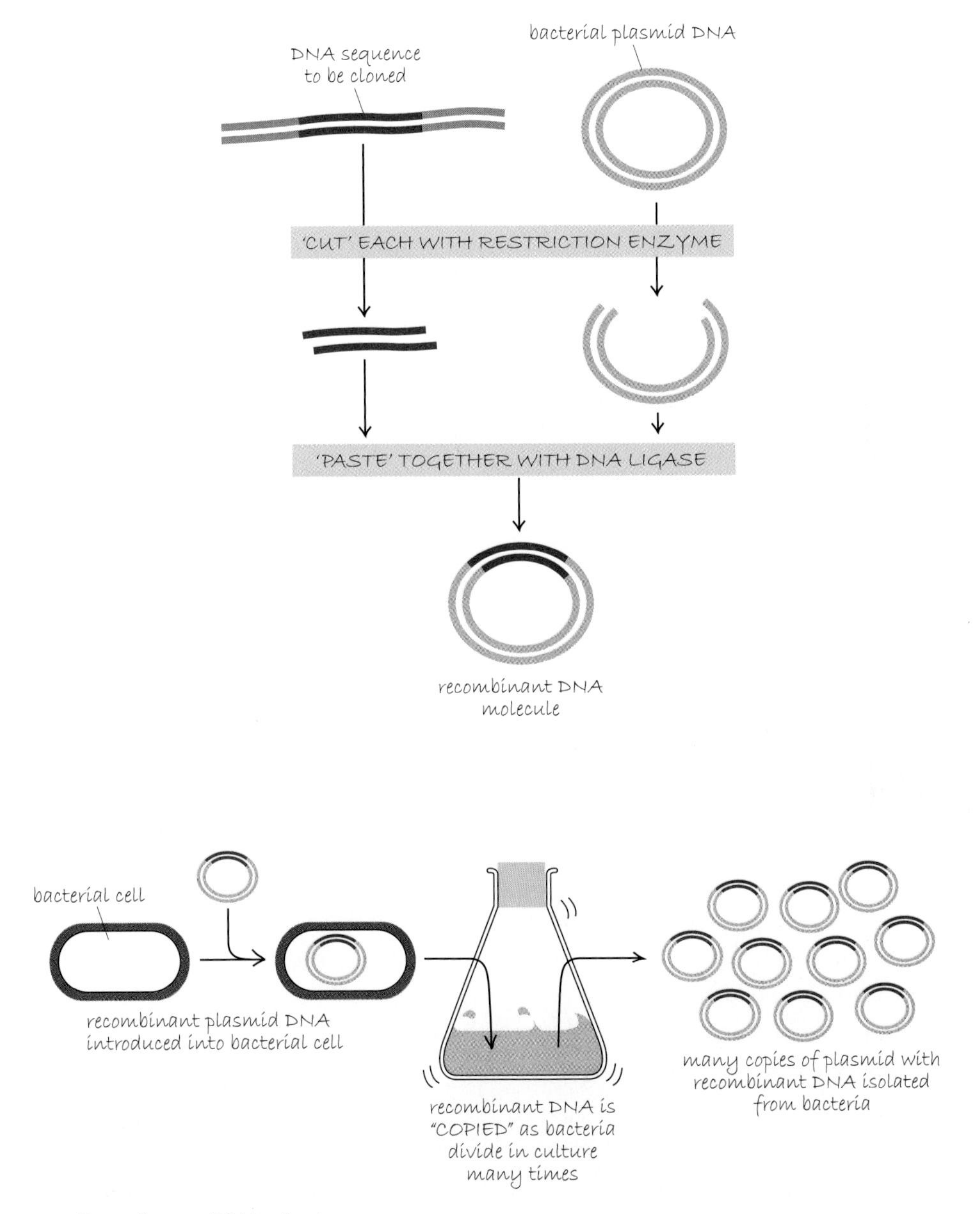

Recombinant DNA: cloning a gene

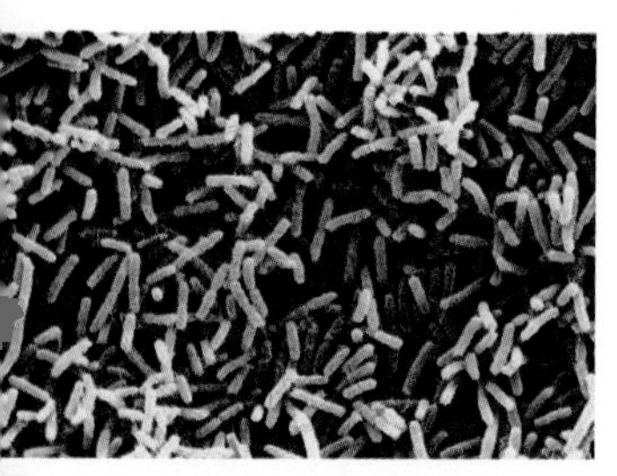

The gut microbe E. coli. *Should you care to look, about 10 million of these can be found in every gram of human feces.*

prising, in retrospect: a piece of DNA after all is finally still DNA, its chemical properties the same irrespective of its source. It was soon clear that Cohen and Boyer's protocols for cloning fragments of plasmid DNA would work just fine with DNA from any and every creature.

Phase 2 of the molecular biology revolution was thus under way. In phase 1 we aimed to describe how DNA works in the cell; now, with recombinant DNA,* we had the tools to intervene, to manipulate DNA. The stage was set for rapid progress, as we spied the chance to "play God." It was intoxicating: the extraordinary potential for delving deep into the mysteries of life and the opportunities for making real progress in the fight against diseases like cancer. But while Cohen and Boyer may indeed have opened our eyes to extraordinary scientific vistas, had they also opened a Pandora's box? Were there undiscovered perils in molecular cloning? Should we go on cheerfully inserting pieces of human DNA into *E. coli,* a species predominant in the microbial jungle in our guts? What if the altered forms should find their way into our bodies? In short, could we in good conscience simply turn a deaf ear to the cry of the alarmists, that we were creating bacterial Frankensteins?

In 1961 a monkey virus called SV40 ("SV" stands for "simian virus") was isolated from rhesus monkey kidneys being used for the preparation of polio vaccine. Although the virus was believed to have no effect on the monkeys in which it naturally occurs, experiments soon showed that it could cause cancer in rodents and, under certain laboratory conditions, even in human cells. Because the polio vaccination program had, since its inception in 1955, infected millions of American children with the virus, this discovery was alarming indeed. Had the polio prevention program inadvertently condemned a gen-

*The term "recombinant DNA" may present a little confusion in light of our encounter with "recombination" in the context of classical genetics. In Mendelian genetics, recombination involved the breaking and re-forming of chromosomes, with the result of a "mixing and matching" of chromosomal segments. In the molecular version, "mixing and matching" occurs on a much smaller scale, *recombining* two stretches of DNA into a single composite molecule.

Paul Berg with his viral Honda

eration to cancer? The answer, fortunately, seems to be "no"; no epidemic of cancer has resulted, and SV40 seems to be no more pernicious in living humans than it is in monkeys. Nevertheless, even as SV40 was becoming a fixture in molecular biology laboratories, there remained doubts about its safety. I was particularly concerned since I was by this time head of the Cold Spring Harbor Laboratory, where growing ranks of young scientists were working with SV40 to probe the genetic basis of cancer.

Meanwhile, at Stanford University Medical School, Paul Berg was more excited by the promise than by the dangers of SV40; he foresaw the possibility of using the virus to introduce pieces of DNA—foreign genes—into mammalian cells. The virus would work as a molecular delivery system in mammals, just as plasmids had been put to work in bacteria by Stanley Cohen. But whereas Cohen used bacteria essentially as copy machines, which could amplify up a particular piece of DNA, Berg saw in SV40 a means to introduce corrective genes into the victims of genetic disease. Berg was ahead of his time. He aspired to carry out what today is called gene therapy: introducing new genetic material into a living person to compensate for inherited genetic flaws.

Berg had come to Stanford as a junior professor in 1959 as part of the package deal that also brought the more eminent Arthur Kornberg there from Washington University in St. Louis. In fact, Berg's connections to Kornberg can be traced all the way back to their common birthplace of Brooklyn, New York, where each in his time was to pass through the same high-school science club run by a Miss Sophie Wolfe. Berg recalled: "She made science fun, she made us share ideas." It was an understatement really: Miss Wolfe's science club at Abraham Lincoln High School would produce three Nobel laureates—Kornberg (1959), Berg (1980), and the crystallographer Jerome Karle (1985)—all of whom have paid tribute to her influence.

While Cohen and Boyer, and by now others, were ironing out the details of how to cut and paste DNA molecules, Berg planned a truly bold experiment: he would see whether SV40, implanted with a piece of DNA not its own, could be made to transport that foreign gene into an animal cell. For convenience he would use as the source of his non-SV40 DNA a readily available bacterial virus, a bacteriophage. The aim was to see whether a composite molecule consisting of SV40 DNA and the bacteriophage DNA could successfully invade an animal cell. If it could, as Berg hoped, then the possibility existed that he could ultimately use this system to insert useful genes into human cells.

At Cold Spring Harbor Laboratory in the summer of 1971, a graduate student of Berg's gave a presentation explaining the planned experiment. One scientist in the audience was alarmed enough to phone Berg straightaway. What if, he asked, things happened to work in reverse? In other words, what if the SV40 virus, rather than taking up the viral DNA and then inserting it into the animal cell, was itself manipulated by the bacteriophage DNA, which might cause the SV40 DNA to be inserted into, say, an *E. coli* bacterial cell? It was not an unrealistic scenario: after all, that is precisely what many bacteriophages are programmed to do—to insert their DNA into bacterial cells. Since *E. coli* is both ubiquitous and intimately associated with humans, as the major component of our gut flora, Berg's well-meaning experiment might result in dangerous colonies of *E. coli* carrying SV40 monkey virus, a potential cancer agent. Berg heeded his colleague's misgivings, though he did not share them: he decided to postpone the experiments until more could be learned about SV40's potential to cause human cancer.

Biohazard anxieties followed hard on the heels of the news of Boyer and Cohen's success with their recombinant DNA procedures. At a scientific conference on nucleic acids in New Hampshire in the summer of 1973, a majority voted to petition the National Academy of Sciences to investigate without delay the dangers of the new technology. A year later a committee appointed by the National Academy and chaired by Paul Berg published its conclusions in a letter to the journal *Science*. I myself signed the letter, as did many of the others—including Cohen and Boyer—who were most active in the relevant research. In what has since come to be known as the "Moratorium Letter" we called upon "scientists throughout the world" to suspend voluntarily all recombinant studies "until the potential hazards of such recombinant DNA molecules have been better evaluated or until adequate methods are developed for preventing their spread." An important element of this statement was the admission that "our concern is based on judgements of potential rather than demonstrated risk since there are few experimental data on the hazards of such DNA molecules."

All too soon, however, I found myself feeling deeply frustrated and regretful of my involvement in the Moratorium Letter. Molecular cloning had the obvious potential to do a fantastic amount of good in the world, but now, having worked so hard and arrived at the brink of a biological revolution, here we were conspiring to draw back. It was a confusing moment. As Michael Rogers wrote in his 1975 report on the subject for *Rolling Stone*, "The molecular biologists had clearly reached the edge of an experimental precipice that may ultimately prove equal to that faced by nuclear physicists in the years prior to the atom bomb." Were we being prudent or chickenhearted? I couldn't quite tell yet, but I was beginning to feel it was the latter.

The "Pandora's Box Congress": that's how Rogers described the February 1975 meeting of 140 scientists from around the world at the Asilomar conference center in Pacific Grove, California. The agenda was to determine once and for all whether recombinant DNA really held more peril than promise. Should the moratorium be permanent? Should we press ahead regardless of potential risk, or wait for the development of certain safeguards? As chair of the organizing committee, Paul Berg was also nominal head of the conference, and so had the almost impossible task of drafting a consensus statement by the end of the meeting.

The press was there, scratching its collective head as scientists bandied about the latest jargon. The lawyers were there, too, just to remind us that there were also legal issues to be addressed: for example, would I, as head of a lab doing recombinant research, be liable if a technician of mine developed cancer? As to the scientists, they were by nature and training averse to hazarding predictions in the absence of knowledge; they rightly suspected that it would be impossible to reach a unanimous decision. Perhaps Berg was equally doubtful; in any case, he opted for freedom of expression over firm leadership from the chair. The resulting debate was therefore something of a free-for-all, with the proceedings not infrequently derailed by some speaker intent only on rambling irrelevantly and at length about the important work going on in his or her lab.

Debating DNA: Maxine Singer, Norton Zinder, Sydney Brenner, and Paul Berg grapple with the issues during the Asilomar conference.

Opinions ranged wildly, from the timid—"prolong the moratorium"—to the gung ho—"the moratorium be damned, let's get on with the science." I was definitely on the latter end of the spectrum. I now felt that it was more irresponsible to defer research on the basis of unknown and unquantified dangers. There were desperately sick people out there, people with cancer or cystic fibrosis—what gave us the right to deny them perhaps their only hope?

Sydney Brenner, then based in the United Kingdom, at Cambridge, offered one of the very few pieces of relevant data. He had collected colonies of the *E. coli* strain known as K-12, the favorite bacterial workhorse for this kind of molecular cloning research. Particular rare strains of *E. coli* occasionally cause outbreaks of food poisoning, but in fact the vast majority of *E. coli* strains are harmless, and Brenner assumed that K-12 was no exception. What interested him was not his own health but K-12's: could it survive outside the laboratory? He stirred the microbes into a glass of milk (they were rather unpalatable served up straight), and went on to quaff the vile mixture. He monitored what came out the other end to see whether any K-12 cells had managed to colonize his intestine. His finding was negative, suggesting that K-12, despite thriving in a petri dish, was not viable in the "natural" world. Still, others questioned the inference: even if the K-12 bacteria were themselves unable to survive, this was no proof they could not exchange plasmids—or other genetic information—

with strains that could live perfectly well in our guts. Thus "genetically engineered" genes could still enter the population of intestine-dwelling bacteria. Brenner then championed the idea that we should develop a K-12 strain that was without question incapable of living outside the laboratory. We could do this by a genetic alteration that would ensure the strain could grow only when supplied with specialized nutrients. And of course we would specify a set of nutrients that would never be available in the natural world; the full complement of nutrients would occur together only in the lab. A K-12 thus modified would be a "safe" bacterium, viable in our controlled research setting, but doomed in the real world.

With Brenner's urging, this middle-ground proposal carried the day. There was plenty of grumbling from both extremes, of course, but the conference ended with coherent recommendations allowing research to continue on disabled, non-disease-causing bacteria and mandating expensive containment facilities for work involving the DNA of mammals. These recommendations would form the basis for a set of guidelines issued a year later by the National Institutes of Health.

I departed feeling despondent, isolated from most of my peers. Stanley Cohen and Herb Boyer found the occasion disheartening as well; they believed, as I did, that many of our colleagues had compromised their better judgment as scientists just to be seen by the assembled press as "good guys" (and not as potential Dr. Frankensteins). In fact, the vast majority had never worked with disease-causing organisms and little understood the implications of the research restrictions they wanted to impose on those of us who did. I was irked by the arbitrariness of much of what had been agreed: DNA from cold-blooded vertebrates was, for instance, deemed acceptable, while mammalian DNA was ruled off-limits for most scientists. Apparently it was safe to work with DNA from a toad but not with DNA from a mouse. Dumbstruck by such nonsense, I offered up a bit of my own: didn't everyone know that toads cause warts? But my facetious objections were in vain.

The guidelines led many participants in the Asilomar conference to expect clear sailing for research based on cloning in "safe bacteria." But anyone who set off under such an impression very soon hit choppy seas. According to the logic peddled by the popular press, if scientists themselves saw cause for concern, then the public at large should *really* be alarmed. These were, after all, still the days, though waning, of the American counterculture. Both the Vietnam War and Richard Nixon's political career had only recently petered out; a suspicious public, ill-equipped to understand complexities that science itself was only beginning to fathom, was only too eager to swallow theories of evil conspiracies perpetrated by the Establishment. For our part, we scientists were quite surprised to see ourselves counted among this elite, to which we had never before imagined we belonged. Even Herb Boyer, the veritable model of a hippie scientist, would find himself named in the special Halloween issue of the *Berkeley Barb,* the Bay Area's underground paper, as one of the region's "ten biggest bogeymen," a distinction otherwise reserved for corrupt pols and union-busting capitalists.

My greatest fear was that this blooming public paranoia about molecular biology would result in draconian legislation. Having experimental dos and don'ts laid down for us in some cumbersome legalese could only be bad for science. Plans for experiments would have to be submitted to politically minded review panels, and the whole hopeless bureaucracy that comes with this kind of territory would take hold like the moths in Grandmother's closet. Meanwhile, our best attempts to assess the real risk potential of our work continued to be dogged by a complete lack of data and by the logical difficulty of proving a negative. No recombinant DNA catastrophe had ever occurred, but the press continued to outdo itself imagining "worst case scenarios." In his account of a meeting in Washington, D.C., in 1977, the biochemist Leon Heppel aptly summed up the absurdities scientists perceived in the controversy.

> I felt the way I would feel if I had been selected for an *ad hoc* committee convened by the Spanish Government to try to evaluate the risks assumed by Christopher Columbus and his sailors, a committee that was supposed to set up guidelines for what to do in case the earth was flat, how far the crew might safely venture to the earth's edge, etc.

Even withering irony, however, could little hinder those hell-bent on countering what they saw as science's Promethean hubris. One such crusader was

Hearings in Cambridge, Massachusetts, that resulted in a citywide ban on recombinant DNA research

Mark Ptashne

Tom Maniatis

BUILD
WISDOM
NOT
CONTAINMENT

He hath
shewed thee
O Man what is
good And what
doth the LORD
require of thee
but to do Justly
and to Love
MERCY and to
humbly
THY

Mayor Vellucci

Matt Meselson

MAYOR ALFRED VELLUCCI

Alfred Vellucci, the mayor of Cambridge, Massachusetts. Vellucci had earned his political chops championing the common man at the expense of his town's elite institutions of learning, namely, MIT and Harvard. The recombinant DNA tempest provided him with a political bonanza. A contemporary account captures nicely what was going on.

> In his cranberry doubleknit jacket and black pants, with his yellow-striped blue shirt struggling to contain a beer belly, right down to his crooked teeth and overstuffed pockets, Al Vellucci is the incarnation of middle-American frustration at these scientists, these technocrats, these smartass Harvard eggheads who think they've got the world by a string and wind up dropping it in a puddle of mud. And who winds up in the puddle? Not the eggheads. No, it's always Al Vellucci and the ordinary working people who are left alone to wipe themselves off.

Whence this heat? Scientists at Harvard had voiced a desire to build an on-campus containment facility for doing recombinant work in strict accordance with the new NIH guidelines. But, seeing his chance and backed by a left-wing Harvard-MIT cabal with its own anti-DNA agenda, Vellucci managed to push through a several months' ban on all recombinant DNA research in Cambridge. The result was a brief but pronounced local brain drain, as Harvard and MIT biologists headed off to less politically charged climes. Vellucci, meanwhile, began to enjoy his newfound prominence as society's scientific watchdog. In 1977 he would write to the president of the National Academy of Sciences:

> In today's edition of the *Boston Herald American,* a Hearst Publication, there are two reports which concern me greatly. In Dover, MA, a "strange, orange-eyed creature" was sighted and in Hollis, New Hampshire, a man and his two sons were confronted by a "hairy, nine foot creature."
>
> I would respectfully ask that your prestigious institution investigate these findings. I would hope as well that you might check to see whether or not these "strange creatures" (should they in fact exist), are in any way connected to recombinant DNA experiments taking place in the New England area.

Though much debated, attempts to enact national legislation regulating recombinant DNA experiments fortunately never came to fruition. Senator Ted Kennedy of Massachusetts entered the fray early on, holding a Senate hearing just a month after Asilomar. In 1976, he wrote President Ford to advise that the federal government should control industrial as well as academic DNA research. In March of '77, I testified before a hearing of the California state legislature. Governor Jerry Brown was in attendance, and so I had the occasion to advise him in person that it would be a mistake to consider any legislative action except in the event of unexplained illnesses among the scientists at Stanford. If those actually handling recombinant DNA remained perfectly healthy, the public would be better served if lawmakers focused on more evident dangers to public health, like bike riding.

As more and more experiments were performed, whether under NIH guidelines or under those imposed by regulators in other countries, it became more and more apparent that recombinant DNA procedures were not creating Frankenbugs (much less—*pace* Mr. Vellucci—"strange orange-eyed crea-

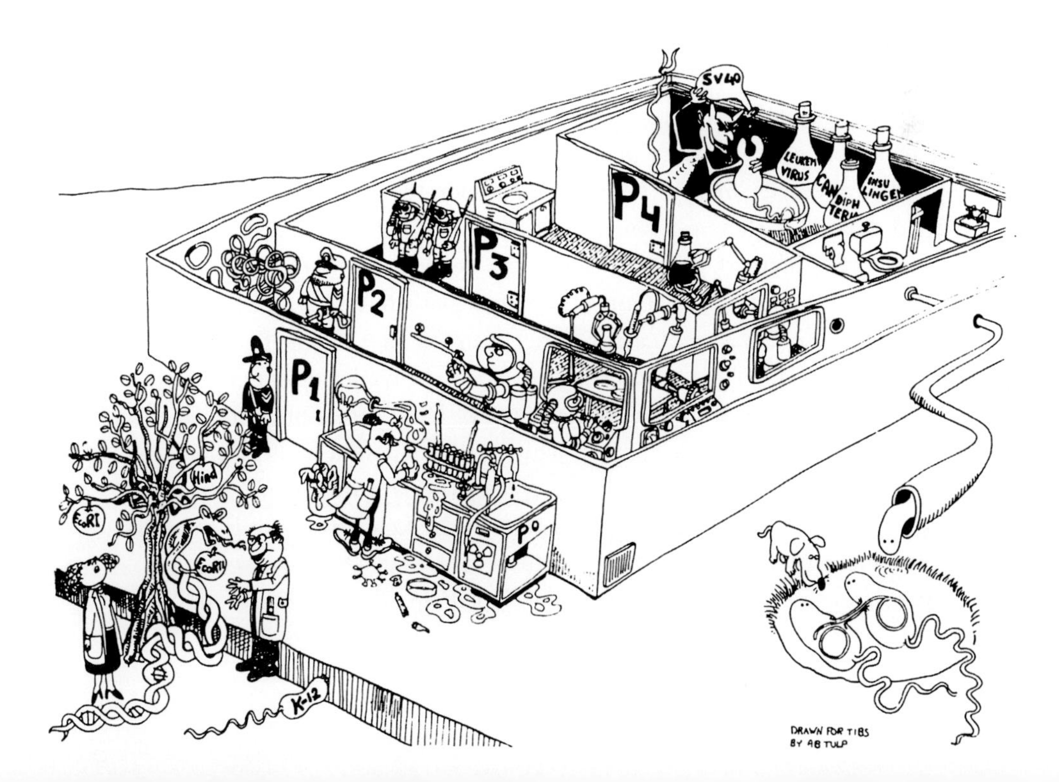

tures"). By 1978 I could write, "Compared to almost any other object that starts with the letter D, DNA is very safe indeed. Far better to worry about daggers, dynamite, dogs, dieldrin, dioxin, or drunken drivers than to draw up Rube Goldberg schemes on how our laboratory-made DNA will lead to the extinction of the human race."

Later that year, in Washington, D.C., the Recombinant DNA Advisory Committee (RAC) of the NIH proposed much less restrictive guidelines that would permit most recombinant work—including tumor virus DNA research—to go forward. And in 1979, Joseph Califano, Secretary of Health, Education, and Welfare, approved the changes, thus ending a period of pointless stagnation for mammalian cancer research.

In practical terms, the outcome of the Asilomar consensus was ultimately nothing more than five sad years of delay in important research, and five frustrating years of disruption in the careers of many young scientists.

As the 1970s ended, the issues raised by Cohen and Boyer's original experiments turned gradually into non-issues. We had been forced to take an unprofitable detour, but at least it showed that molecular scientists wanted to be socially responsible.

Molecular biology during the second half of the 1970s, however, was not completely derailed by politics; these years did in fact see a number of important advances, most of them building upon the still controversial Boyer-Cohen molecular cloning technology. The most significant breakthrough was the invention of methods for reading the sequence of DNA. Sequencing depends on having a large quantity of the particular stretch of DNA that you are interested in, so it was not feasible—except in the case of small viral DNA—until cloning technologies had been developed. As we have seen, cloning, in essence, involves inserting the desired piece of DNA into a plasmid, which is then itself inserted into a bacterium. The bacteria, allowed to divide and grow, will then produce a vast number of copies of the DNA fragment. Once harvested from the bacteria, this large quantity of the DNA fragment is then ripe for sequencing.

Two sequencing techniques were developed simultaneously, one by Wally

Gilbert in Cambridge, Massachusetts (Harvard), and the other by Fred Sanger in Cambridge, England. Gilbert's interest in sequencing DNA stemmed from his having isolated the repressor protein in the *E. coli* beta-galactosidase gene regulation system. As we have seen, he had shown that the repressor binds to the DNA close to the gene, preventing its transcription into RNA chains. Now he wanted to know the sequence of that DNA region. A fortuitous meeting with the brilliant Soviet chemist Andrei Mirzabekov suggested to Gilbert a way—using certain potent combinations of chemicals—to break DNA chains at just the desired, base-specific sites.

Wally Gilbert (top) and Fred Sanger, sequence kings

As a high-school senior in Washington, D.C., Gilbert used to cut class to read up on physics at the Library of Congress. He was then pursuing the Holy Grail of all high-school science prodigies: a prize in the Westinghouse Talent Search.* He duly won his prize in 1949. (Years later, in 1980, he would receive a call from the Swedish Academy in Stockholm, adding to the statistical evidence that winning the Westinghouse is one of the best predictors of a future Nobel.) Gilbert stuck with physics as an undergraduate and graduate student, and a year after I arrived at Harvard in 1956 he joined the physics faculty. But once I got him interested in my lab's work on RNA, he abandoned his field for mine. Thoughtful and unrelenting, Gilbert has ever since been at the forefront of molecular biology.

Of the two sequencing methods, however, it is Sanger's that has better withstood the test of time. Some of the DNA-breaking chemicals required by Gilbert's are difficult to work with; given half a chance, they will start breaking up the researcher's own DNA. Sanger's method, on the other hand, uses the

*In 1998, as the Old Economy gave way to the New, the honor was renamed the Intel Prize.

same enzyme that copies DNA naturally in cells, DNA polymerase. His trick involves making the copy out of base pairs that have been slightly altered. Instead of using only the normal "deoxy" bases (As, Ts, Gs, and Cs) found naturally in DNA (deoxyribonucleic acid), Sanger also added some so-called "dideoxy bases." Dideoxy bases have a peculiar property: DNA polymerase will happily incorporate them into the growing DNA chain (i.e., the copy being assembled as the complement of the template strand), but it cannot then add any further bases to the chain. In other words, the duplicate chain cannot be extended beyond a dideoxy base.

Imagine a template strand whose sequence is GGCCTAGTA. There are many, many copies of that strand in the experiment. Now imagine that the strand is being copied using DNA polymerase, in the presence of a mixture of normal A, T, G, and C plus some dideoxy A. The enzyme will copy along, adding first a C (to correspond to the initial G), then another C, then a G, and another G. But when the enzyme reaches the first T, there are two possibilities: either it can add a normal A to the growing chain, or it can add a dideoxy A. If it picks up a dideoxy A, then the strand can grow no further, and the result is a short chain that ends in a dideoxy A (ddA): CCGGddA. If it happens to add a normal A, however, then DNA polymerase can continue adding bases: T, C, etc. The next chance for a dideoxy "stop" of this kind will not come until the enzyme reaches the next T. Here again it may add either a normal A or a ddA. If it adds a ddA, the result is another truncated chain, though a slightly longer one: this chain has a sequence of CCGGATCddA. And so it goes every time the enzyme encounters a T (i.e., has occasion to add an A to the chain); if by chance it selects a normal A, the chain continues, but in the case of a ddA the chain terminates there.

Where does this leave us? At the end of this experiment, we have a whole slew of chains of varying lengths copied from the template DNA; what do they all have in common? They all end with a ddA.

Now, imagine the same process carried out for each of the other three bases: in the case of T, for instance, we use a mix of normal A, T, G, and C plus ddT; the resultant molecules will be either CCGGAddT or CCGGATCAddT.

Having staged the reaction all four ways—once with ddA, once with ddT, once with ddG, and once with ddC—we have four sets of DNA chains: one consists of chains ending in ddA, one with chains ending with ddT, and so on.

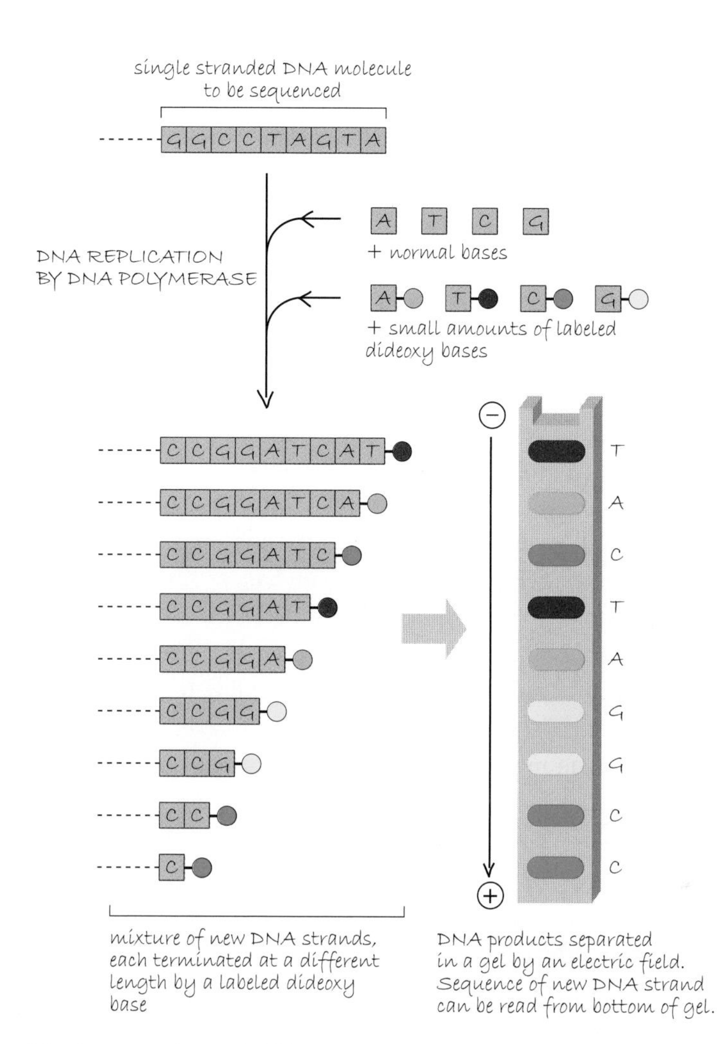

The Sanger method of DNA sequencing

Now if we could only sort all these mini-chains according to their respective, slightly varying lengths, we could infer the sequence. How? A moment, please. First, let's see how we could do the sorting. We can place all the DNA fragments on a plate full of a special gel, and place the plate of gel in an electric field. In the pull of the electric field the DNA molecules will be forced to migrate through the gel, and the speed with which a particular mini-chain will travel is a function of its size: short chains travel faster than long ones. Within a fixed interval of time, the smallest mini-chain, in our case a simple ddC, will travel furthest; the next smallest, CddC, will travel a slightly shorter distance;

and the next one, CCddG, a slightly shorter one still. Now Sanger's trick should be clear: by reading off the relative positions of all these mini-chains after a timed race through our gel, we can infer the sequence of our piece of DNA: first is a C, then another C, then a G, and so on.

In 1980, Sanger shared the Nobel Prize in Chemistry with Gilbert and with Paul Berg, who was recognized for his contribution to the development of the recombinant DNA technologies. (Inexplicably neither Stanley Cohen nor Herb Boyer has been so honored.)

For Sanger, this was his second Nobel.* He had received the chemistry prize in 1958 for inventing the method by which proteins are sequenced—that is, by which their amino acid sequence is determined—and applying it to human insulin. But there is absolutely no relation between Sanger's method for protein sequencing and the one he devised for sequencing DNA; neither technically nor imaginatively did the one give rise to the other. He invented both from scratch, and should perhaps be regarded as the presiding technical genius of the early history of molecular biology.

Sanger is not what you might expect of a double Nobel laureate. Born to a Quaker family, he became a socialist and was a conscientious objector during the Second World War. More improbably, he does not advertise his achievements, preferring to keep the evidence of his Nobel honors in storage: "You get a nice gold medal, which is in the bank. And you get a certificate, which is in the loft." He has even turned down a knighthood: "A knighthood makes you different, doesn't it? And I don't want to be different." Having retired, Sanger is content these days to tend his garden outside Cambridge, though he still makes the occasional self-effacing and cheerful appearance at the Sanger Centre, the genome-sequencing facility near Cambridge that opened in 1993.

Sequencing would confirm one of the most remarkable findings of the 1970s. We already knew that genes were linear chains of As, Ts, Gs, and

*As a double Nobelist, Sanger is in exalted company. Marie Curie received the prize in physics (1903) and then in chemistry (1911); John Bardeen received the physics prize twice, for the discovery of transistors (1956) and for superconductivity (1972); and Linus Pauling received the chemistry prize (1954) and the peace prize (1962).

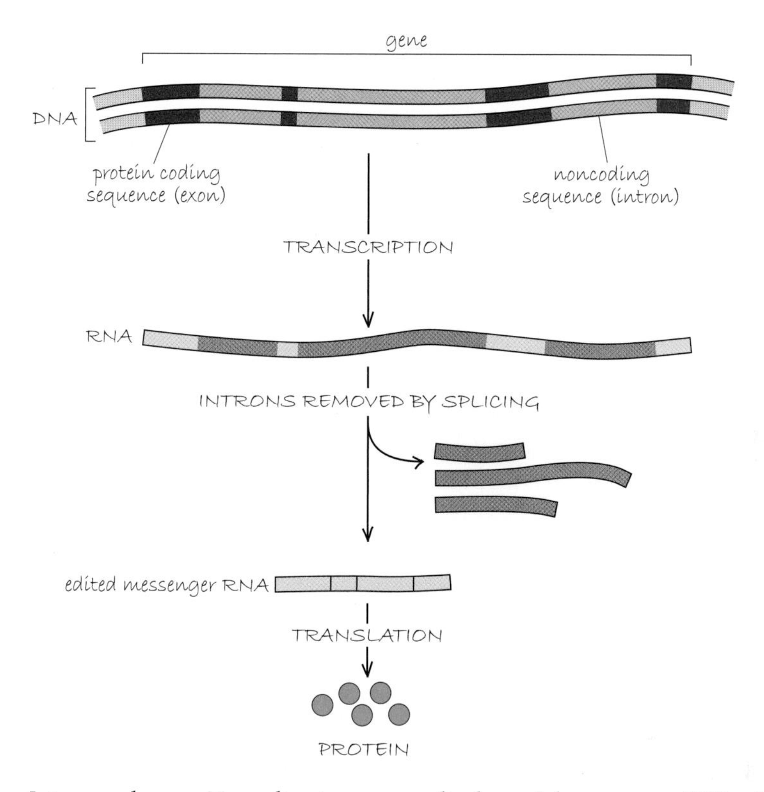

Introns and exons. Noncoding introns are edited out of the messenger RNA prior to protein production.

Cs, and that these bases were translated three at a time, in accordance with the genetic code, to create the linear chains of amino acids we call proteins. But remarkable research by Richard Roberts, Phil Sharp, and others revealed that, in many organisms, genes actually exist in pieces, with the vital coding DNA broken up by chunks of irrelevant DNA. Only once the messenger RNA has been transcribed is the mess sorted out by an "editing" process that eliminates the irrelevant parts. It would be as though this book contained occasional extraneous paragraphs, apparently tossed in at random, about baseball or the history of the Roman Empire. Wally Gilbert dubbed the intrusive sequences "introns" and the ones responsible for actual protein-coding (i.e., functionally part of the gene) he named "exons." It turns out that introns are principally a feature of sophisticated organisms; they do not appear in bacteria.

Some genes are extraordinarily intron-rich. For example, in humans, the gene for blood clotting factor VIII (which may be mutated in people with hemophilia) has twenty-five introns. Factor VIII is a large protein, some two thousand amino acids long, but the exons that code for it constitute a mere 4 percent of the total length of the gene. The remaining 96 percent of the gene is made up of introns.

Why, then, do introns exist? Obviously their presence vastly complicates cellular processes, since they always have to be edited out to form the messenger RNA; and that editing seems a tricky business, especially when you consider that a single error in excising an intron from the messenger RNA for, say, clotting factor VIII would likely result in a frameshift mutation that would render the resulting protein useless. One theory holds that these molecular intruders are merely vestigial, an evolutionary heirloom, left over from the early days of life on earth. Still it remains a much-debated issue how introns came to be and what if any use they may have in life's great code.

Once we became aware of the general nature of genes in eukaryotes (organisms whose cells contain a compartment, the nucleus, specialized for storing the genetic material; prokaryotes, such as bacteria, lack nuclei), a scientific gold rush was launched. Teams of eager scientists armed with the latest technology raced to be the first to isolate (clone) and characterize key genes. Among the earliest treasures to be found were genes in which mutations give rise to cancers in mammals. Once scientists had completed the DNA sequencing of several well-studied tumor viruses, SV40 for one, they could then pinpoint the exact cancer-causing genes. These genes were capable of transforming normal cells into cells with cancerlike properties, with for instance a propensity for the kind of uncontrolled growth and cell division that results in tumors. It was not long until molecular biologists began to isolate genes from human cancer cells, finally confirming that human cancer arises because of changes at the DNA level and not from simple nongenetic accidents of growth, as had been supposed. We found genes that accelerate or promote cancer growth and we found genes that slow or inhibit it. Like an automobile, a cell, it seems, needs both an accelerator and a brake to function properly.

The treasure hunt for genes took over molecular biology. In 1981, Cold Spring Harbor Laboratory started an advanced summer course that taught gene-cloning techniques. *Molecular Cloning,* the lab manual that was developed out of this course, sold more than eighty thousand copies over the following three years. The first phase of the DNA revolution (1953–72)—the early excitement that grew out of the discovery of the double helix and led to the genetic code—eventually involved some three thousand scientists. But the second phase, inaugurated by recombinant DNA and DNA sequencing technologies, would see those ranks swell a hundredfold in little more than a decade.

Part of this expansion reflected the birth of a brand new industry: biotechnology. After 1975, DNA was no longer solely the concern of biologists trying to understand the molecular underpinnings of life. The molecule moved beyond the academic cloisters inhabited by white-coated scientists into a very different world populated largely by men in silk ties and sharp suits. The name Francis Crick had given his home in Cambridge, the Golden Helix, now had a whole new meaning.

MARCH 9, 1981 $1.50

TIME

Prince Charles Picks a Bride

Shaping Life in the Lab

The Boom In Genetic Engineering

Genentech's Herbert Boyer

A. MAGEE

0 10090 10
724404

CHAPTER FIVE

DNA, DOLLARS, AND DRUGS: BIOTECHNOLOGY

Herb Boyer has a way with meetings. We have seen how his 1972 chat with Stanley Cohen in a Waikiki deli led to the experiments that made recombinant DNA a reality. In 1976, lightning struck a second time: the scene was San Francisco, the meeting was with a venture capitalist named Bob Swanson, and the result was a whole new industry that would come to be called biotechnology.

Only twenty-seven when he took the initiative and contacted Boyer, Swanson was already making a name for himself in high-stakes finance. He was looking for a new business opportunity, and with his background in science he sensed one in the newly minted technology of recombinant DNA. Trouble was, everyone Swanson spoke to told him that he was jumping the gun. Even Stanley Cohen suggested that commercial applications were at least several years away. As for Boyer himself, he disliked distractions, especially when they involved men in suits, who always look out of place in the jeans-and-T-shirt world of academic science. Somehow, though, Swanson cajoled him into sparing ten minutes of his time one Friday afternoon.

Ten minutes turned into several hours, and then several beers when the meeting was adjourned to nearby Churchill's Bar, where Swanson discovered he had succeeded in rousing a latent entrepreneur. It was in Derry Borough High School's 1954 yearbook that class president Boyer had first declared his ambition "to become a successful businessman."

The basic proposition was extraordinarily simple: find a way to use the

Time *magazine marks the birth of the biotechnology business (and looks forward to a royal wedding).*

Cohen-Boyer technology to produce proteins that are marketable. A gene for a "useful" protein—say, one with therapeutic value, such as human insulin—could be inserted into a bacterium, which in turn would start manufacturing the protein. Then it would just be a matter of scaling up production, from petri dishes in the laboratory to vast industrial-size vats, and harvesting the protein as it was produced. Simple in principle, but not so simple in practice. Nevertheless, Boyer and Swanson were optimistic: each plunked down $500 to form a partnership dedicated to exploiting the new technology. In April 1976 they formed the world's first biotech company. Swanson's suggestion that they call the firm "Her-Bob," a combination of their first names, was mercifully rejected by Boyer, who offered instead "Genentech," short for "genetic engineering technology."

Insulin was an obvious commercial first target for Genentech. Diabetics require regular injections of this protein since their bodies naturally produce either too little of it (Type II diabetes) or none at all (Type I). Before the discovery in 1921 of insulin's role in regulating blood-sugar levels, Type I diabetes was lethal. Since then, the production of insulin for use by diabetics has become a major industry. Because blood-sugar levels are regulated much the same way in all mammals, it is possible to use insulin from domestic animals, mainly pigs and cows. Pig and cow insulins differ slightly from the human version: pig insulin by 1 amino acid in the 51-amino-acid protein chain, and cow insulin by 3. These differences can occasionally cause adverse effects in patients; diabetics sometimes develop allergies to the "foreign" protein. The biotech way around these allergy problems would be to provide diabetics with the real McCoy, human insulin.

With an estimated 8 million diabetics in the United States, insulin promised a biotech gold mine. Boyer and Swanson, however, were not alone in recognizing its potential. A group of Boyer's colleagues at the University of California, San Francisco (UCSF), as well as Wally Gilbert at Harvard, had also realized that cloning human insulin would prove both scientifically and commercially valuable. In May 1978, the stakes were raised when Gilbert and several others from the United States and Europe formed their own company, Biogen. The contrasting origins of Biogen and Genentech show just how fast things were moving: Genentech was envisioned by a twenty-seven-year-old willing to work the phones; Biogen was put together by a consortium of seasoned venture capitalists who head-hunted top scientists. Genentech was born in a San Francisco bar, Biogen in a fancy European hotel. Both companies, however, shared the same vision, and insulin was part of it. The race was on.

Inducing a bacterium to produce a human protein is tricky. Particularly awkward is the presence of introns, those noncoding segments of DNA found in human genes. Since bacteria have no introns, they have no means for dealing with them. While the human cell carefully "edits" the messenger RNA to remove these noncoding segments, bacteria, with no such capacity, cannot produce a protein from a human gene. And so, if *E. coli* were really going to be harnessed to produce human proteins from human genes, the intron obstacle needed to be overcome first.

The rival start-ups approached the problem in different ways. Genentech's strategy was to chemically synthesize the intron-free portions of the gene, which could then be inserted into a plasmid. They would in effect be cloning an artificial copy of the original gene. Nowadays, this cumbersome method is seldom used, but at the time Genentech's was a smart strategy. The Asilomar biohazard meeting had occurred only a short time earlier, and genetic cloning, particularly when it involved human genes, was still viewed with great suspicion and fell under heavy regulation. However, by using an artificial copy of the gene, rather than one actually extracted from a human being, Genentech had found a loophole. The company's insulin hunt could proceed unimpeded by the new rules.

Genentech's competitors followed an alternative approach—the one generally used today—but, working with DNA taken from actual human cells, they would soon find themselves stumbling into a regulatory nightmare. Their method employed one of molecular biology's most surprising discoveries to date: that the central dogma governing the flow of genetic information—the rule that DNA begets RNA, which in turn begets protein—could occasionally be violated. In the 1950s scientists had discovered a group of viruses that contain RNA but lack DNA. HIV, the virus that causes AIDS, is a member of this group. Subsequent research showed that these viruses could nevertheless

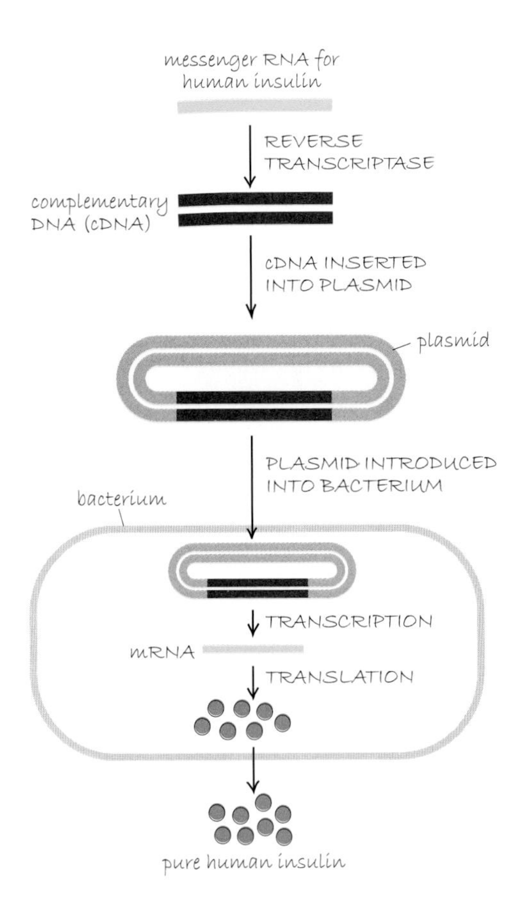

Cloning a gene without its introns using reverse transcriptase

convert their RNA into DNA after inserting it into a host cell. These viruses thus defy the central dogma with their backward RNA → DNA path. The critical trick is performed by an enzyme, reverse transcriptase, that converts RNA to DNA. Its discovery in 1970 earned Howard Temin and David Baltimore the 1975 Nobel Prize in Physiology or Medicine.

Reverse transcriptase suggested to Biogen and others an elegant way to create their own intron-free human insulin gene for insertion in bacteria. The first step was to isolate the messenger RNA produced by the insulin gene. Because of the editing process, the messenger RNA lacks the introns in the DNA from which it is copied. The RNA itself is not especially useful because RNA, unlike DNA, is a delicate molecule liable to degrade rapidly; also the Cohen-Boyer system calls for inserting DNA—not RNA—into bacterial cells. The goal, therefore, was to make DNA from the edited messenger RNA molecule using reverse transcriptase. The result would be a piece of DNA without the introns but with all the information that bacteria would require to make the human insulin protein—a cleaned-up insulin gene.

In the end Genentech would win the race, but just barely. Using the reverse transcriptase method, Gilbert's team had succeeded in cloning the rat gene for insulin and then coaxing a bacterium into producing the rat protein. All that remained was to repeat the process with the human gene. Here, however, is where Biogen met its regulatory Waterloo. To clone human DNA, Gilbert's team had to find a P4 containment facility—one with the highest level of containment, the sort required for work on such unpleasant beasts as the Ebola virus. They managed to persuade the British military to grant them access to Porton Down, a biological warfare laboratory in the south of England.

In his book about the race to clone insulin, Stephen Hall records the almost surreal indignities suffered by Gilbert and his colleagues.

> Merely *entering* the P4 lab was an ordeal. After removing all clothing, each researcher donned government-issue white boxer shorts, black rubber boots, blue pajama-like garments, a tan hospital-style gown open in the back, two pairs of gloves, and a blue plastic hat resembling a shower cap. Everything then passed through a quick formaldehyde wash. Everything. All the gear, all the bottles, all the glassware, all the equipment. All the scientific recipes, written down on paper, had to pass through the wash; so the researchers slipped the instructions, one sheet at a time, inside plastic Ziploc bags, hoping that formaldehyde would not leak in and turn the paper into a brown, crinkly, parchment-like mess. Any document exposed to lab air would ultimately have to be destroyed, so the Harvard group could not even bring in their lab notebooks to make entries. After stepping through a basin of formaldehyde, the workers descended a short flight of steps into the P4 lab itself. The same hygienic rigmarole, including a shower, had to be repeated whenever anyone left the lab.

All this for the simple privilege of cloning a piece of human DNA. Today, in our less paranoid and better informed times, the same procedure is often performed in rudimentary labs by undergraduates taking introductory molecular biology. The whole episode was a bust for Gilbert and his team as they failed to clone the insulin gene. Not surprisingly they blamed their P4 nightmare.

The Genentech team faced no such regulatory hurdles, but their technical challenges in inducing *E. coli* to produce insulin from their chemically synthesized gene were considerable all the same. For Swanson the businessman, the problems were not merely scientific. Since 1923, the U.S. insulin market had been dominated by a single producer, Eli Lilly, which by the late seventies was a $3 billion company with an 85 percent share of the insulin market. Swanson knew Genentech was in no position to compete with the 800-pound gorilla, even with a genetically engineered *human* insulin, a product patently superior to Lilly's farm-animal version. He decided to cut a deal and approached Lilly, offering an exclusive license to Genentech's insulin. And so as his scientist partners beavered away in the lab, Swanson hustled away in the boardroom.

Lilly, he was sure, would agree; even such a giant could ill afford to miss out on what recombinant DNA technology represented, namely the very future of pharmaceutical production.

But Swanson wasn't the only one with a proposal, and Lilly was actually funding one of the competing efforts. A Lilly official had even been dispatched to Strasbourg, France, to oversee a promising attempt to clone the insulin gene using methods similar to Gilbert's. However, when the news came through that Genentech had gotten there first, Lilly's attention was instantly diverted to California. Genentech and Lilly signed an agreement on August 25, 1978, one day after the final experimental confirmation. The biotech business was no longer just a dream. Genentech would go public in September 1980. Within minutes its shares rose from a starting price of $35 to $89. At the time, this was the most rapid escalation in value in the history of Wall Street. Boyer and Swanson suddenly found themselves worth some $66 million apiece.

Traditionally in academic biology, all that mattered was precedence: who made the discovery first. One was rewarded in kudos, not cash. There were exceptions—the Nobel Prize, for instance, does come with a hefty financial award—but in general we did biology because we loved it. Our meager academic salaries certainly did not offer much of an inducement.

With the advent of biotechnology, all that changed. The 1980s would see changes in the relationship of science and commerce that were unimaginable a decade before. Biology was now a big-money game, and with the money came a whole new mind-set, and new complications.

For one thing, the founders of biotech companies were typically university professors, and not surprisingly the research underpinning their companies' commercial prospects typically originated in their university labs. It was in his Zurich University lab, for instance, that Charles Weissmann, one of Biogen's founders, cloned human interferon, which, as a treatment for multiple sclerosis, has since become the company's biggest moneymaker. And Harvard University hosted Wally Gilbert's ultimately unsuccessful attempt to add recombinant insulin to Biogen's roster of products. Certain questions were soon bound to be asked: Should professors be permitted to enrich themselves on the basis of work done in their university's facilities? Would the commercialization of academic science create irreconcilable conflicts of interest? And the prospect of a new era of industrial-scale molecular biology fanned the still-glowing embers of the safety debate: with big money at stake, just how far would the captains of this new industry push the safety envelope?

Harvard's initial response was to form a biotech company of its own. With plenty of venture capital and the intellectual capital of two of the university's star molecular biologists, Mark Ptashne and Tom Maniatis, the business plan seemed a sure thing; a major player was about to enter the biotech game. In the fall of 1980, however, the plan fell apart. When the measure was put to a vote, the faculty refused to allow Fair Harvard to dip its lily-white academic toes into the murky waters of commerce. There were concerns that the enterprise would create conflicts of interest within the biology department: with a profit center in place, would faculty continue to be hired strictly on the basis of academic merit or would their potential to contribute to the firm now come into consideration? Ultimately, Harvard was forced to withdraw, giving up its 20 percent stake in the company. Sixteen years later, the cost of that call would become apparent when the firm was sold to the pharmaceutical giant Wyeth for $1.25 billion. And to this day, Harvard's Department of Molecular and Cellular Biology lacks a designated endowment to support research above the cost of salaries.

The decision of Ptashne and Maniatis to press on regardless precipitated a fresh set of obstacles. Mayor Vellucci's moratorium on recombinant DNA research in Cambridge was a thing of the past, but anti-DNA sentiment lingered on. Carefully avoiding a flashy high-tech name like Genentech or Biogen, Ptashne and Maniatis named their company Genetics Institute, hoping to evoke the less threatening fruit fly era of biology, rather than the brave new world of DNA. In the same spirit, the fledgling company decided to hang its shingle not in Cambridge but in the neighboring city of Somerville. A stormy hearing in Somerville City Hall, however, demonstrated that the Vellucci effect extended beyond the Cambridge city limits: Genetics Institute was denied a license to operate. Fortunately the city of Boston, just across the Charles River from Cambridge, proved more receptive, and the new firm set up shop in an empty hospital building in Boston's Mission Hill district. As it became more and more

apparent that recombinant methods posed no health or environmental risk, the Vellucci brand of antibiotech fanaticism could not endure. Within a few years, Genetics Institute would move to North Cambridge, just down the road from the university parent that had abandoned it at birth.

Over the past twenty years, the suspicion and sanctimoniousness attending the early days of the relationship between academic and commercial molecular biology has given way to something approaching a productive symbiosis. For their part universities now actively encourage their faculty to cultivate commercial interests. Learning from Harvard's mistake with Genetics Institute, they have developed ways to cash in on the lucrative applications of technology invented on campus. New codes of practice aim to prevent conflicts of interest for professors straddling both worlds. In the early days of biotech, academic scientists were all too often accused of "selling out" when they became involved with a company. Now involvement in commercial biotech is a standard part of a hotshot DNA career. The money is handy, and there are intellectual rewards as well because, for good business reasons, biotech is invariably on the scientific cutting edge.

Stanley Cohen proved himself a forerunner not only in technology but also in the evolution from a purely academic mind-set to one adapted to the age of big-bucks biology. He had known from the beginning that recombinant DNA had potential for commercial applications, but it had never occurred to him that the Cohen-Boyer cloning method should be patented. It was Niels Reimers in Stanford's technology licensing office who suggested that a patent might be in order when he read on the front page of the *New York Times* about the home team's big win. At first Cohen was dubious; the breakthrough in question, he argued, was dependent on generations of earlier research that had been freely shared, and so it seemed inappropriate to patent what was merely the latest development. But every invention builds on ones that have come before (the steam locomotive could only come after the steam engine); and patents rightly belong to those innovators who extend the achievements of the past in decisive and influential ways. In 1980, six years after Stanford first submitted the application, the Cohen-Boyer process was granted its patent.

In principle the patenting of methods could stifle innovation by restricting the application of important technologies, but Stanford handled the matter wisely, and there were no such negative consequences. Cohen and Boyer (and their institutions) were rewarded for their commercially significant contribution, but not at the expense of academic progress. In the first place, the patent ensured that only corporate entities would be charged for use of the technology; academic researchers could use it free of charge. Second, Stanford resisted the temptation to impose a very high licensing fee, which would have prevented all but the wealthiest companies and institutions from using recombinant DNA. For a relatively modest $10,000 a year with a maximum 3 percent royalty on the sales of products based on the technology, the Cohen-Boyer method was available to anyone who wanted to use it. This strategy, good for science, proved to be good for business as well: the patent has contributed some quarter of a billion dollars to the coffers of UCSF and Stanford. And both Boyer and Cohen generously donated part of their shares of the proceeds to their universities.

It was only a matter of time before organisms genetically altered by technology would themselves be patented. The test case had in fact originated in 1972; it involved a bacterium that had been modified using not recombinant DNA technology but traditional genetic methods. The implications for the biotech business were clear nevertheless: if bacteria modified with conventional techniques were patentable, then those modified by the new recombinant methods would be too.

In 1972, Ananda Chakrabarty, a research scientist at General Electric, applied for a patent on a *Pseudomonas* bacteria strain he had developed as an all-in-one oil-slick degrader. Before this, the most efficient way to break down an oil spill was to use a number of different bacteria, each of which degraded a different component of the oil. By combining different plasmids, each coding for a different degradation pathway, he managed to produce a superdegrader strain of *Pseudomonas*. Chakrabarty's initial patent application was turned down, but after wending its way through the legal system for eight years it was finally granted in 1980, when the Supreme Court ruled five to four in his favor, concluding that "a live, human-made micro-organism is patentable subject matter" if, as in this case, it "is the result of human ingenuity and research."

Despite the clarification supplied by the Chakrabarty case, the early encounters between biotechnology and the law were inevitably messy. The stakes were high and—as we shall see in the case of DNA fingerprinting in chapter 10—lawyers, juries, and scientists too often speak different languages. By 1983, both Genentech and Genetics Institute had successfully cloned the gene for tissue plasminogen activator (t-PA), which is an important weapon against the blood clots that cause strokes and heart attacks. Genetics Institute did not, however, apply for a patent, deeming the science underlying the cloning of t-PA "obvious"—in other words, unpatentable. Genentech, however, applied for and was granted a patent, on which, by definition, Genetics Institute had infringed.

The case first came to court in England. The presiding judge, Mr. Justice Whitford, sat behind a large stack of books for much of the trial, appearing to be asleep. The basic question was whether the first party to clone a gene should be granted all subsequent rights over the production and use of the protein. In finding for Genetics Institute and its backers, the drug company Wellcome, Justice Whitford concluded that Genentech could justify a narrow claim for the limited process used by them to clone t-PA but could not justify broad claims for the protein product. Genentech appealed. In England when such esoteric technical cases are appealed they are heard by three specialist judges, who are led through the issues by an independent expert—in this instance, Sydney Brenner. The judges turned down Genentech's appeal, agreeing with Genetics Institute that the "discovery" was indeed obvious, and therefore the Genentech patent was invalid.

In the United States, such cases are argued in front of a jury. Genentech's lawyers ensured that no member of the jury had a college education. Thus what might be obvious to a scientist or to legal experts trained in science was not obvious to members of that jury. The jury found against Genetics Institute, deeming the broad-based Genentech patent valid. Not, perhaps, American justice's finest hour, but the case did nevertheless establish a precedent: from then on, people applied for patents on their products regardless of whether or not the science was "obvious." In future disputes, all that would matter was who cloned the gene first.

Good patents, I would suggest, strike a balance: they recognize and reward innovative work and protect it from being ripped off, but they also make new technology available to do the most good. Unfortunately, Stanford's wise example has not been followed in every case of important new DNA methodology. The polymerase chain reaction (PCR), for instance, is an invaluable technique for amplifying small quantities of DNA. Invented in 1983 at the Cetus Corporation, PCR—about which we shall hear more in chapter 7, in connection with the Human Genome Project—quickly became one of the workhorses of academic molecular biology. Its commercial applications, however, have been much more limited. After granting one commercial license to Kodak, Cetus sold PCR for $300 million to the Swiss giant Hoffmann-LaRoche, makers of chemical, pharmaceutical, and medical diagnostic products. Hoffmann-LaRoche in turn decided that, rather than granting further licenses, the way to maximize the return on their investment was to establish a monopoly on PCR-based diagnostic testing. As part of this strategy, it cornered the AIDS testing business. And only as the patent expiration date drew near did the firm grant any licenses for the technology; those granted have generally been to other major diagnostic companies that can afford the commensurably large fees. To create a subsidiary revenue stream from the same patent, Hoffmann-LaRoche has also levied hefty charges on producers of machines that carry out PCR. And so, to market a simple device for schoolchildren to use, the Cold Spring Harbor Dolan DNA Learning Center must pay the company a 15 percent royalty.

An even more pernicious effect on the productive availability of new technologies has been exerted by lawyers moving aggressively to patent not only new inventions but also the general ideas underpinning them. The patent on a genetically altered mouse created by Phil Leder is a case in point. In the course of their cancer research, Leder's group at Harvard produced a strain of mouse that was particularly prone to developing breast cancer. They did this using established techniques for inserting a genetically engineered cancer gene into a fertilized mouse egg cell. Because the factors inducing cancer in mice may be similar to those at work in humans, this "onco-mouse" was expected to help us understand human cancer. But instead of applying for a patent limited to the specific mouse Leder's team had produced, Harvard's lawyers sought one that covered all cancer-prone transgenic animals—they didn't even draw the line at mice. This umbrella patent was granted in 1988, and so was born the cancerous little rodent dubbed the "Harvard

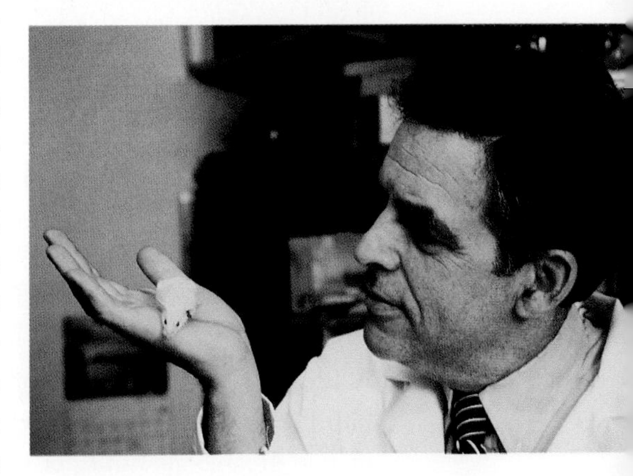

Phil Leder with his "Harvard" onco-mouse

mouse." In fact, because the work in Leder's laboratory was underwritten by Du Pont, the commercial rights resided not with the university but with the chemical giant. The "Harvard mouse" might have been more aptly called the "Du Pont mouse." But whatever its name, the impact of the patent on cancer research has been profound and counterproductive.

Companies interested in developing new forms of cancer-prone mice have been put off by the fees demanded by Du Pont, and those keen to use existing cancer mouse strains to screen experimental drugs have likewise curtailed their programs. Du Pont has begun demanding that academic institutions disclose what experiments are being performed using the company's patented oncomice. This represents an unprecedented, and unacceptable, intrusion of big business into academic laboratories. UCSF, MIT's Whitehead Institute, and Cold Spring Harbor Laboratory, among other research institutions, have refused to cooperate.

When patents involve "enabling technologies" that are fundamental to carrying out the necessary molecular manipulations, the patent holders can literally hold an entire area of research for ransom. And while every patent application should be treated on its particular merits, there are nevertheless some general rules that should be observed. Patents on methods clearly vital to scientific progress should follow the precedent set by the Cohen-Boyer case: the technology should be generally available (not controlled by a single licensee) and should be reasonably priced. These limitations by no means go against the ethic of free enterprise. If a new method is a genuine step forward, then it will be extensively used and even a modest royalty will result in substantial revenue. Patents on *products,* however—drugs, transgenic organisms—should be limited to the specific product created, not the entire range of additional products the new one might suggest.

Genentech's insulin triumph put biotechnology on the map. A quarter of a century later, genetic engineering with recombinant DNA technology is a routine part of the drug-discovery industry. These procedures permit the production in large quantities of human proteins, which are otherwise difficult to acquire. In many cases, the genetically engineered proteins are safer for therapeutic and diagnostic uses than their predecessors. Extreme short stature, dwarfism, often stems from a lack of human growth hormone (HGH). In 1959, doctors first started treating dwarfism with HGH, which then could be obtained only from the brains of cadavers. The treatment worked fine, but it was later recognized to carry the risk of a terrible infection: patients sometimes developed Creutzfeldt-Jakob disease, a ghastly brain-wasting affliction, similar to so-called mad cow disease. In 1985, the FDA banned the use of HGH derived from cadavers. By happy coincidence, Genentech's recombinant HGH—which carries no risk of infection—was approved for use that same year.

During the biotech industry's first phase, most companies focused on proteins of known function. Cloned human insulin was bound to succeed; after all, people had already been injecting themselves with some form of insulin for more than fifty years when Genentech introduced its product. Another example was epoetin alpha (EPO), a protein that stimulates the body to produce red blood cells. The target population for EPO is patients undergoing kidney dialysis who suffer from anemia caused by loss of red blood cells. To meet the need for this product, Amgen, based in Southern California, and Genetics Institute both developed a recombinant form of EPO. That EPO was a useful and commercially viable product was a given; the only unknown was which company would come to dominate the market. Despite being trained in the arcane subtleties of physical chemistry, Amgen CEO George Rathmann has adapted well to the rough and tumble of the business world. Competition brings out a decidedly unsubtle side in him: negotiating with him is like wrestling with a large bear whose twinkling eye assures you that it is only mauling you because it is obliged to. Amgen and its backer, Johnson & Johnson, duly won the court battle with Genetics Institute, and EPO is now worth $2 billion a year to Amgen alone. Amgen is accordingly today the biggest player in the biotech stakes, worth some $64 billion.

After biotech's pioneers had rounded up the "obvious" products, proteins with known physiological function like insulin, t-PA, HGH, and EPO, a second, more speculative phase in the industry got under way. Having run out of surefire winners, companies hungry for further bonanzas began to back possible contenders, even long shots. From knowing that something worked, they went to merely *hoping* that a potential product would work. Unfortunately, the combination of longer odds, technical challenges, and regulatory hurdles to be cleared before a drug is approved by the FDA has taken its toll on many a bright-eyed biotech start-up.

The discovery of growth factors—proteins that promote cell proliferation and survival—provoked a proliferation of new biotech companies. Among them, both New York–based Regeneron and Synergen, located in Colorado, hoped to find a treatment for ALS (amyotrophic lateral sclerosis or Lou Gehrig's disease), the awful degenerative affliction of nerve cells. Their idea was fine in principle, but in practice there was simply too little known at the time about how nerve growth factors act for these efforts to be anything more than shots in the dark. Trials on two groups of ALS patients failed, and the disease remains untreatable today. The experiments did, however, reveal an interesting side effect: those taking the drugs lost weight. In a twist that illustrates just how serendipitous the biotech business can be, Regeneron is today developing a modified version of its drug as a weight-loss therapy.

Another initially speculative enterprise that has seen more than its fair share of dashed commercial hopes is monoclonal antibody (MAb) technology. When they were invented in the mid-1970s at the MRC Laboratory of Molecular Biology at Cambridge University by César Milstein and Georges Köhler, MAbs were hailed as the silver bullets that would quickly change the face of medicine. Nevertheless, in an oversight that would today be unthinkable, the MRC failed to patent them. Silver bullets they proved not to be, but, after decades of disappointment, they are just now coming into their own.

Antibodies are molecules produced by the immune system to bind to and identify invading organisms. Derived from a single line of antibody-producing cells, MAbs are antibodies programmed to bind to a unique target. They can be readily produced in mice by injecting animals with the target material, inducing an immune response, and culturing the blood cells from the mouse that produced the MAb. Because MAbs can recognize and bind to specific molecules, it was hoped that they could be used with pinpoint accuracy against any number of pernicious intruders—tumor cells, for instance. Such optimism prompted the founding of a slew of MAb-based companies, but they quickly ran into obstacles. Ironically, the most significant of these was the human body's own immune system, which identified the mouse MAbs as foreign and duly destroyed them before they could act on their targets. A variety of methods have since been devised to "humanize" MAbs—to replace as much as possible of the mouse antibodies with human components. And the latest generation of MAbs represents the biggest growth area in biotech today.

Centocor, based near Philadelphia, now owned by Johnson & Johnson, has developed ReoPro, an MAb specific to a protein on the surface of platelets, which promote the formation of blood clots. By preventing platelets from sticking together, ReoPro reduces the chance of lethal clot formation in patients undergoing angioplasty, for instance. Genentech, never one to lag in the biotech stakes, now markets Herceptin, an MAb that targets certain forms of breast cancer. Immunex in Seattle produces an MAb-based drug called Enbrel, which fights rheumatoid arthritis, a condition associated with the presence of excessive amounts of a particular protein, tumor necrosis factor (TNF), involved in regulating the immune system. Enbrel works by capturing the excess TNF molecules, preventing them from provoking an immune reaction against the tissue in our joints.

Still other biotech companies are interested in cloning genes whose protein products are potential targets for new pharmaceuticals. Among the most eagerly sought are the genes for proteins usually found on cell surfaces that serve as receptors for neurotransmitters, hormones, and growth factors. It is through such chemical messengers that the human body coordinates the actions of any individual cell with the actions of trillions of others. Drugs developed blindly in the past through trial and error have recently been found to operate by affecting these receptors. And that same new molecular understanding has also explained why so many of these drugs have side effects. Receptors often belong to large families of similar proteins. A drug may indeed effectively target a receptor relevant to the disease in question, but it also may wind up inadvertently targeting similar receptors, thus producing side effects. Intelligent drug design should permit more specific targeting of the receptors so that *only* the relevant one is blocked. However, as with MAbs, what seems a great idea on paper is too often hard to apply in practice, and even harder to make big bucks from.

This depressing lesson was learned by SIBIA, a San Diego start-up associated with the Salk Institute. The discovery of membrane receptors for the neurotransmitter nicotinic acid promised a breakthrough treatment for Parkinson disease, but as so often in biotech a good idea was only the beginning of a long scientific process. Ultimately, after giving promising results in monkeys, SIBIA's drug candidate failed in humans.

Like the unexpected weight loss associated with Regeneron's nerve growth factor, breakthroughs in this area too are often born of pure luck rather than the scientific calculus of rational drug design. In 1991, for instance, a Seattle-based company, ICOS, led by George Rathmann of Amgen fame, was working with a class of enzymes called "phosphodiesterases," which degrade cell-signaling molecules. Their quarry was new drugs to lower blood pressure, but one of their test drugs had a surprising side effect. They had stumbled onto a Viagra-like therapy for erectile dysfunction, which may well yield a bigger jackpot than any they previously dreamed of.*

The market for easier erections notwithstanding, the search for cancer therapies has, not surprisingly, become the single greatest driving force for the biotech industry. The classic "cell-killing" approach to attacking cancer, using radiation or chemotherapy, invariably also kills healthy normal cells, typically with dreadful side effects. With developing DNA methodologies researchers are finally closing in on drugs that can target only those key proteins—many of them growth factors and their receptors on the cell surface—that promote cancer cell growth and division. Developing a drug that inhibits a desired target without disabling other vital proteins is a formidable challenge even for the best of medicinal chemists. And the uncertain journey from a successfully cloned drug target gene to the widespread availability of an FDA-approved pharmaceutical is a veritable odyssey that seldom takes less than ten years.

Success stories are hard to come by, but will, I am sure, become more common. Discovered by chemists at the Swiss company Novartis, Gleevec works against a blood cancer called chronic myeloid leukemia (CML) by specifically blocking the growth-stimulating activity of membrane receptor proteins that are overproduced by cancerous cells of this type. If given early in the course of CML, Gleevec generally leads to long disease-free remissions, and hopefully in many cases to true cures. For some unlucky individuals, though, the disease reappears when new mutations in the gene encoding the membrane-receptor proteins render Gleevec ineffective.

One of the most important anticancer-drug target proteins may be the receptor for epidermal growth factor (EGFR). This receptor frequently shows up in

*Viagra itself has a similar history. Also originally developed to combat high blood pressure, trials on male medical students convinced researchers that it had other properties.

much higher quantities in cancer cells (particularly in breast and lung cancers) than in normal ones, which suggests that it may well be a winner as a drug target. Several potent drugs that specifically block EGFR action are now in late-stage clinical testing. But while the arrival of target-specific drugs will certainly introduce big new guns in the war against cancer, the likelihood is that, after initial remission, many patients will suffer a relapse as resistance to the new drugs evolves among the cancer cells colonizing the body.

For this reason, many have come to believe that a better long-term way of fighting cancer cells may involve targeting their nutritional lifelines. They, like all cells in the body, need nutrients to grow, and they receive these nutrients from blood vessels that grow near them. If you block the growth of blood vessels into tumors, you can eventually starve to death the cancer cells they serve. The idea that small tumors become dangerous only once they are infiltrated by newly formed blood vessels (a process called "angiogenesis") first occurred to Judah Folkman in the early 1960s while he was doing his military service in the Naval Medical Research Institute outside Washington, D.C. The precocious son of an Ohio rabbi, Folkman was the first graduate of Ohio State University to enter Harvard Medical School. By the time he went to high school he had already assisted in surgery on a dog, and in college he invented a surgical device to cool the liver when its blood supply was temporarily cut off. At thirty-four, he became the youngest professor of surgery in the history of Harvard University. Folkman's anti-angiogenesis ideas could not, however, be explored therapeutically until the recent discovery of three specific growth factors that play vital roles in the growth of "endothelial" cells, those that line blood vessels. Inhibitors developed against these growth factors—anti-angiogenesis drugs—might very well prove effective against many forms of cancer. Some forty years after Folkman's original insight, we may at last be able in the foreseeable future to cure most cancers, including those that have become resistant to the best conventional anticancer drugs.

Already Sugen, a firm outside San Francisco, has developed two highly specific small-molecule drugs that work against distinct angiogenesis growth factors and inhibit tumors in model animal systems. Neither drug given separately has yet proved effective against advanced human cancers. However, preliminary data from experiments with cancer-prone mice done by Doug Hanahan at

UCSF suggest the Sugen drugs might have worked had they been administered in tandem. Unfortunately the future of onco-mouse experiments at UCSF and elsewhere is jeopardized by the ongoing dispute provoked by Du Pont's aggressive onco-mouse licensing policies.

Blood vessel infiltration into mouse tumors has also been prevented by a newly discovered group of proteins that are likely naturally occurring inhibitors of blood vessel formation. Two such proteins, angiostatin and endostatin, isolated by Michael O'Reilly in Judah Folkman's lab, are currently in clinical trials. While neither is present in blood in amounts large enough to be extracted for human testing, recombinant DNA procedures permit both proteins to be made in yeast cells in quantities sufficient for clinical use. And while neither angiostatin nor endostatin alone has yet demonstrated miracle-like anticancer effects in humans, mouse experiments suggest that, as with Sugen's drugs, an efficacious combination of the two may soon be discovered. Over the next decade, a virtual armada of small-molecule and protein inhibitors will probably be ready to sail through the systems of cancer sufferers, thwarting blood vessel formation before tumors have a chance to become lethal. And if tumor growth can indeed be curtailed in this way, we may come to regard cancer as we do diabetes, as a disease that can be controlled rather than completely cured outright.

Since recombinant technologies allow us to harness cells to produce virtually any protein, the question has logically arisen: Why limit ourselves to pharmaceuticals? Consider the example of spider silk. So-called dragline silk, which forms the radiating spokes of a spider web, is an extraordinarily tough fiber. By weight, it is five times as strong as steel. Though there are ways spiders can be coaxed to spin more than their immediate needs require, unfortunately, attempts to create spider farms have foundered because the creatures are too territorial to be reared en masse. Now, however, the silk-protein-producing genes have been isolated and can be inserted into other organisms, which can thus serve as spider-silk factories. This very line of research is being funded by the Pentagon, which sees Spiderman in the U.S. Army's future: soldiers may one day be clad in protective suits of spider-silk body armor.

Another exciting new frontier in biotechnology involves improving on natural proteins. Why be content with nature's design, arrived at by sometimes arbitrary and now irrelevant evolutionary pressures, when a little manipulation might yield something more useful? Starting with an existing protein, we now have the ability to make slight alterations in its amino acid sequence. The limitation, unfortunately, is in our knowledge of what effect altering even a single amino acid in the chain is likely to have on the protein's properties.

Here we can return to nature's example for a solution: a procedure known as "directed molecular evolution" effectively mimics natural selection. In natural selection new variants are generated at random by mutation and then winnowed by competition among individuals; successful—better adapted—variants are more likely to live and contribute to the next generation. Directed molecular evolution stages this process in the test tube. After using biochemical tricks to introduce random mutations into the gene for a protein, we can then mimic genetic recombination to shuffle the mutations to create new sequences. From among the resulting new proteins our system selects the ones that perform best under the conditions specified. The whole cycle is repeated several times, each time with the "successful" molecules from the previous cycle competing in the next.

For a nice example of how directed molecular evolution can work, we need look no farther than the laundry room. Here disasters occur when a single colored item finds its way accidentally into a load of whites: some of the dye inevitably leaches out of that red T-shirt and before you know it every sheet in the house is a pale pink. It so happens that a peroxidase enzyme naturally produced by a toadstool—the inkcap mushroom, to be specific—has the property of decolorizing the dyes that have leached out of clothing. The problem, however, is that the enzyme cannot function in the hot soapy environment of a washing machine. By using directed molecular evolution, however, it has been possible to improve the enzyme's capacity for coping with these conditions: one specially "evolved" enzyme, for instance, demonstrated an ability to withstand high temperatures 174 times greater than that of the toadstool's own enzyme. And such useful "evolutions" do not take long. Natural selection takes eons, but directed molecular evolution in the test tube does the job in just hours or days.

Genetic engineers realized early that their technologies could also have a positive impact on agriculture. As the biotech world now knows all too well, the resulting genetically modified (GM) plants are now at the center of a firestorm of controversy. So it's interesting to note that an earlier contribution to agriculture—one that increased milk production—also led to an outcry.

Bovine growth hormone (BGH) is similar in many ways to human growth hormone, but it has an agriculturally valuable side effect: it increases milk production in cows. Monsanto, the St. Louis–based agricultural chemical company, cloned the BGH gene and produced recombinant BGH. Cows naturally produce the hormone, but, with injections of Monsanto's BGH, their milk yields increased by about 10 percent. In late 1993 the FDA approved the use of BGH, and by 1997 some 20 percent of the nation's 10 million cows were receiving BGH supplements. The milk produced is indistinguishable from that produced by nonsupplemented cows: they both contain the same small amounts of BGH. In fact, a major argument against labeling milk as "non-BGH-supplemented" versus "BGH-supplemented" is that it is impossible to distinguish between milk from supplemented and nonsupplemented cows, so there is no way to determine whether or not such advertising is fraudulent. Because BGH permits farmers to reach their milk production targets with fewer cattle, it is in principle beneficial to the environment because it could result in a reduction in the size of dairy herds. Because methane gas produced by cattle contributes significantly to the greenhouse effect, herd reduction may actually have a long-term effect on global warming. Methane is twenty-five times more effective at retaining heat than carbon dioxide, and on average a grazing cow produces six hundred flatulent liters of the stuff a day—enough to inflate forty party balloons.

At the time I was surprised that BGH provoked such an outburst from the anti-DNA lobby. Now, as the GM food controversy drags on, I have learned that professional polemicists can make an issue out of anything. Jeremy Rifkin, biotechnology's most obsessive foe, was launched on his career in naysaying by the U.S. Bicentennial in 1976. He objected. After that he moved on to objecting to DNA. His response in the mid-1980s to the suggestion that BGH would not likely inflame the public was, "I'll *make* it an issue! I'll find something! It's the first product of biotechnology out the door, and I'm going to fight it." Fight it he did. "It's unnatural" (but it's indistinguishable from "natural" milk). "It contains proteins that cause cancer" (it doesn't, and in any case proteins are broken

Jeremy Rifkin, professional naysayer: you name it, he has tried to stop it.

down during digestion). "It'll drive the small farmer out of business" (but, unlike with many new technologies, there are no up-front capital costs, so the small farmer is not being discriminated against). "It'll hurt the cows" (nearly nine years of commercial experience on millions of cows has proved this not to be the case). In the end, rather like the Asilomar-era objections to recombinant techniques, the issue petered out when it became clear that none of Rifkin's gloom-and-doom scenarios were realistic.

The spat over BGH was a taste of what was to come. For Rifkin and like-minded DNA-phobes, BGH was merely the appetizer: genetically modified foods would be the protesters' main course.

GM FOOD Are you buying Frankenstein food?

Where you're most likely to find GM food and drinks as you shop.

As you navigate your way around a supermarket or smaller food shops, it's helpful to know what categories of food are most likely to conceal products that may have been genetically modified so that you can be on your guard. Not all shopping zones are the same in this respect. In certain categories of food — fresh meat, for example — you can afford to be relaxed. While in others — such as sandwiches — extreme vigilance is called for.

Use the guide below to identify categories or types of food and drink where you need to pay special attention.

THE RED ZONE

It is quite likely that the food and drinks in this zone will contain GM ingredients.

- Lo-cal, 'lite' and reduced-fat foods
- Processed foods imported from the US
- Prepared sandwiches
- Prepared ready meals, both meat and vegetarian (especially ethnic-themed ones, e.g. Mexican, Chinese)
- Vegetarian meals containing soya
- Vegetarian products containing soya milk or tofu
- Prepared pasta and noodles

THE AMBER ZONE

It is possible that the foods and drinks in this zone will contain GM ingredients.

- Bread and rolls
- Sweet and savoury breads (e.g. croissants, muffins, scones, pikelets)
- Artificial flavourings (e.g. vanilla or peppermint essence)
- Flavoured teas, fruit teas and herbal infusions
- Tinned fruit
- Tinned vegetables
- Tinned fish packed in oil
- Any food fortified with vitamin B2 riboflavin (breakfast cereals, baby food and soft drinks)

THE GREEN ZONE

Use this section to get more information about 'branded' foods. These are products carrying brand names of individual companies as opposed to those that carry the supermarket's own label. This section looks at familiar branded foods that could be genetically modified or could become so in the near future and gives each brand a red/amber/green rating*.

RED The company sees some potential benefits in GM food and is prepared to use GM ingredients and derivatives. Its foods are more likely to contain GM ingredients and derivatives.

AMBER The company's foods could be genetically modified. Companies in this category have no clear stance for or against genetic modification. They either cannot guarantee all their products or can give only a qualified GM-free guarantee 'for the time being'.

GREEN The company never knowingl sources genetically modified ingredients or derivatives. Its foods are highly unlikely to contain GM ingredients or derivatives and would only do so through unintentional contamination.

*This rating is based on policy and statements of intent supplied by the company or its parent company. It is not possible to verify these claims

THE GREEN BRANDS

FRANKENSTEIN FOOD FIASCO

By DAVID DERBYSHIRE
Science Correspondent

Looks familiar: these cloned sheep at the Roslin Institute in Edinburgh are seen primarily as a means to genetic engineering

We modify their genes at our peril

Food technology: A new breed of eco-warrior is challenging the big corporations by taking direct action

Wheatfields turn into war zones

BY CHARLES ARTHUR
Technology Editor

ONE OF the most intense commercial battles of recent years found a focus yesterday at the Royal Show in Berkshire, in a patch of wheat covering an area the size of two table-tennis tables. It looked no different from any other wheat, except that it had its own 24-hour security guard.

What the guards are protecting is actually invisible. The wheat has been genetically modified (GM) to contain extra gluten genes – so bread made from it should be lighter and bouncier. But that makes it a prime target for the groups aiming to uproot such crops. So far this year GM plants have been torn up at almost 30 of 200 experimental sites around Britain. To keep the wheat

in ten food products contain soya or soya oils or soya lecithin. From September, the food will have labels: "Does not contain genetically modified elements", or: "Contains genetically modified elements".

The Laboratory of the Government Chemist (LGC) is one of two UK organisations offering DNA tests to identify whether foods contain GM elements. Recently it hosted a crowded industry session with speakers from the supermarket Sainsbury's and Rank Hovis McDougall, which makes

…ly-modified crops such as oilseed rape are the latest target for eco-protest groups

Ian Duncan

…iddle-class activists…

Giant killers

How Europe's eco-warriors humbled

Finance

News Analysis Agribusiness is running scared from GM foods

Test fields of conflict

Julia Finch

Consumers are concerned, eco-warriors are on the warpath and British agribusiness is on the run. The scare over genetically modified food which blew up a year ago shows no sign of abating and companies the length of the food supply chain are responding.

The issue is threatening to explode into a transatlantic trade war which could dwarf the recent EU-US conflagrations over bananas and hormone-treated beef.

The GM food scare, prompted by research which claimed to have found a link between tumours in mice and genetically modified potatoes, and since discredited, is swiftly becoming a huge corporate liability. Fresh concerns and protests about the potential health and environmental impact of GM crops are reported almost weekly.

Within the past 10 days:

- Lord Melchett, director of Greenpeace, was imprisoned after an attack on a genetically engineered cornfield in Norfolk.

However, this week AstraZeneca's group chief executive, Tom McKillop, made it clear that the GM business was a sideline it could do without: "There isn't the management resource to devote to everything."

The supermarkets were the first to react to mounting consumer concern. "We are in the firing line," said a senior executive from one of the big four chains. "Consumers make their feelings known to us much more than to the food producers and far more than to companies like Monsanto which are at the forefront of this research. When the tabloids started talking about 'Frankenstein foods' our postbag just took off. It was bordering on hysteria, though we compounded the problem by seizing on the issue and using it for competitive advantage."

The big four groups, along with other retailers, rushed to be able to declare that their own-brand foods were GM-free and, although so far only Sainsbury among them has achieved it, Safeway, Tesco and Asda are expected to follow suit shortly.

"The international food producers are

Environmental protesters make

'The main risk is loss

CHAPTER SIX

. . .

TEMPEST IN A CEREAL BOX: GENETICALLY MODIFIED AGRICULTURE

In June 1962, Rachel Carson's book *Silent Spring* created a sensation when it was serialized in *The New Yorker.* Her terrifying claim was that pesticides were poisoning the environment, contaminating even our food. At that time I was a consultant to John Kennedy's President's Scientific Advisory Committee (PSAC). My main brief was to look over the military's biological warfare program, so I was only too glad to be diverted by an invitation to serve on a subcommittee that would formulate the administration's response to Carson's concerns. Carson herself gave evidence, and I was impressed by her careful exposition and circumspect approach to the issues. In person, too, she was nothing like the hysterical ecofreak she was portrayed as by the pesticide industry's vested interests. An executive of the American Cyanamid Company, for instance, insisted that "if man were to faithfully follow the teachings of Miss Carson, we would return to the Dark Ages, and the insects and diseases and vermin would once again inherit the earth." Monsanto, another giant pesticide producer, published a rebuttal of *Silent Spring,* called *The Desolate Year,* and distributed five thousand copies free to the media.

My most direct experience of the world Carson described, however, came a year later when I headed a PSAC panel looking into the threat posed to the nation's cotton crop by herbivorous insects, especially the boll weevil. Touring the cotton fields of the Mississippi Delta, West Texas, and the Central Valley of California, one could hardly fail to notice the utter dependence of cotton growers on chemical pesticides. En route to an insect research laboratory near

The British press made a meal of the genetically modified foods issue

The Famous Wartime Insecticide Discovery

Now Available to Civilians

D D T is a powder used as a basic ingredient for various types of insecticides, powdered and liquid. There are various mixtures for different uses:

- **Compounded with other powders, for dusting purposes**
- **Solutions, containing DDT mixed with water**
- **DDT mixed with other ingredients and volatile solvents, for spraying or brushing**

Prices will vary according to type and quality of the product.

There Will Be Varying Types and Qualities of D D T Sprays. For Your Protection, Get Facts About Each Before You Buy!

MADACO DDT Base Insecticide

is a liquid compound of D D T, PYRETHRUM and other ingredients. (CONTAINS NO KEROSENE) making it most effective against household insects. It is to be used as a spray for killing, as well as repelling, all household pests. Used according to directions MADACO DDT BASE INSECTICIDE will kill them promptly and act to repel them for a period of from sixty to ninety days.

55 Gal. Drums, per gal. **$3.10**
30 Gal. Drums, per gal. **$3.25**
5 Gallon Containers, per gal. **$3.50**
2 Gallon Containers, per gal. **$3.75**
1 Gallon Containers, per gal. **$3.85**

Out-of-town shipments prepaid on orders of five (5) gallons or over. Check or money order must accompany one and two gallon mail orders. Deliveries will be made promptly. Add 25c per gal. to cover shipping charges on one or two gallon orders out of Jacksonville.

Proper Precautions Necessary for Best Results

All containers are labeled with proper instructions and precautions conforming with bulletins issued by the War Food Administration, Offices of Marketing Services, Washington, D. C.

Marion H. Davis & Co.

1212 Mary St. **Phone 9-6636**

Rachel Carson testifying in 1962 before a congressional subcommittee appointed to look into her claims about the dangers posed by pesticides. Before she rang the alarm, DDT (right) was seen as everyone's best friend.

Brownsville, Texas, our car was inadvertently doused from above by a crop duster. Here billboards featured not the familiar Burma-Shave ads but pitches for the latest and greatest insect-killing compounds. Poisonous chemicals seemed to be a major part of life in cotton country.

Whether Carson had gauged the threat accurately or not, there had to be a better way to deal with the cotton crop's six-legged enemies than drenching huge tracts of country with chemicals. One possibility promoted by the U.S. Department of Agriculture scientists in Brownsville was to mobilize the insects' own enemies—the polyhedral virus, for instance, which attacks the bollworm (soon to become a greater threat to cotton than the boll weevil)—but such strategies proved impracticable. Back then, I could not have conceived of a solution that would involve creating plants with built-in resistance to pest insects: such an idea would simply have seemed too good to be true. But these days that is exactly how farmers are beating the pests while at the same time reducing dependence on noxious chemicals.

Genetic engineering has produced crop plants with onboard pest resistance. The environment is the big winner because pesticide use is decreased, and yet paradoxically organizations dedicated to protecting the environment have been the most vociferous in opposing the introduction of these so-called genetically modified (GM) plants.

As with genetic engineering in animals, the tricky first step in plant biotechnology is to get your desired piece of DNA (the helpful gene) into the plant cell, and afterwards into the plant's genome. As molecular biologists frequently discover, nature had devised a mechanism for doing this eons before biologists even thought about it.

Crown gall disease results in the formation of an unattractive lumpy "tumor," known as a gall, on the plant stem. It is caused by a common soil bacterium called *Agrobacterium tumefaciens,* which opportunistically infects plants where they are damaged by, say, the nibbling of a herbivorous insect. How the bacterial parasite carries out the attack is remarkable. It constructs a tunnel through which it delivers a parcel of its own genetic material into the plant cell. The parcel consists of a stretch of DNA that is carefully excised from a special plasmid and then wrapped in a protective protein coat before being shipped off through the tunnel. Once the DNA parcel is delivered, it becomes integrated, as a virus's DNA would be, into the host cell's DNA. Unlike a virus, however, this stretch of DNA, once lodged, does not crank out more copies of itself. Instead, it produces both plant growth hormones and specialized proteins, which serve as nutrients for the bacterium. These promote simultaneous plant cell division and bacterial growth by creating a positive feedback loop: the growth hormones cause the plant

A plant with crown gall disease, caused by Agrobacterium tumefaciens. *The lumpy tumor is the bacterium's ingenious way of ensuring that the plant produces plenty of what the bacterium needs.*

cells to multiply more rapidly, with the invasive bacterial DNA being copied at each cell division along with the host cell's, so that more and more bacterial nutrients *and* plant growth hormones are produced.

For the plant the result of this frenzy of uncontrolled growth is a lumpy cell mass, the gall, which for the bacterium serves as a kind of factory in which the plant is coerced into producing precisely what the bacterium needs, and in ever greater quantities. As parasitic strategies go, *Agrobacterium*'s is brilliant: it has raised the exploitation of plants to an art form.

The details of *Agrobacterium's* parasitism were worked out during the 1970s by Mary-Dell Chilton at the University of Washington in Seattle and by Marc van Montagu and Jeff Schell at the Free University of Ghent, Belgium. At the time the recombinant DNA debate was raging at Asilomar and elsewhere. Chilton and her Seattle colleagues later noted ironically that, in transferring DNA from one species to another without the protection of a P4 containment facility, *Agrobacterium* was "operating outside the National Institutes of Health guidelines."

Chilton, van Montagu, and Schell soon were not alone in their fascination with *Agrobacterium*. In the early eighties Monsanto, the same company that had condemned Rachel Carson's attack on pesticides, realized that *Agrobacterium* was more than just a biological oddity. Its bizarre parasitic lifestyle might hold the key to getting genes into plants. When Chilton moved from Seattle to Washington University, St. Louis, Monsanto's hometown, she found that her new neighbors took a more than passing interest in her work. Monsanto may have made its entry late in the *Agrobacterium* stakes, but it had the money and other resources to catch up fast. Before long both the Chilton and the van Montagu/Schell laboratories were being funded by the chemical giant in return for a promise to share their findings with their benefactor.

Monsanto's success was built on the scientific acumen of three men, Rob Horsch, Steve Rogers, and Robb Fraley, all of whom joined the company in the early eighties. Over the next two decades they would engineer an agricultural revolution. Horsch always "loved the smell of [the soil], the heat of it" and, even as a boy, wanted "always to grow things better than what I could find at the grocery store." He instantly saw a job at Monsanto as an opportunity to follow that dream on an enormous scale. By contrast, Rogers, a molecular biologist at Indiana University, initially discarded the company's letter of invitation, viewing the prospect of such work as "selling out" to industry. Upon visiting, however, he discovered not only a vigorous research environment but also an abundance of one key element that was always in short supply in academic research: money. He was converted. Fraley was possessed early on by a vision for agricultural biotechnology. He came to the company after approaching Ernie Jaworski, the executive whose bold vision had started Monsanto's biotechnology program. Jaworski proved not only a visionary but also an affable employer. He was unfazed by his first encounter with the new man when they were both passing through Boston's Logan Airport: Fraley announced that one of his goals was to take over Jaworski's job.

All three *Agrobacterium* groups—Chilton's, van Montagu and Schell's, and Monsanto's—saw the bacterium's strategy as an invitation to manipulate the genetics of plants. By then it wasn't hard to imagine using the standard cut-and-paste tools of molecular biology to perform the relatively simple act of inserting into *Agrobacterium*'s plasmid a gene of one's choice to be transferred to the plant cell. Thereafter, when the genetically modified bacterium infected a host, it would insert the chosen gene into the plant cell's chromosome. *Agrobacterium* is a ready-made delivery system for getting foreign DNA into plants; it is a natural genetic engineer. In January 1983, at a watershed conference in Miami, Chilton, Horsch (for Monsanto), and Schell all presented independent results confirming that *Agrobacterium* was up to the task. And by this time, each of the three groups had also applied for patents on *Agrobacterium*-based methods of genetic alteration. Schell's was recognized in Europe, but in the United States, a falling-out between Chilton and Monsanto would rumble through the courts until 2000, when a patent was finally awarded to Chilton and her new employer, Syngenta. But having now seen a bit of the Wild West show that is intellectual property patents, one shouldn't be surprised to hear that the story does not end so neatly there: as I write, Syngenta is in court suing Monsanto for patent infringement.

At first *Agrobacterium* was thought to work its devious magic only on certain plants. Among these, we could not, alas, count the agriculturally important group that includes cereals such as corn, wheat, and rice. However, in the years since it gave birth to plant genetic engineering, *Agrobacterium* has itself been the focus of genetic engineers, and technical advances have extended its empire to even the most recalcitrant crop species. Before these innovations, we had to rely upon a rather more haphazard, but no less effective, way of getting our DNA selection into a corn, wheat, or rice cell. The desired gene is affixed to tiny gold or tungsten pellets, which are literally fired like bullets into the cell. The trick is to fire the pellets with enough force to enter the cell, but not so much that they will exit the other side! The method lacks *Agrobacterium*'s finesse, but it does get the job done.

This "gene gun" was developed during the early 1980s by John Sanford at Cornell's Agricultural Research Station. Sanford chose to experiment with onions because of their conveniently large cells; he recalls that the combination of blasted onions and gunpowder made his lab smell like a McDonald's franchise on a firing range. Initial reactions to his concept were incredulous, but in 1987 Sanford unveiled his botanical firearm in the pages of *Nature*. By 1990, scientists had succeeded in using the gun to shoot new genes into corn, America's most important food crop, worth $19 billion in 2001 alone.

A "gene gun" for shooting DNA into plant cells

Corn is not only a valuable food crop; unique among major American crops, it also has long been a valuable seed crop. The seed business has traditionally been something of a financial dead-end: a farmer buys your seed, but then for subsequent plantings he can take seed from the crop he has just grown, so he never needs to buy your seed again. American corn seed companies solved the problem of nonrepeat business in the twenties by marketing hybrid corn, each hybrid the product of a cross between two particular genetic lines of corn. The hybrid's characteristic high yield makes it attractive to farmers. Because of the Mendelian mechanics of breeding, the strategy of using seed from the crop itself (i.e., the product of a hybrid × hybrid cross) fails because most of the seed will lack those high-yield characteristics of the original hybrid. Farmers therefore must return to the seed company every year for a new batch of high-yield hybrid seed.

Hybrid corn companies have for years hired an army of "detasselers" to remove the male flowers, tassels, from corn plants. This prevents self-pollination, ensuring that the seeds produced are indeed hybrid—the product of the cross between two separate strains.

America's biggest hybrid corn seed company, Pioneer Hi-Bred International (now owned by Du Pont), has long been a midwestern institution. Today it controls about 40 percent of the U.S. corn seed market, with $1 billion in annual sales. Founded in 1926 by Henry Wallace, who went on to become Franklin D. Roosevelt's vice president, the company used to hire as many as forty thousand high-schoolers every summer to ensure the hybridity of its hybrid corn. The two parental strains were grown in neighboring stands, and then these "detasselers" removed by hand the male pollen-producing flowers (tassels) before they became mature from one of the two strains. Therefore, only the other strain could serve as a possible source of pollen, so all the seed produced by the detasseled strain was sure to be hybrid. Even today, detasseling provides summer work for thousands: in July 2002, Pioneer hired thirty-five thousand temps for the job.

One of Pioneer's earliest customers was Roswell Garst, an Iowa farmer who, impressed by Wallace's hybrids, bought a license to sell Pioneer seed corn. On September 23, 1959, in one of the less frigid moments of the Cold War, the Soviet leader Nikita Khrushchev visited Garst's farm to learn more about the American agricultural miracle and the hybrid corn behind it. The nation Khrushchev had inherited from Stalin had neglected agriculture in the drive

The Cold War Corn Summit: Soviet leader Khrushchev with Iowa farmer Roswell Garst (right) in 1959

toward industrialization, and the new premier was keen to make amends. In 1961, the incoming Kennedy administration approved the sale to the Soviets of corn seed, agricultural equipment, and fertilizer, all of which contributed to the doubling of Soviet corn production in just two years.

As the GM food debate swirls around us, it is important to appreciate that our custom of eating food that has been genetically modified is actually thousands of years old. In fact, both our domesticated animals, the source of our meat, and the crop plants that furnish our grains, fruits, and vegetables, are very far removed genetically from their wild forebears.

Agriculture did not suddenly arise, fully fledged, ten thousand years ago. Many of the wild ancestors of crop plants, for example, offered relatively little to the early farmers: they were low-yield and hard to grow. Modification was necessary if agriculture was to succeed. Early farmers understood that modification must be bred in ("genetic," we would say) if desirable characteristics were to be maintained from generation to generation. Thus began our agrarian ancestors' enormous program of genetic modification. And in the absence of

The effect of eons of artificial selection: corn and its wild ancestor, teosinte (left)

gene guns and the like, this activity depended on some form of artificial selection, whereby farmers bred only those individuals exhibiting the desired traits—the cows with the highest milk yield, for example. In effect, the farmers were doing what nature does in the course of natural selection: picking and choosing from among the range of available genetic variants to ensure that the next generation would be enriched with those best adapted for consumption, in the case of farmers; for survival, in the case of nature. Biotechnology has given us a way to generate the desired variants, so that we do not have to wait for them to arise naturally; as such, it is but the latest in a long line of methods that have been used to *genetically modify* our food.

Weeds are difficult to eliminate. Like the crop whose growth they inhibit, they are plants too. How do you kill weeds without killing your crop? Ideally, there would be some kind of pass-over system whereby every plant lacking a "protective mark"—the weeds, in this case—would be killed, while those possessing the mark—the crop—would be spared. Genetic engineering has furnished farmers and gardeners just such a system in the form of Monsanto's "Roundup Ready" technology. "Roundup" is a broad-spectrum herbicide that can kill almost any plant. But through genetic alteration Monsanto scientists have also produced "Roundup Ready" crops that possess built-in resistance to the herbicide, and do just fine as all the weeds around them are biting the dust. Of course, it suits the company's commercial interests that farmers who buy Monsanto's adapted seed will buy Monsanto's herbicide as well. But such an approach is also actually beneficial to the environment. Normally a farmer must use a range of different weed killers, each one toxic to a particular group of weeds but safe for the crop. There are many potential weed groups to guard against. Using a single herbicide for all the weeds in creation actually reduces the environmental levels of such chemicals, and Roundup itself is rapidly degraded in the soil.

Unfortunately, the rise of agriculture was a boon not only to our ancestors but to herbivorous insects as well. Imagine being an insect that eats wheat and related wild grasses. Once upon a time, thousands of years ago, you had to forage far and wide for your dinner. Then along came agriculture, and humans conveniently started laying out dinner in enormous stands. It is not surprising that crops have to be defended against insect attack. From the elimination point of view at least, insects pose less of a problem than weeds because it is possible to devise poisons that target animals, not plants. The trouble is that humans and other creatures we value are animals as well.

The full extent of the risks involved with the use of pesticides was not widely apparent until Rachel Carson first documented them. The impact on the environment of long-lived chlorine-containing pesticides like DDT (banned in Europe and North America since 1972) has been devastating. In addition, there is a danger that residues from these pesticides will wind up in our food. While these chemicals at low dosage may not be lethal—they were, after all, designed to kill animals at a considerable evolutionary remove from us—there remain concerns about possible mutagenic effects, resulting in human cancers and birth defects. An alternative to DDT came in the form of a group of organophosphate pesticides, like parathion. In their favor, they decompose rapidly once applied and do not linger in the environment. On the other hand, they are even more acutely toxic than DDT; the sarin nerve gas used in the terrorist attack on the Tokyo subway system in 1995, for instance, is a member of the organophosphate group.

Even solutions using nature's own chemicals have produced a backlash. In the mid-1960s, chemical companies began developing synthetic versions of a natural insecticide, pyrethrin, derived from a small daisylike chrysanthemum. These helped keep farm pests in check for more than a decade until, not surprisingly, their widespread use led to the emergence of resistant insect populations. Even more troubling, however, pyrethrin, though natural, is not necessarily good for humans; in fact, like many plant-derived substances it can be quite toxic. Pyrethrin experiments with rats have produced Parkinson-like symptoms, and epidemiologists have noted that this disease has a higher incidence in rural environments than in urban ones. Overall—and there is a dearth of reliable data—the Environmental Protection Agency estimates that there may be as many as 300,000 pesticide-related illnesses among U.S. farmworkers every year.

Organic farmers have always had their tricks for avoiding pesticides. One ingenious organic method relies on a toxin derived from a bacterium—or, often, the bacterium itself—to protect plants from insect attack. *Bacillus thuringiensis* (Bt) naturally assaults the cells of insect intestines, feasting upon the nutrients released by the damaged cells. The guts of the insects exposed to the bacterium are paralyzed, causing the creatures to die from the combined effects of starvation and tissue damage. Originally identified in 1901, when it decimated Japan's silkworm population, *Bacillus thuringiensis* was not so named until 1911, during an outbreak among flour moths in the German province of Thuringia. First used as a pesticide in France in 1938, the bacterium was originally thought to work only against lepidopteran (moth/butterfly) caterpillars, but different strains have subsequently proved effective against the larvae of beetles and flies. Best of all, the bacterium is insect-specific: most animal intestines are acidic—that is, low pH—but the insect larval gut is highly alkaline—high pH—just the environment in which the pernicious Bt toxin is activated.

In the age of recombinant DNA technology the success of *Bacillus thuringiensis* as a pesticide has inspired genetic engineers. What if, instead of applying the bacterium scattershot to crops, the gene for the Bt toxin were engineered into the genome of crop plants? The farmer would never again need to dust his crops because every mouthful of the plant would be lethal to the insect ingesting it (and harmless to us). The method has at least two clear advantages over the traditional dumping of pesticides on crops. First, only insects that actually eat the crop will be exposed to the pesticide; non-pests are not harmed, as they would be with external application. Second, implanting the Bt toxin gene into the plant genome causes it to be produced by every cell of the plant; traditional pesticides are typically applied only to the leaf and stem. And so bugs that feed on the roots or that bore inside plant tissues, formerly immune to externally applied pesticides, are now also condemned to a Bt death.

Today we have a whole range of Bt designer crops, including "Bt corn," "Bt potato," "Bt cotton," and "Bt soybean," and the net effect has been a massive

Bt cotton: cotton genetically engineered to produce insecticidal Bt toxin (right) thrives while a non-Bt crop is trashed by pest insects.

reduction in the use of pesticides. In 1995 cotton farmers in the Mississippi Delta sprayed their fields an average of 4.5 times per season. Just one year later, as Bt cotton caught on, that average—for all farms, including those planting non-Bt cotton varieties—dropped to 2.5 times. It is estimated that since 1996 the use of Bt crops has resulted in an annual reduction of 2 million gallons of pesticides in the United States. I have not visited cotton country lately but I would wager that billboards there are no longer hawking chemical insect-killers; in fact, I suspect that Burma-Shave ads are more likely to make a comeback than ones for pesticides. And other countries are starting to benefit as well: in China in 1999 the planting of Bt cotton reduced pesticide use by an estimated 1,300 *tons*.

Biotechnology has also fortified plants against other traditional enemies in a surprising form of disease prevention superficially similar to vaccination. We inject our children with mild forms of various pathogens to induce an immune response that will protect them against infection when they are subsequently exposed to the disease. Remarkably when a plant, which has no immune system properly speaking, has been exposed to a particular virus, it often becomes resistant to other strains of the same virus. Roger Beachy at Washington University, in St. Louis, realized that this phenomenon of "cross-protection" might allow genetic engineers to "immunize" plants against threatening diseases. He tried inserting the gene for the virus's protein coat into the plants to see whether this might induce cross-protection without exposure to the virus itself. It did indeed. Somehow the presence in the cell of the viral coat protein prevents the cell from being taken over by invading viruses.

Beachy's method saved the Hawaiian papaya business. Between 1993 and 1997, production declined by 40 percent thanks to an invasion of the papaya ringspot virus; one of the islands' major industries was thus threatened with extinction. By inserting a gene for just part of the virus's coat protein into the papaya's genome, scientists were able to create plants resistant to attacks by the virus. Hawaii's papayas lived to fight another day.

Scientists at Monsanto later applied the same harmless method to combat a common disease caused by potato virus X. (Potato viruses are unimaginatively named. There is also a potato virus Y.) Unfortunately, McDonald's and other major players in the burger business feared the use of such modified spuds would lead to boycotts organized by the anti-GM food partisans. Consequently, the fries they now serve cost more than they should.

Nature conceived onboard defense systems hundreds of millions of years before human genetic engineers started inserting Bt genes into crop plants. Biochemists recognize a whole class of plant substances, so-called secondary products, that are not involved in the general metabolism of the plant. Rather, they are produced to protect against herbivores and other would-be attackers. The average plant is, in fact, stuffed full of chemical toxins developed by evolution. Over the ages, natural selection has understandably favored those plants containing the nastiest range of secondary products because they are less vulnerable to damage by herbivores. In fact, many of the substances that humans have learned to extract from plants for use as medicine (digitalis from the foxglove plant, used in precise doses, can treat heart patients), stimulants (cocaine from the coca plant), or pesticides (pyrethrin from chrysanthemums) belong to this class of secondary products. Poisonous to the plant's natural enemies, these substances constitute the plant's meticulously evolved defensive response.

Bruce Ames, who devised the Ames test, a procedure widely relied upon for determining whether or not a particular substance is carcinogenic, has noted that the natural chemicals in our food are every bit as lethal as the noxious chemicals we worry about. Referring to tests on rats, he takes coffee as an example:

> There are more rodent carcinogens in one cup of coffee than pesticide residues you get in a year. And there's still a thousand chemicals left to test in a cup of coffee. So it just shows our double standard: If it's synthetic we really freak out, and if it's natural we forget about it.

One ingenious set of chemical defenses in plants involves furanocoumarins, a group of chemicals that become toxic only when directly exposed to ultraviolet light. By this natural adaptation, the toxins are activated only when a herbivore starts munching on the plants, breaking open the cells and exposing their contents to sunlight. Furanocoumarins present in the peel of limes were responsible for a bizarre plague that struck a Club Med resort in the Caribbean. The guests who found themselves afflicted with ugly rashes on their thighs had all participated in a game that involved passing a lime from one person to the next without using hands, feet, arms, or head. In the bright Caribbean sunlight the activated furanocoumarins in the humiliated lime had wreaked a terrible revenge on numerous thighs.

Plants and herbivores are involved in an evolutionary arms race: nature selects plants to be ever more toxic, and herbivores to be ever more efficient at detoxifying the plant's defensive substances while metabolizing the nutritious ones. In the face of furanocoumarins, some herbivores have evolved clever countermeasures. Some caterpillars, for example, roll up a leaf before starting to munch. Sunlight does not penetrate the shady confines of their leaf roll, and thus the furanocoumarins are not activated.

Adding a particular Bt gene to crop plants is merely one way the human species as an interested party can give plants a leg up in this evolutionary arms race. We should not be surprised, however, to see pest insects eventually evolve resistance to that particular toxin. Such a response, after all, is the next stage in the ancient conflict. When it happens, farmers will likely find that the multiplicity of available Bt toxin strains can furnish them yet another exit from the vicious evolutionary cycle: as resistance to one type becomes common, they can simply plant crops with an alternative strain of Bt toxin onboard.

In addition to defending a plant against its enemies, biotechnology can also help bring a more desirable product to market. Unfortunately, however, sometimes the cleverest biotechnologists can fail to see the forest for the trees (or the crop for the fruits). So it was with Calgene, an innovative California-based company. In 1994 Calgene earned the distinction of producing the very first GM product to reach supermarket shelves. Calgene had solved a major problem of tomato growing: how to bring ripe fruit to market instead of picking them when green, as is customary. But in their technical triumph they forgot fundamentals: their rather unfortunately named "Flavr-Savr" tomato was neither tasty nor cheap enough to succeed. And so it was that the tomato had the added distinction of being one of the first GM products to disappear from supermarket shelves.

Still, the technology was ingenious. Tomato ripening is naturally accompanied by softening, thanks to the gene encoding an enzyme called polygalacturonase (PG), which softens the fruit by breaking down the cell walls. Because soft tomatoes do not travel well, the fruit are typically picked when they are still green (and firm) and then reddened using ethene gas, a ripening agent. Calgene researchers figured that knocking out the PG gene would result in fruit that stayed firm longer, even after ripening on the vine. They inserted an inverted copy of the PG gene, which, owing to the affinities between complementary base pairs, had the effect of causing the RNA produced by the PG gene proper to become "bound up" with the RNA produced by the inverted gene, thus neutralizing the former's capacity to create the softening enzyme. The lack of PG function meant that the tomato stayed firmer, and so it was now possible in principle to deliver fresher, riper tomatoes to supermarket shelves. But Calgene, triumphant in its molecular wizardry, underestimated the trickiness of basic tomato farming. (As one grower hired by the company commented, "Put a molecular biologist out on a farm, and he'd starve to death.") The strain of tomato Calgene had chosen to enhance was a particularly bland and tasteless one: there simply was not much "flavr" to save, let alone savor. The tomato was a technological triumph but a commercial failure.

Overall, plant technology's most potentially important contribution to human well-being may involve enhancing the nutrient profile of crop plants, compensating for their natural shortcomings as sources of nourishment. Because plants are typically low in amino acids essential for human life, those who eat a purely vegetarian diet, among whom we may count most of the developing world, may suffer from amino acid deficiencies. Genetic engineering can ensure that crops contain a fuller array of nutrients, including amino acids, than the unmodified versions that would otherwise be grown and eaten in these parts of the world.

To take an example, in 1992 UNICEF estimated that some 124 million children around the world were dangerously deficient in vitamin A. The annual result is some half million cases of childhood blindness; many of these children will even die for want of the vitamin. Since rice does not contain vitamin A or its biochemical precursors, these deficient populations are concentrated in parts of the world where rice is the staple diet.

An international effort, funded largely by the Rockefeller Foundation (a nonprofit organization and therefore protected from the charges of commercialism or exploitation often leveled at producers of GM foods), has developed what has come to be called "golden rice." Though this rice doesn't contain vitamin A per se, it yields a critical precursor, beta-carotene (which gives carrots their bright orange color and golden rice the fainter orange tint that inspired its name). As those involved in humanitarian relief have learned, however, malnutrition can be more complex than a single deficiency: the absorption of vitamin A precursors in the gut works best in the presence of fat, but the malnourished whom the golden rice was designed to help often have little or no fat in their diet. Nevertheless golden rice represents at least one step in the right direction. It is here that we see the broader promise of GM agriculture to diminish human suffering.

We are merely at the beginning of a great GM plant revolution, only starting to see the astonishing range of potential applications. Apart from delivering nutrients where they are wanting, plants may also one day hold the key to distributing orally administered vaccine proteins. By simply engineering a banana that produces, say, the polio vaccine protein—which would remain intact in the fruit, which travels well and is most often eaten uncooked—we could one day distribute the vaccine to parts of the world that lack public health infrastructure. Plants may also serve less vital but still immensely helpful purposes. One company, for example, has succeeded in inducing cotton plants to produce a form of polyester, thereby creating a natural cotton-polyester blend. With such potential to reduce our dependence on chemical manufacturing processes (of which polyester fabrication is but one) and their polluting by-products, plant engineering will provide ways as yet unimagined to preserve the environment.

Monsanto was definitely the leader of the GM food pack, but naturally its primacy was challenged. The German pharmaceutical company Hoechst developed its own Roundup equivalent, an herbicide called Basta (or Liberty in the United States), with which they marketed "LibertyLink" crops genetically engineered for resistance. Another European pharmaceutical giant, Aventis, produced a version of Bt corn called "Starlink."

But Monsanto, aiming to capitalize on being biggest and first, aggressively lobbied the big seed companies, notably Pioneer, to license Monsanto's products. But Pioneer was still wed to its long-established hybrid corn methods so its response to the heated courtship was frustratingly lukewarm and, in deals made in 1992 and 1993, Monsanto looked inept when it was able to exact from the seed giant only a paltry $500,000 for rights to Roundup Ready soybeans and $38 million for Bt corn. When he became CEO of Monsanto in 1995, Robert Shapiro aimed to redress this defeat by positioning the company for all-out domination of the seed market. For a start, he broadened the attack on the old seed-business problem of farmers who replant using seed from last year's crop rather than paying the seed company a second time. The hybrid solution that worked so well for corn was unworkable for other crops. Shapiro, therefore, proposed that farmers using Bt seed sign a "technology agreement" with Monsanto, obliging them both to pay for use of the gene and to refrain from replanting with seed generated by their own crops. What Shapiro had engineered was a hugely effective way to make Monsanto anathema in the farming community.

Shapiro was an unlikely CEO for a midwestern agrichemical company. Working as a lawyer at the pharmaceutical outfit Searle, he had the marketing equivalent of science's "Eureka!" moment. By compelling Pepsi and Coca-Cola to put the name of Searle's brand of chemical sweetener on their diet soft drink containers, Shapiro made NutraSweet synonymous with a low-calorie lifestyle. In 1985, Monsanto acquired Searle and Shapiro started to make his way up the parent company's corporate ladder. Naturally, once he was appointed CEO, Mr. NutraSweet had to prove he was no one-trick pony.

In an $8 billion spending spree in 1997–98, Monsanto bought a number of major seed companies, including Pioneer's biggest rival, Dekalb, as Shapiro schemed to make Monsanto into the Microsoft of seeds. One of his intended purchases, the Delta and Pine Land Company, controlled 70 percent of the U.S. cottonseed market. Delta and Pine also owned the rights to an interesting biotech innovation invented in a U.S. Department of Agriculture research lab in Lubbock, Texas: a technique for preventing a crop from producing any fertile seeds. The ingenious molecular trick involves flipping a set of genetic switches in the seed before it is sold to the farmer. The crop develops normally but produces seeds incapable of germinating. Here was the real key to making money in the seed business! Farmers would *have* to come back every year to the seed company.

Though it might seem in principle counterproductive and something of an oxymoron, nongerminating seed is actually of general benefit to agriculture in the long run. If farmers buy seed every year (as they do anyway, in the case of hybrid corn), then the improved economics of seed production promote the development of new (and better) varieties. Ordinary (germinating) forms would always be available for those who wished them. Farmers would buy the nongerminating kind only if it were superior in yield and other characteristics farmers care about. In short, nongerminating technology, while closing off one option, provides farmers with more and ever improved seed choices.

For Monsanto, however, this technology precipitated a public relations disaster. Activists dubbed it the "terminator gene." They evoked visions of the downtrodden third world farmer, accustomed by tradition to relying on his last crop to provide seeds to sow for the new one. Suddenly finding his own seeds useless, he would have no choice but to return to the greedy multinational and, like Oliver Twist, beg pathetically for more. Monsanto backed off, a humiliated Shapiro publicly disavowed the technology, and the terminator gene remains out of commission to this day. Through the public relations fallout, its only real impact to date has been the termination of Monsanto's grandiose ambitions of the late 1990s.

Much of the hostility to GM foods, as we saw in the last chapter with bovine growth hormone, has been orchestrated by professional alarmists like Jeremy Rifkin. His counterpart in the United Kingdom, Lord Peter Melchett, was equally effective until he lost credibility in the environmental movement by quitting Greenpeace to join a public relations firm that has in the past worked for Monsanto. Rifkin, the son of a self-made plastic-bag manufacturer from Chicago, may differ in style from Melchett, a former Eton boy from a grand family, but they share a vision of corporate America as conspiratorial juggernaut pitted against the helpless common man.

Nor has the reception of GM foods been aided by the knee-jerk, politically craven attitudes and even scientific incompetence typical of governmental regulatory agencies—in this country the Food and Drug Administration (FDA) and the Environmental Protection Agency (EPA)—when they have been confronted with these new technologies. Roger Beachy, who first identified the "cross-protection" phenomenon that saved Hawaii's papaya farmers from ruin, remembers how the EPA responded to his breakthrough:

> I naively thought that developing virus-resistant plants in order to reduce the use of insecticides would be viewed as a positive advance. However the EPA basically said, "If you use a gene that protects the plant from a virus, which is a pest, that gene must be considered a pesticide." Thus the EPA considered the genetically transformed plants to be pesticidal. The point of the story is that as genetic sciences and biotech developed, the federal agencies were taken somewhat by surprise. The agencies did not have the background or expertise to regulate the new varieties of crop plants that were developed, and they did not have the background to regulate the environmental impacts of transgenic crops in agriculture.

An even more glaring instance of the government regulators' ineptitude came in the so-called Starlink episode. Starlink, a Bt corn variety produced by the European multinational Aventis, had run afoul of the EPA when its Bt protein was found not to degrade as readily as other Bt proteins in an acidic environment, one like that of the human stomach. In principle, therefore, eating Starlink corn *might* cause an allergic reaction, though there was never any evidence that it actually would. The EPA dithered. Eventually it decided to approve Starlink for use in cattle feed, but not for human consumption. And so under EPA "zero-tolerance" regulations, the presence of a single molecule of Starlink in a food product constituted illegal contamination. Farmers were growing Starlink and non-Starlink corn side by side, and non-Starlink crops inevitably became contaminated: even a single Starlink plant that had inadvertently found its way into the harvest from whole fields of non-Starlink was enough. Not surprisingly, Starlink began to show up in food products. The absolute quantities were tiny, but genetic testing to detect the presence of Starlink is supersensitive. In late September 2000, Kraft Foods launched a recall of taco shells deemed to be tainted with Starlink, and a week later Aventis began a buy-back program to recover Starlink seed from the farmers who had bought it. The estimated cost of this "cleanup" program: $100 million.

Blame for this debacle can only be laid at the door of an overzealous and irrational EPA. Permitting the use of corn for one purpose (animal feed) and not another (human consumption), and then mandating absolute purity in food is, as is now amply apparent, absurd. Let us be clear that if "contamination" is defined as the presence of a single molecule of a foreign substance, then every morsel of our food is contaminated! With lead, with DDT, with bacterial toxins, and a host of other scary things. What matters, from the point of view of public health, is the concentration levels of these substances, which can range from the negligible to the lethal. It should also be considered a reasonable requirement in labeling something a contaminant that there be at least minimal evidence of demonstrable detriment to health. Starlink has never been shown to harm anyone, not even a laboratory rat. The only positive outcome of this whole sorry episode has been a change in EPA policy abolishing "split" permits: an agricultural product will hereafter be approved for all food-related uses or not.

That the anti-GM food lobby is most powerful in Europe is no accident. Europeans, the British in particular, have good reason both to be suspicious about what is in their food and to distrust what they are told about it. In 1984, a farmer in the south of England first noticed that one of his cows was behaving strangely; by 1993, 100,000 British cattle had died from a new brain disease, bovine spongiform encephalopathy (BSE), commonly known as mad cow disease. Government ministers scrambled to assure the public that the disease, probably transmitted in cow fodder derived from remnants of slaughtered animals, was not transmissible to humans. By February 2002, 106 Britons had died from the human form of BSE. They had been infected by eating BSE-contaminated meat.

The insecurity and distrust generated by BSE has spilled over into the discussion of GM foods, dubbed by the British press "Frankenfoods." As Friends of the Earth announced in a press release in April 1997, "After BSE, you'd think the food industry would know better than to slip 'hidden' ingredients down people's throats." But that, more or less, is exactly what Monsanto was planning to do in Europe. Certain the anti-GM food campaign was merely a passing distraction, management pressed ahead with its plans to bring GM products to European supermarket shelves. It was to prove a major miscalculation: through 1998, the consumer backlash gained momentum. Headline writers at the British tabloids had a field day: "GM Foods Are Playing Games with Nature: If Cancer Is the Only Side-Effect We Will Be Lucky"; "Astonishing Deceit of GM Food Giant"; "Mutant Crops." Prime Minister Tony Blair's halfhearted defense merely provoked tabloid scorn: "The Prime Monster; Fury As Blair Says: I Eat Frankenstein Food and It's Safe." In March 1999, the British supermarket chain Marks and Spencer announced that it would not carry GM food products, and soon Monsanto's European biotech dreams were in jeopardy. Not surprisingly other food retailers took similar actions: it made good sense to show supersensitivity to consumer concerns, and no sense at all to stick one's neck out in support of an unpopular American multinational.

It was around this time of the Frankenfood maelstrom in Europe that news of the terminator gene and Monsanto's plans to dominate the global seed market began to circulate on the home front. With much of the opposition orchestrated by environmental groups, the company's attempts to defend itself were hamstrung by its own past. Having started out as a producer of pesticides, Monsanto was loath to incur the liability of explicitly renouncing these chemicals as environmental hazards. Yet one of the greatest virtues of both Roundup Ready and Bt technologies is the extent to which they reduce the need for herbicides and insecticides. The official industry line since the 1950s had been that proper use of the right pesticides harmed neither the environment nor the farmer applying them: Monsanto still could not now admit that Rachel Carson had been right all along. Unable to simultaneously condemn pesticides and sell them, the company could not make use of one of the most compelling of arguments in defense of the use of biotechnology on the farm.

Monsanto was never able to reverse this unfortunate momentum. In April 2000, the company effected a merger but its partner, the pharmaceutical giant Pharmacia & Upjohn, was primarily interested in acquiring Monsanto's drug division, Searle. The agricultural business, later spun off as an independent entity, still exists today under the name Monsanto. Gone, however, are the company's pioneering bravado and aura of invincibility.

The GM foods debate has conflated two distinct sets of issues. First, there have been the purely scientific questions of whether GM foods pose a threat to our health or to the environment. Second, there are economic and political questions centered on the practices of aggressive multinational companies and the effects of globalization. Much of the rhetoric has focused on agribusiness, Monsanto in particular. Having seemed throughout the 1990s to view the technology as little more than a means of dominating the world food supply, the company may indeed have harbored unwholesome dreams of becoming the Microsoft of the food industry, but since its stunning reversal of fortunes, this aspect of the controversy has been rendered largely baseless. It is not likely that another company with as much to lose will stumble into the same minefield. A meaningful evaluation of GM food should be based on scientific considerations, not political or economic ones. Let us therefore review some of the common claims.

It ain't natural. Virtually no human being, save the very few remaining genuine hunter-gatherers, eats a strictly "natural" diet. *Pace* Prince Charles, who famously declared in 1998 that "this kind of genetic modification takes mankind into realms that belong to God," our ancestors have in fact been fiddling in these realms for eons.

Early plant breeders often crossed different species, bringing into existence entirely new ones with no direct counterparts in nature. Wheat, for example, is the product of a whole series of crosses. Einkorn wheat, a naturally occurring progenitor, crossed with a species of goat grass, produced emmer wheat. And the bread wheat we know was produced by a subsequent crossing of emmer with yet another goat grass. Our wheat is thus a combination—perhaps one nature would have never devised—of the characteristics of all these ancestors.

Furthermore, crossing plants in this way results in the wholesale generation of genetic novelty: every gene is affected, often with unforeseeable effects. Biotechnology, by contrast, allows us to be much more precise in introducing new genetic material into a plant species, one gene at a time. It is the difference

Detail of Brueghel's painting The Harvesters *shows wheat as it was in the sixteenth century—five feet high. Artificial selection has since halved its height, making it easier to harvest; because the plant puts less energy into growing its stem, its seed heads are larger and more nutritious.*

between traditional agriculture's genetic sledgehammer and biotech's genetic tweezers.

It will result in allergens and toxins in our food. Again, the great advantage of today's transgenic technologies is the precision they allow us in determining how we change the plant. Aware that certain substances tend to provoke allergic reactions, we can accordingly avoid them. But this concern persists, stemming to some degree from an oft-told tale about the addition of a Brazil nut protein to soybeans. It was a well-intentioned undertaking: the West African diet is often deficient in methionine, an amino acid abundant in a protein produced by Brazil nuts. It seemed a sensible solution to insert the gene for the protein into West Africa's soybean, but then someone remembered that there is a common allergic reaction to Brazil nut proteins that can have serious consequences, and so the project was shelved. Obviously the scientists involved had no intention of unleashing a new food that would promptly send thousands of people into anaphylactic shock; they halted the project once the serious drawbacks were appreciated. But for most commentators it was an instance of molecular engineers playing with fire, heedless of the consequences. In principle, genetic engineering can actually *reduce* the instance of allergens in food: perhaps the Brazil nut itself will one day be available free of the protein that was deemed unsafe to import into the soybean.

It is indiscriminate, and will result in harm to nontarget species. In 1999 a now-famous study showed that monarch butterfly caterpillars feeding on leaves heavily dusted with pollen from Bt corn were prone to perish. This was scarcely surprising: Bt pollen contains the Bt gene, and therefore the Bt toxin, and the toxin is intentionally lethal to insects. But everyone loves butterflies, and so environmentalists opposed to GM foods had found an icon. Would the monarch, they wondered, be but the first of many inadvertent victims of GM technology? Upon examination, the experimental conditions under which the caterpillars were tested were found to be so extreme—the levels of the Bt pollen so high—as to tell us virtually nothing of practical value about the likely mortality of caterpillar populations in nature. Indeed, further study has suggested that the impact of Bt plants on the monarch (and other nontarget insects) is trivial. But even if it were not, we should ask how it might compare

Reports of the impact of Bt corn pollen on the caterpillars of monarch butterflies galvanized opponents of agricultural biotechnology. In 2000, this protester dressed as a monarch attracted the interest of Boston's finest.

with the effects of the traditional non-GM alternative: pesticides. As we have seen, in the absence of GM methods, these substances must be applied liberally if we are to have agriculture that is as productive as modern society requires. Whereas the toxin built into Bt plants affects only those insects that actually feed off the plant tissue (and to some lesser degree, insects exposed to Bt pollen), pesticides unambiguously affect all insects exposed, pest and non-pest alike. The monarch butterfly, were it capable of weighing in on the debate, would assuredly cast its vote in favor of Bt corn.

It will lead to an environmental meltdown with the rise of "superweeds." The worry here is that genes for herbicide resistance (like those in Roundup Ready plants) will migrate out of the crop genome into that of the weed population through interspecies hybridization. This is not inconceivable, but it is unlikely to occur on a wide scale for the following reason: interspecies hybrids tend to be feeble creations, not well equipped for survival. This is especially true when one of the species is a domesticated variety bred to thrive only when mollycoddled by a farmer. But let us suppose, for argument's sake, that the resistance gene does enter the weed population and is sustained there. It would not actually be the end of the world, or even of agriculture, but rather an instance of something that has occurred frequently in the history of farming: resistance arising in pest species in response to attempts to eradicate them. The most famous example is the evolution of resistance to DDT in pest insects. In applying a pesticide, a farmer is exerting strong natural selection in favor of resistance, and evolution, we know, is a subtle and able foe: resistance arises readily. The result is that the scientists have to go back to the drawing board and come up with a new pesticide or herbicide, one to which the target species is not resistant; the whole evolutionary cycle will then run its course before culminating once more in the evolution of resistance in the target species. The acquisition of resistance, therefore, is the potential undoing of virtually all attempts to control pests; it is by no means peculiar to GM strategies. It's simply the bell that signals the next round, and summons human ingenuity to invent anew.

Despite her concern about the impact of multinational corporations on farmers in countries like India, Suman Sahai of the New Delhi–based Gene Campaign has pointed out that the GM food controversy is a feature of societies for which food is not a life-and-death issue. In India, where people literally starve to death, as Sahai points out, up to 60 percent of fruit grown in hill regions rots before it reaches market. Just imagine the potential good of a technology that delays ripening, like the one used to create the Flavr-Savr tomato. The most important role of GM foods may lie in the salvation they offer developing regions, where surging birthrates and the pressure to produce on the limited available arable land lead to an overuse of pesticides and herbicides with devastating effects upon both the environment and the farmers applying them; where nutritional deficiencies are a way of life and, too often, of death; and where the destruction of one crop by a pest can be a literal death sentence for farmers and their families.

As we have seen, the invention of recombinant DNA methods in the early

1970s resulted in a round of controversy and soul-searching centered on the Asilomar conference. Now it is happening all over again. At the time of Asilomar, it may at least be said, we were facing several major unknowns: we could not then say for certain that manipulating the genetic makeup of the human gut bacterium, *E. coli,* would not result in new strains of disease-causing bacteria. But our quest to understand and our pursuit of potential for good proceeded, however haltingly. In the case of the present controversy, anxieties persist despite our much greater understanding of what we are actually doing. While a considerable proportion of Asilomar's participants urged caution, today one would be hard-pressed to find a scientist opposed in principle to GM foods. Recognizing the power of GM technologies to benefit both our species and the natural world, even the renowned environmentalist E. O. Wilson has endorsed them: "Where genetically engineered crop strains prove nutritionally and environmentally safe upon careful research and regulation . . . they should be employed."

The opposition to GM foods is largely a sociopolitical movement whose arguments, though couched in the language of science, are typically unscientific. Indeed, some of the anti-GM pseudoscience propagated by the media—whether in the interests of sensationalism or out of misguided but well-intentioned concern—would be actually amusing were it not evident that such gibberish is in fact an effective weapon in the propaganda war. Monsanto's Rob Horsch has had his fair share of run-ins with protesters:

> I was once accused of bribing farmers by an activist at a press conference in Washington, D.C. I asked what they meant. The activist answered that by giving farmers a better performing product at a cheaper price those farmers profited from using our products. I just looked at them with my mouth hanging open.

Let me be utterly plain in stating my belief that it is nothing less than an absurdity to deprive ourselves of the benefits of GM foods by demonizing them; and, with the need for them so great in the developing world, it is nothing less than a crime to be governed by the irrational suppositions of Prince Charles and others.

Experimental plots vandalized at Cold Spring Harbor Laboratory, 2000

In fact, a few years from now, when the West inevitably regains its senses and throws off the shackles of Luddite paranoia, it may find itself seriously lagging in agricultural technology. Food production in Europe and the United States will come to be more expensive and less efficient than elsewhere in the world. Meanwhile, countries like China, which can ill afford to entertain illogical misgivings, will forge ahead. The Chinese attitude is entirely pragmatic: With 23 percent of the world's population but only 7 percent of its arable land, China *needs* the increased yields and added nutritional value of GM crops if it is to feed its population.

On reflection, we erred too much on the side of caution at Asilomar, quailing before unquantified (indeed, unquantifiable) concerns about unknown and unforeseeable perils. But after a needless and costly delay, we resumed our pursuit of science's highest moral obligation: to apply what is known for the greatest possible benefit of humankind. In the current controversy, as our society delays in sanctimonious ignorance, we would do well to remember how much is at stake: the health of hungry people and the preservation of our most precious legacy, the environment.

In July 2000 anti-GM-food protesters vandalized a field of experimental corn at Cold Spring Harbor Lab. In fact there were no GM plants in the field; all the vandals managed to destroy was two years' hard work on the part of two young scientists at the lab. But the story is instructive all the same. At a time in which the destruction of GM crops has become positively fashionable in parts of Europe, when even the pursuit of knowledge on that continent and this one can come under attack, those in the vanguard of the cause might do well to ask themselves: what are we fighting *for*?

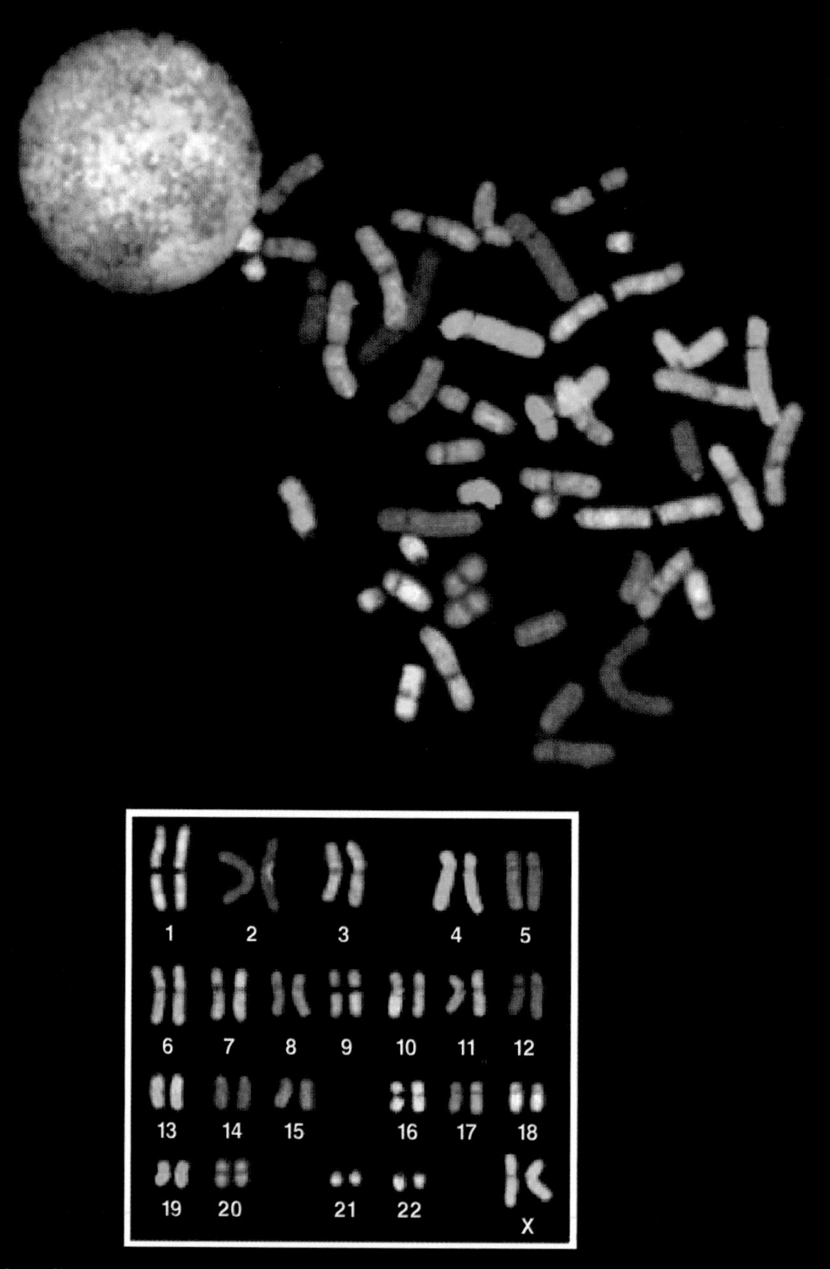

The full complement of human chromosomes highlighted by chromosome-specific fluorescent stains. The total number of chromosomes in each cell's nucleus is 46—two full sets, one from each parent. The genome is one set: twenty-three chromosomes—twenty-three very long DNA molecules.

CHAPTER SEVEN

. . .

THE HUMAN GENOME: LIFE'S SCREENPLAY

The human body is bewilderingly complex. Traditionally biologists have focused on one small part and tried to understand it in detail. This basic approach did not change with the advent of molecular biology. Scientists for the most part still specialize on one gene or on the genes involved in one biochemical pathway. But the parts of any machine do not operate independently. If I were to study the carburetor of my car engine, even in exquisite detail, I would still have no idea about the overall function of the engine, much less the entire car. To understand what an engine is for, and how it works, I'd need to study the whole thing—I'd need to place the carburetor in context, as one functioning part among many. The same is true of genes. To understand the genetic processes underpinning life, we need more than a detailed knowledge of particular genes or pathways; we need to place that knowledge in the context of the entire system—the genome.

The genome is the entire set of genetic instructions in the nucleus of every cell. (In fact, each cell contains *two* genomes, one derived from each parent: the two copies of each chromosome we inherit furnish us with two copies of each gene, and therefore two copies of the genome.) Genome sizes vary from species to species. From measurements of the amount of DNA in a single cell, we have been able to estimate that the human genome—half the DNA contents of a single nucleus—contains some 3.1 billion base pairs: 3,100,000,000 As, Ts, Gs, and Cs.

Genes figure in our every success and woe, even the ultimate one: they are implicated to some extent in all causes of mortality except accidents. In the most obvious cases, diseases like cystic fibrosis and Tay-Sachs are caused directly by mutations. But there are many other genes whose work is just as deadly, if more oblique, influencing our susceptibility to common killers like cancer and heart disease, both of which may run in families. Even our response to infectious diseases like measles and the common cold has a genetic component since the immune system is governed by our DNA. And aging is largely a genetic phenomenon as well: the effects we associate with getting older are to some extent a reflection of the lifelong accumulation of mutations in our genes. Thus, if we are to understand fully, and ultimately come to grips with, these life-or-death genetic factors, we must have a complete inventory of all the genetic players in the human body.

Above all, the human genome contains the key to our humanity. The freshly fertilized egg of a human and that of a chimpanzee are, superficially at least, indistinguishable, but one contains the human genome and the other the chimp genome. In each, it is the DNA that oversees the extraordinary transformation from a relatively simple single cell to the stunningly complex adult of the species, comprised, in the human instance, of 100 trillion cells. But only the chimp genome can make a chimp, and only the human genome a human. The human genome is the great set of assembly instructions that governs the development of every one of us. Human nature itself is inscribed in that book.

Understanding what is at stake, one might imagine that to champion a project seeking to sequence all the human genome's DNA would be no more controversial than sticking up for Mom and apple pie. Who in his right mind would object? In the mid-1980s, however, when the possibility of sequencing the genome was first discussed, this was viewed by some as a decidedly dubious idea. To others it simply seemed too preposterously ambitious. It was like suggesting to a Victorian balloonist that we attempt to put a man on the moon.

It was a telescope, of all things, that inadvertently helped inaugurate the Human Genome Project (HGP). In the early 1980s, astronomers at the University of California proposed to build the biggest, most powerful telescope in the world, with a projected cost of some $75 million. When the Max Hoffman

Foundation pledged $36 million, a grateful UC agreed to name the project for its generous benefactor. Unfortunately, this way of saying thank-you complicated the business of raising the remaining money. Other potential donors were reluctant to put up funds for a telescope already named for someone else, so the project stalled. Eventually, a second, much wealthier California philanthropy, the W. M. Keck Foundation, stepped in with a pledge to underwrite the entire project. UC was happy to accept, Hoffman or no. (The new Keck telescope, on the summit of Mauna Kea in Hawaii, would be fully operational by May 1993.) Unprepared to play second fiddle to Keck, the Hoffman Foundation withdrew its pledge, and UC administrators sensed a $36 million opportunity. In particular, Robert Sinsheimer, chancellor of UC Santa Cruz, realized that the Hoffman money could bankroll a major project that would "put Santa Cruz on the map."

Sinsheimer, a biologist by training, was keen to see his field enter the major leagues of big-money sciences. Physicists had their pricey supercolliders, astronomers their $75 million telescopes and satellites; why shouldn't biologists have their own high-profile, big-money project? So he suggested that Santa Cruz build an institute dedicated to sequencing the human genome; in May 1985, a conference convened at Santa Cruz to discuss Sinsheimer's idea. Overall it was deemed too ambitious and the participants agreed that the initial emphasis should instead be on exploring particular regions of the genome that were of medical importance. In the end, the discussion was moot because the Hoffman money did not actually make its way into the University of California's coffers. However, the Santa Cruz meeting had sown the seed.

The next step toward the Human Genome Project also came from deep in left field: the U.S. Department of Energy (DOE). Though its brief naturally concentrated on the nation's energy needs, the DOE did have at least one biological mandate: to assess the health risks of nuclear energy. In this connection, it had funded monitoring of long-term genetic damage in survivors of the atomic blasts at Nagasaki and Hiroshima and their descendants. What could be more useful in identifying mutations caused by radiation than a full reference sequence of the human genome? In the fall of 1985, the DOE's Charles DeLisi called a meeting to discuss his agency's genome initiative. The biological establishment was skeptical at best: Stanford geneticist David Botstein condemned the project as "DOE's program for unemployed bomb-makers," and James Wyngaarden, then head of the National Institutes of Health (NIH), likened the idea to "the National Bureau of Standards proposing to build the B-2 bomber." Not surprisingly, the NIH itself was eventually to become the most prominent member of the Human Genome Project coalition; nevertheless, the DOE played a significant role throughout the project, and, in the final reckoning, would be responsible for some 11 percent of the sequencing.

By 1986 the genome buzz was getting stronger. That June, I organized a special session to discuss the project during a major meeting on human genetics at Cold Spring Harbor Laboratory. Wally Gilbert, who had attended Sinsheimer's meeting the year before in California, took the lead by making a daunting cost projection: 3 billion base pairs, 3 billion dollars. This was big-money science for sure. It was an inconceivable sum to imagine without public funding, and some at the meeting were naturally concerned that the megaproject, whose success was hardly assured, would inevitably suck funds away from other critical research. The Human Genome Project, it was feared, would become scientific research's ultimate money pit. And at the level of the individual scientific ego, there was, even in the best case, relatively little career bang for the buck. While the HGP promised technical challenges aplenty, it failed to offer much in the way of intellectual thrill or fame to those who actually met them. Even an important breakthrough would be dwarfed by the size of the undertaking as a whole and who was going to dedicate his life to the endless tedium of sequencing, sequencing, sequencing? Stanford's David Botstein, in particular, demanded extreme caution: "It means changing the structure of science in such a way as to indenture us all, especially the young people, to this enormous thing like the Space Shuttle."

Despite the less than overwhelming endorsement, that meeting at Cold Spring Harbor Laboratory convinced me that sequencing the human genome was destined soon to become an international scientific priority, and that, when it did, the NIH should be a major player. I persuaded the James S. McDonnell Foundation to fund an in-depth study of the relevant issues under the aegis of the National Academy of Sciences (NAS). With Bruce Alberts of UC San Francisco chairing the committee, I felt assured that all ideas would be subject to the fiercest scrutiny. Not long before, Alberts had published an article warning that the rise of "big science" threatened to swamp traditional research's vast

Genesis of the genome project: Wally Gilbert and David Botstein at loggerheads at Cold Spring Harbor Laboratory, 1986

archipelago of innovative contributions from individual labs the world over. Without knowing for sure what our group would find, I took my place, along with Wally Gilbert, Sydney Brenner, and David Botstein, on the fifteen-member committee that during 1987 would hammer out the details of a potential genome project.

In those early days, Gilbert was the Human Genome Project's most forceful proponent. He rightly called it "an incomparable tool for the investigation of every aspect of human function." But having discovered the allure of the heady biotech mix of science and business at Biogen, the company he had helped found, Gilbert saw in the genome an extraordinary new business opportunity. And so, after serving briefly, he ceded his spot on the committee to Washington University's Maynard Olson to avoid any possible conflict of interest. Molecular biology had already proved its potential as big business, and Gilbert saw no need to go begging at the public trough. He reasoned that a private company with its own enormous sequencing laboratory could do the job and then sell genome information to pharmaceutical manufacturers and other interested parties. In spring 1987, Gilbert announced his plan to form Genome Corporation. Deaf to the howls of complaint at the prospect of genome data coming under private ownership (thus possibly limiting its application for the general good), Gilbert set about trying to raise venture capital. Unfortunately, he was handicapped at the outset by his own less-than-golden track record as a CEO. Following his resignation in 1982 from the Harvard faculty to take the reins of

Biogen, the company promptly lost $11.6 million in 1983 and $13 million in 1984. Understandably, Gilbert took refuge behind ivy-covered walls, returning to Harvard in December 1984, but Biogen continued to lose money after his departure. It was hardly the stuff of a mouth-watering investment prospectus, but ultimately Gilbert's grand plan foundered owing more to circumstances beyond his control than to any managerial shortcoming: the stock market crash of October 1987 abruptly terminated Genome Corp.'s gestation.

In fact, Gilbert was guilty of nothing as much as being ahead of his time. His plan was not so different from the one Celera Genomics would implement so successfully a full ten years after Genome Corp. was stillborn. And the concerns his venture provoked about the private ownership of DNA sequence data would come into ever sharper focus as the HGP progressed.

The plan our Gilbert-less NAS committee devised under Alberts made sense at the time—and indeed the Human Genome Project has been carried out more or less according to its prescriptions. Our cost and timing projections have also proved respectably close to the mark. Knowing, as any PC owner has learned, that over time technology gets both better and cheaper, we recommended that the lion's share of actual sequencing work be put off until the techniques reached a sensibly cost-effective level. In the meanwhile, the improvement of sequencing technologies should have high priority. In part toward this end, we recommended that the (smaller) genomes of simpler organisms be sequenced as well. The knowledge gained thereby would be valuable both intrinsically (as a basis for enlightening comparisons with the eventual human sequence) and as a means for honing our methods before attacking the big enchilada. (Of course the obvious nonhuman candidates were the geneticists' old flames: *E. coli,* baker's yeast, *C. elegans* [the nematode worm popularized for research by Sydney Brenner], and the fruit fly.)

Meanwhile, we should concentrate on *mapping* the genome as accurately as possible. Mapping would be both genetic and physical. Genetic mapping entails determining relative positions, the order of genetic landmarks along the chromosomes, just as Morgan's boys had originally done for the chromosomes of fruit flies. Physical mapping entails actually identifying the absolute positions of those genetic landmarks on the chromosome. (Genetic mapping tells you that gene 2, say, lies between genes 1 and 3; physical mapping tells you that gene 2 is 1 million base pairs from gene 1, and gene 3 is located 2 million base pairs further along the chromosome.) Genetic mapping would lay out the basic structure of the genome; physical mapping would provide the sequencers, when eventually they were let loose on the genome, with fixed positional anchors along the chromosomes. The location on a chromosome of each separate chunk of sequence could then be determined by reference to those anchors.

We estimated that the entire project would take about fifteen years and cost about $200 million per year. We did a lot more fancy arithmetic, but there was no getting away from Gilbert's $1 per base pair estimate. Each space shuttle mission costs some $470 million. The Human Genome Project would cost six space shuttle launches.

Our report was published in February 1988. The rough draft of the genome was published in 2001. The gaps continue to be filled in by sequencing labs around the world as I write, and in 2003—the fiftieth anniversary of the discovery of the double helix and the fifteenth of the committee's report—we will see the completion of the sequence.

While the NAS committee was still deliberating, I went to see key members of the House and Senate subcommittees on health that oversee the NIH's budget. James Wyngaarden, head of NIH, was in favor of the genome project "from the very start," as he put it, but less farsighted individuals at NIH were opposed. In my pitch for $30 million to get NIH on the genome track, I emphasized the medical implications of knowing the genome sequence. Lawmakers, like the rest of us, have all too often lost loved ones to diseases like cancer that have genetic roots, and could appreciate how knowing the sequence of the human genome would facilitate our fight against such diseases. In the end we got $18 million.

Meanwhile the DOE was able to secure $12 million for its own effort, mainly by playing up the project as a technological feat. This, one must remember, was the era of Japanese dominance in manufacturing technology; Detroit was in peril of being run over by Japan's automobile industry, and many feared the American edge in high-tech would be the next domino to fall. Rumor had it that three giant Japanese conglomerates (Matsui, Fuji, and Seiko) had combined forces to produce a machine capable of sequencing 1 million base pairs a day. It turned out to be a false alarm, but such anxieties ensured that the U.S. genome initiative would be pursued with the sort of fervor that put Americans on the moon before the Soviets.

In May 1988 Wyngaarden asked me to run NIH's side of the project. When I expressed reluctance to forsake the directorship of the Cold Spring Harbor Laboratory, he was able to arrange for me to do the NIH job on a part-time basis. I couldn't say no. Eighteen months later, with the HGP fast becoming an irresistible force, NIH's genome office was upgraded to the National Center for Human Genome Research; I was appointed its first director.

It was my job both to pry the cash away from Congress and to ensure that it was wisely spent. A major concern of mine was that the HGP's budget be separate from that of the rest of NIH. I thought it vitally important that the Human Genome Project not jeopardize the livelihood of non-HGP science; we had no right to succeed if by our success other scientists could legitimately charge that their research was being sacrificed on the altar of the megaproject. At the same time, I felt that we, the scientists embarking on this unprecedented enterprise, ought to signal somehow our awareness of its profundity. The Human Genome Project is much more than a vast roll call of As, Ts, Gs, and Cs: it is as precious a body of knowledge as humankind will ever acquire, with a potential to speak to our most basic philosophical questions about human nature, for purposes of good and mischief alike. I decided that 3 percent of our total budget (a small proportion, but a large sum nevertheless) should be dedicated to exploring the ethical, legal, and social implications of the Human Genome Project. Later at Senator Al Gore's urging, this was increased to 5 percent.

It was during these early days of the project that a pattern of international collaboration was established. The United States was directing the effort and carrying out more than half the work; the rest would be done mainly in the United Kingdom, France, Germany, and Japan. Despite a long tradition in genetics and molecular biology, the U.K.'s Medical Research Council was only a minor contributor. Like the whole of British science, it was suffering from Mrs. Thatcher's myopically stingy funding policies. Fortunately, the Wellcome Trust, a private biomedical charity, came to the rescue: in 1992 it established a purpose-built sequencing facility outside Cambridge—the Sanger Centre named, as we have seen, for Fred Sanger. In managing the international effort,

I decided to assign distinct parts of the genome to different nations. In this way, I figured, a participating nation would feel that it was invested in something concrete, say, a particular chromosome arm, rather than laboring on a nameless collection of anonymous clones. The Japanese effort, for example, focused largely on chromosome 21. Sad to say, in the rush to finish, this tidy order broke down, and it proved to be not so easy after all to superimpose the genome map on a map of the world.

From the start I was certain that the Human Genome Project could not be accomplished through a large number of small efforts—a combination of many, many contributing labs. The logistics would be hopelessly messy, and the benefits of scale and automation would be lost. Early on, therefore, genome mapping centers were established at Washington University in St. Louis, Stanford and UCSF in California, the University of Michigan at Ann Arbor, MIT in Cambridge, and Baylor College of Medicine in Houston. The DOE's operations, first centered at their Los Alamos and Livermore National Laboratories, in time came to be centralized in Walnut Creek, California.

The next order of business was to investigate and develop alternative sequencing technologies with a view to reducing overall cost to about 50 cents a base pair. Several pilot projects were launched. Ironically, the method that eventually paid off, fluorescent dye-based automated sequencing, did not fare especially well during this phase. In retrospect, the pilot automated machine effort should have been carried out by Craig Venter, an NIH staff researcher who had already proved adept at getting the most out of the procedure. He had applied to do it, but Lee Hood, as the technology's original developer, was preferred. This early rebuff of Venter was to have repercussions later.

In the end, the HGP did not involve the wholesale invention of new methods of analyzing DNA; rather, it was the improvement and automation of familiar methods that ultimately enabled a progressive scaling up from hundreds to thousands and then to millions of base pairs of sequence. Critical to the project, however, was a revolutionary technique for generating large quantities of particular DNA segments (you need large quantities of a given segment, or gene, if you are going to sequence it). Until the mid-eighties, amplifying a

Kary Mullis, inventor of PCR

particular DNA region depended on the Cohen-Boyer method of molecular cloning: you would cut out your piece of DNA, insert it into a plasmid, and then insert the modified plasmid into a bacterial cell. The cell would then replicate, duplicating each time your inserted DNA segment. Once sufficient bacterial growth had occurred, you would purify your DNA segment out from the total mass of DNA in the bacterial population. This procedure, though refined since Boyer and Cohen's original experiments, was still cumbersome and time-consuming. The development of the polymerase chain reaction (PCR) was therefore a great leap forward: it achieves the same goal, selective amplification of your piece of DNA, within a couple of hours, and without any need to mess around with bacteria.

PCR was invented by Kary Mullis, then an employee of Cetus Corporation. By his own account, "The revelation came to me one Friday night in April, 1983, as I gripped the steering wheel of my car and snaked along a moonlit mountain road into northern California's redwood country." It is remarkable that he should have been inspired in the face of such peril. Not that the roads in Northern California are particularly treacherous, but as a friend—who once saw the daredevil Mullis in Aspen skiing down the center of an icy road through speeding two-way traffic—explained to the *New York Times:* "Mullis had a vision that he would die by crashing his head against a redwood tree. Hence he is fearless wherever there are no redwoods." Mullis received the Nobel Prize in Chemistry for his invention in 1993 and has since become ever more eccentric. His advocacy of the revisionist theory that AIDS is not caused by HIV has damaged both his credibility and public health efforts.

PCR is an exquisitely simple process. By chemical methods, we synthesize two primers—short stretches of single-stranded DNA, usually about twenty base pairs in length—that correspond in sequence to regions flanking the piece of DNA we are interested in. These primers bracket our gene. We add the

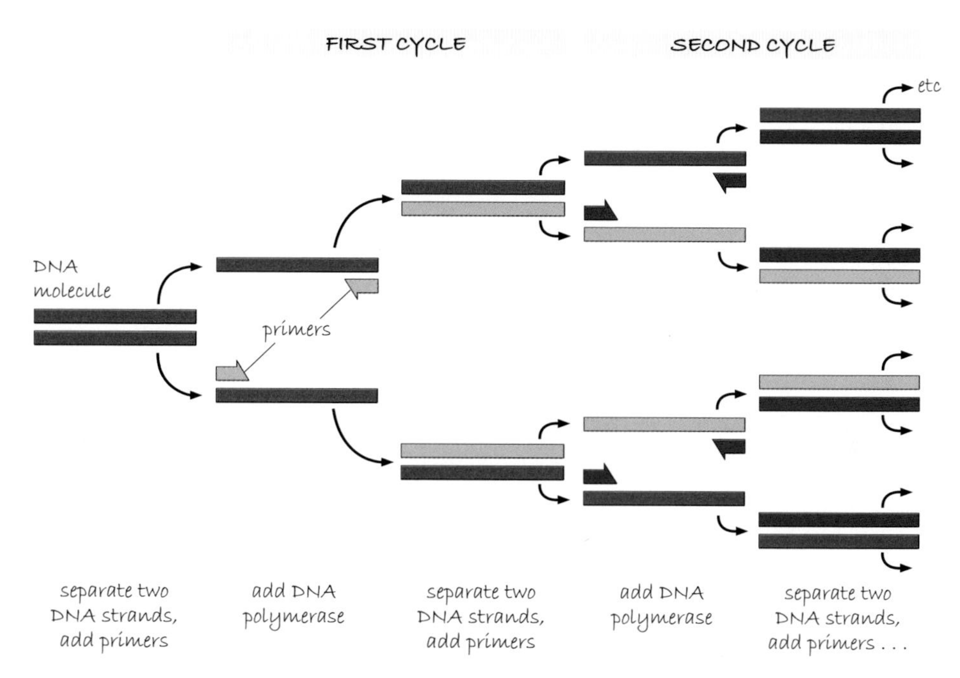

Amplifying the DNA region you're interested in: the polymerase chain reaction

primers to our template DNA, which has been extracted from a sample of tissue. The template effectively consists of the entire genome, and the goal is to massively enrich our sample for the target region. When DNA is heated up to 95°C, the two strands come apart. This allows each primer to bond to the twenty-base-pair stretches of template whose sequences are complementary to the primer's. We have thus formed two small twenty-base-pair islands of double-stranded DNA along the single strands of the template DNA. DNA polymerase—the enzyme that copies DNA by incorporating new base pairs in complementary positions along a DNA strand—will only start at a site where the DNA is already double-stranded. DNA polymerase therefore starts its work at the double-stranded island made by the union of the primer and the complementary template region. The polymerase makes a complementary copy of the template DNA starting from each primer, and therefore copying the target region. At the end of this process, the total amount of target DNA will have doubled. Now we repeat the heating step, and the whole process occurs again; once more, the number of copies of the DNA bracketed by the two primers doubles. Each cycle of this process results in a doubling of the target region. After twenty-five cycles of PCR—which means in less than two hours—we have a 2^{25} (about a 34 million–fold) increase in the amount of our target DNA. In effect, the resulting solution, which started off as a mixture of template DNA, primers, DNA polymerase enzyme, and free As, Ts, Gs, and Cs, is a concentrated solution of the target DNA region.

A major early problem with PCR is that DNA polymerase, the enzyme that does the work, is destroyed at 95°C. It was therefore necessary to add it afresh in each of the process's twenty-five cycles. Polymerase is expensive, and so it was soon apparent that PCR, for all its potential, would not be an economically practical tool if it involved literally burning huge quantities of the stuff. Happily Mother Nature came to the rescue. Plenty of organisms live at temperatures much higher than the 37°C that is optimal for *E. coli,* the original source of the enzyme; and these creatures' proteins—including enzymes like DNA polymerase—have adapted over eons of natural selection to cope with serious heat. Today PCR is typically performed using a form of DNA polymerase derived from *Thermus aquaticus,* a bacterium that lives in the hot springs of Yellowstone National Park.

PCR quickly became a major workhorse of the Human Genome Project. The process is basically the same as that developed by Mullis, but it has been automated. No longer dependent on legions of bleary-eyed graduate students to effect the painstaking transfer of tiny quantities of fluid into plastic tubes, a state-of-the-art genome lab features robot-controlled production lines. PCR robots engaged in a project on the scale of sequencing the human genome inevitably churn through vast quantities of the heat-resistant polymerase enzyme. HGP scientists therefore especially resented the unnecessarily hefty royalties added to the cost of the enzyme by the owner of the PCR patent, the European industrial-pharmaceutical giant Hoffmann-LaRoche.

The other workhorse was the DNA sequencing method itself. Again, the underlying chemistry was not new: the HGP used the same method worked out by Fred Sanger in the mid-seventies. Innovation came as a matter of scale, through the mechanization of sequencing.

Sequencing automation was initially developed in Lee Hood's Caltech lab. As a high-school quarterback in Montana, Hood led his team to successive state championships; he would carry the lesson of teamwork into his academic career. Peopled by an eclectic mixture of chemists, biologists, and engineers, Hood's lab became a leader in technological innovation.

Automated sequencing was actually the brainchild of Lloyd Smith and Mike Hunkapiller. Then in Hood's lab, Hunkapiller approached Smith about a sequencing method using a different colored dye for each base type. In principle the idea promised to make the Sanger process four times more efficient: instead of four separate sets of sequencing reactions, each run in a separate gel lane, color-coding would make it possible to do everything with a single set of reactions, and run the result in a single gel lane. Smith was initially pessimistic, fearing the quantities of dye implied by the method would be too small to detect. But being an expert in laser applications, he soon conceived a solution using special dyes that fluoresce under a laser.

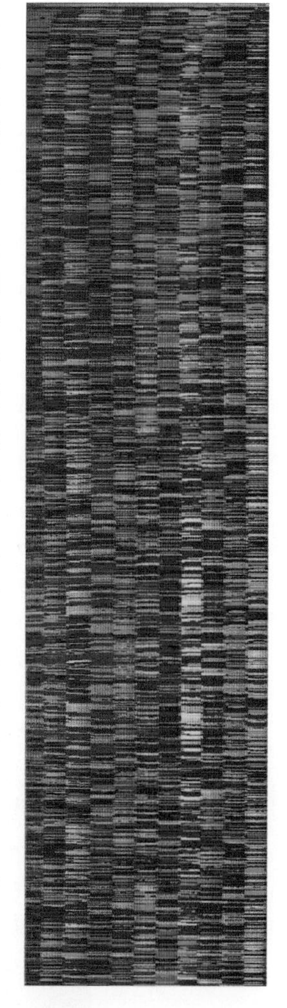

Read the fine print: DNA sequence output from an automated sequencing machine. Each color represents one of the four bases.

Following the standard Sanger method, a procession of DNA fragments would be created and sorted by the gel according to size. Each fragment would be tagged with a fluorescent dye corresponding to its chain-terminating dideoxy nucleotide (see page 106); the color emitted by that fragment would thereby indicate the identity of that base. A laser would then scan across the bottom of the gel, activating the fluorescence, and an electric eye would be in place to detect the color being emitted by each piece of DNA. This information would be fed straight into a computer, obviating the excruciating data-entry process that dogged manual sequencing.

Hunkapiller left Hood's lab in 1983 to join a recently formed instrument manufacturer, Applied Biosystems, Inc. (ABI). It was ABI that produced the first commercial Smith-Hunkapiller sequencing machine. Since then, the efficiency of the process has been enormously improved: gels—unwieldy and slow—have been discarded and replaced with high-throughput capillary systems—thin tubes in which the DNA fragments are size-sorted very rapidly. Today, the latest generation of ABI's sequencing machines is phenomenally fast, some thousand times speedier than the prototype. With minimal human intervention (about fifteen minutes every twenty-four hours), these machines can produce as much as half a million base pairs of sequence per day. It was ultimately this technology that made the genome project doable.

While DNA sequencing strategies were being optimized during the first part of the Human Genome Project, the mapping phase forged ahead. The immediate goal was a rough outline of the entire genome that would guide us in determining where each block of eventual sequence was located. The genome had to be broken up into manageable chunks, and it would be those chunks that would be mapped. Initially we pursued this objective using yeast artificial chromosomes (YACs), a means devised by Maynard Olson of importing large pieces of human DNA into yeast cells. Once implanted, YACs are replicated together with the normal yeast chromosomes. But attempts to load up to a million base pairs of human DNA into a single YAC exposed methodological problems. Segments, it was discovered, were getting shuffled, and since mapping is all about the order of genes along the chromosome, this shuffling of sequences was just about the worst thing that could happen. BACs (bacterial artificial chromosomes), developed by Pieter de Jong in Buffalo, came to the rescue. These are smaller, just 100,000 to 200,000 base pairs long, and much less prone to shuffling.

The team at the heart of France's contribution to the genome project. Jean Weissenbach is third from left and Daniel Cohen is on the right. Next to Cohen is Jean Dausset, the visionary immunologist who launched the effort.

For those attacking the human genome map head on—groups in Boston, Iowa, Utah, and France—the critical first steps involved finding genetic markers—locations where the same stretch of DNA drawn from two different individuals differed by one or more base pairs. These sites of variation would serve as landmarks for orienting our efforts throughout the genome. In short order the French effort, under Daniel Cohen and Jean Weissenbach, produced excellent maps at Généthon, a factorylike genomic research institute funded by the French Muscular Dystrophy Association. Like the Wellcome Trust across the English Channel, the French charity took up some of the slack created by insufficient government support. When, in the final push, detailed physical mapping of BACs became necessary, John McPherson's program at the genome center at Washington University was the major contributor.

As the HGP lurched into high gear, the debate persisted about the best way to proceed. Some pointed out that a large portion of the human genome is what we in the trade call "junk," stretches of DNA that apparently don't code for anything. Indeed those stretches that encode proteins—genes—constitute only a small fraction of the total. Why therefore, these critics asked, should we sequence the entire genome—why bother with the junk? There is actually an extremely quick-and-dirty way to secure a snapshot of all the coding genes in the genome, using the reverse transcriptase technology described in chapter 5. Purify a sample of messenger RNA from any type of tissue; if your source is the brain, you will have a sample of RNA for all the genes expressed in the brain. Using reverse transcriptase, you can then create DNA copies (known as cDNAs) of these genes and the cDNAs can then be sequenced.

This quick and dirty approach, however, was no substitute for doing the whole thing. As we now know, many of the most interesting parts of the genome lie outside genes, constituting the control mechanisms that switch the genes on and off. And so, in the case of the cDNA analysis of brain tissue just described, you will have an overview of the genes switched on in the brain but no idea how they are switched on: the hugely important regulatory regions of DNA are not transcribed into RNA by the RNA polymerase enzyme that copies the DNA strand into messenger RNA.

Working at the relatively cash-strapped Medical Research Council (MRC) in Britain, Sydney Brenner pioneered this cDNA-based approach to large-scale gene discovery. With limited research funds, he figured that sequencing cDNAs was the most cost-effective way of using what little money he had. Keen to reap the commercial benefits of the sequences, the MRC prevented Brenner from publishing them until British pharmaceutical firms had a chance to position themselves to profit from them.

On a visit to Sydney Brenner's lab, Craig Venter was impressed by this cDNA strategy. He could hardly wait to return to his NIH lab outside Washington, D.C., where he would apply the technique himself to produce a treasure trove of new genes. By sequencing even a small part of each one, Venter could determine whether or not it was new to science. In June 1991 an NIH official urged him to apply for patents on 337 of these new genes, although he had, in many instances, no clue about their function. A year later, having applied the technique more broadly, Venter added 2,421 sequences to the list submitted to the patent office. In my judgment, the very notion of blindly patenting sequences without knowledge of what they do was outrageous: what precisely was one protecting? This conduct could only be seen as a preemptive financial claim on a truly meaningful discovery someone else might yet make. I expounded my objections to the higher-ups at NIH, but to no avail. And the agency's persistence in endorsing the practice—a policy that was later reversed—spelled the beginning of the end of my career as a government bureaucrat. I had mixed feelings when Bernadine Healy, head of NIH, forced me to resign in 1992. Four years in the Washington pressure cooker had been enough. But what really mattered to me was that by the time of my departure, the Human Genome Project was undeflectably on course.

Venter's taste of the commercial possibilities of patenting chunks of the genome whetted his appetite for more. But he wanted it both ways: to remain a part of the academic community, in which information was freely shared and salaries were small; and also to enter the business arena, in which his discoveries could be kept under wraps until the patent cleared and he could cash in. With the help of a fairy godfather, venture capitalist Wallace

Steinberg (the inventor of the Reach toothbrush), Venter got his wish in 1992. Steinberg supplied $70 million to set up not one but two organizations: a nonprofit, The Institute for Genomic Research (TIGR, pronounced "tiger"), to be headed by Venter, and a sister company, Human Genome Sciences (HGS), to be headed by commercially inclined molecular biologist William Haseltine. It would work this way: TIGR, the research engine, would crank out cDNA sequences, and HGS, the business arm, would market the discoveries. HGS would always have six months to review TIGR's data prior to publication, except when the findings indicated potential to develop a drug, in which case HGS would have a year.

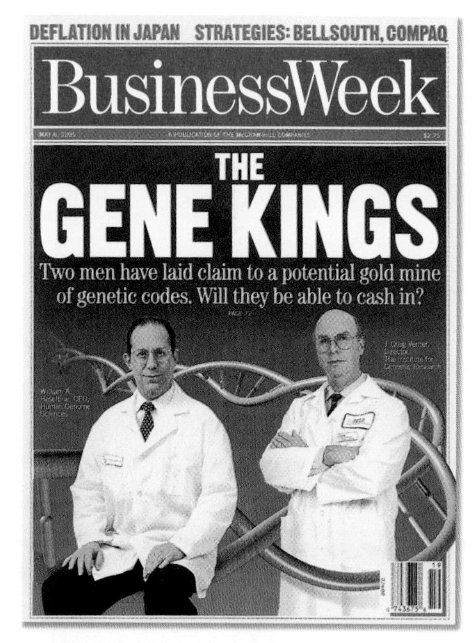

The genome project goes commercial: William Haseltine and Craig Venter.

Having grown up in California, Venter initially chose surfing over higher education. But a traumatic yearlong tour as a medical assistant in Vietnam during the war seemed to focus his mind, and on his return to the United States he acquired in short order an undergraduate degree and a Ph.D. in physiology and pharmacology from the University of California, San Diego. His migration from academia into the commercial sector made sense viewed in relation to his personal finances: by his own reckoning, he had $2,000 in the bank when he founded TIGR. But he was quick to turn his fortunes around: early in 1993 the British pharmaceutical company SmithKline Beecham, anxious for a stake in the genome gold rush, paid $125 million for the exclusive commercial rights to Venter's growing list of new genes. And a year later, the *New York Times* revealed that Venter's 10 percent share of HGS was itself worth $13.4 million. Not afraid to spend it, he dropped $4 million on an eighty-two-foot racing yacht, whose spinnaker he adorned with a twenty-foot image of himself.

In the 1970s William Haseltine had been at Harvard as a graduate student under the joint direction of Wally Gilbert and myself. Afterward, he would run an innovative HIV research center at the medical school's Dana Farber Cancer Center. But it was his marriage to the multimillionaire socialite Gale Hayman (creator of the 1980s must-have perfume Giorgio Beverly Hills) that gave him the most visibility and ensured Haseltine had rather more than $2,000 in the bank when he set up HGS. Even before he went corporate, his jet-setting had provoked comment from members of his Harvard Medical School laboratory. "What's the difference between Bill Haseltine and God?" Answer: "God is everywhere; Haseltine is everywhere but Boston, where he's supposed to be."

Precious little skill or ingenuity was involved in Venter and Haseltine's scramble to patent every human gene they could find on the basis of cDNA sequencing. TIGR and HGS were simply the biotech equivalent of the kids who round up all the toys at the playground just so no other kid can play with them.

In 1995, HGS filed a patent for a gene called CCR5. HGS's preliminary sequence analysis suggested that the gene encoded a cell-surface protein in the immune system, and was therefore worth "owning" since such proteins may potentially serve as targets for drugs affecting the immune system. CCR5 was one of a batch of 140 patents for similar genes that HGS applied for. But in 1996 researchers discovered the role of CCR5 in the pathway by which HIV, the virus that causes AIDS, invades the immune system's T cells. They also found that mutations in CCR5 were responsible for AIDS resistance: it had been observed that some gay men—who turned out to have mutated CCR5 genes—never contracted the disease despite repeated exposure to HIV. Thus CCR5 was and remains clearly destined to play an important part in our assault on HIV. But although it made no contribution whatsoever to the hard work and solid science that determined CCR5's central role in AIDS infection, HGS stands to profit enormously from simply having got its hands on the gene first; and by exacting a fee for every attempted application of the knowledge, its CCR5 patent will sorely tax an area of medical research that desperately needs every penny it has. Haseltine's response is by turns unapologetic—"If somebody uses this gene in a drug discovery program after the patent has been issued . . . and does it for commercial purposes, they have infringed the patent"—and indignant: "We'd be entitled not just to damages, but to double and triple damages."

This kind of speculative gene patenting creates a terrible drag on medical research and development, leading in the long run to fewer and poorer treatment options. The trouble is that the speculators are in effect patenting potential drug targets—the proteins upon which any drug or treatment yet to be invented might act. For most big pharmaceutical firms gene patents on drug targets, filed by biotech companies with little or no biological information on function, become a poison pill. The large royalties demanded by gene-finding monopolies tip the economic balance against drug development; cloning a drug target is at most 1 percent of the way to an approved drug. Furthermore, if a company produces a drug with a particular target for which it also holds the patent on the underlying gene, that company has no immediate incentive to develop better drugs for that target. Why invest in R&D when your patent makes it prohibitively costly—if not simply illegal—for other companies to get in on the act?

The prospect of the TIGR/HGS/SmithKline Beecham triumvirate having a commercial stranglehold on human gene sequences alarmed both the academic and commercial molecular biology communities. In 1994 Merck, one of SmithKline Beecham's traditional rivals in the pharmaceutical business, provided the genome center at Washington University with $10 million to sequence human cDNAs and publish them openly, thus delivering an open-access riposte to HGS.

At about the time that TIGR and HGS were taking their first steps to commercialize the genome, Francis Collins was appointed to succeed me as the director of the NIH's genome effort. Collins was an excellent choice. He had proven himself a top-notch gene mapper, with several major disease genes under his belt, including the ones for cystic fibrosis, neurofibromatosis (the so-called Elephant Man disease), and, as part of a multipronged effort, Huntington disease. Had prizes been awarded in the early matches of the HGP tournament—those contests for the mapping and characterization of important genes—the palm would surely have gone to Collins. He did himself keep score after his own fashion: a Honda Nighthawk motorcycle being his preferred mode of transport, his colleagues added a decal to his helmet every time a new gene was mapped in his lab.

Collins was raised in Virginia's Shenandoah Valley on a ninety-five-acre farm without plumbing. Initially home-schooled by his parents, a drama professor and a playwright, he wrote and directed his own stage production of *The Wizard of Oz* at age seven. The wicked witch of science, however, dragged Collins away from a career in the theater; after completing a Ph.D. in physical chemistry at Yale, he went to medical school and from there into a research career in medical genetics. Collins is a member of a rare species, the devoutly religious scientist. In college, he recalls, "I was a pretty obnoxious atheist," but that changed in medical school, when "I watched people in terrible medical circumstances who were engaged in battles for survival, which many of them lost. I watched how some people leaned on their faith and saw what strength it gave them." To the Human Genome Project Collins brought scientific excellence as well as a spiritual dimension singularly lacking in his predecessor.

By the mid-nineties, with the initial mapping of the human genome accomplished and sequencing technologies fast developing, it was time to get down to the nitty-gritty of As, Ts, Gs, and Cs—time to start sequencing. Sticking to the game plan outlined at the outset by our NAS committee, we would first attack an array of model organisms: bacteria to start with, and then on to more complicated creatures (with more complicated genomes). The lowly nematode worm, *C. elegans,* was the first big nonbacterial challenge, and as the joint achievement of John Sulston at Britain's Sanger Centre and Bob Waterston at Washington University it provided an excellent model of international collaboration. The worm's sequence was published in December 1998, all 97 million base pairs of it. No bigger than a comma on this page, and comprising a fixed number of cells, just 959, the worm nevertheless has some 20,000 genes.

At first sight, Sulston appeared ill-suited to a leadership role in Big Science. He had spent most of his professional life staring down a microscope in order to produce in astonishing detail a complete description, cell by cell, of the development of the worm. Bearded and avuncular, he is the son of a Church of England vicar and also a lifelong socialist who believes passionately that business and the human genome should have nothing in common. Like Francis Collins, he is a motorcycle enthusiast; he used to commute on his 550cc machine from

International collaboration (right): British and American scientists were the first to complete the sequencing of the genome of a complex organism, the nematode, C. elegans. *The project's leaders (below), Bob Waterston and John Sulston, still found time to relax.*

his home outside Cambridge to the Sanger Centre until, just as the HGP was gathering speed, an accident left him severely injured and his bike, as he put it, "little more than nuts and bolts." The Wellcome Trust, which was funding the Sanger Centre, was horrified to learn that the project's scientific leader was taking his life in his hands every time he came to work: "After we'd invested all that money in this bloke!" complained Bridget Ogilvie, then the trust's director.

Sulston's U.S. partner, Waterston, was an engineering major at Princeton and imported plenty of engineering savvy to the big sequencing center he ran at Washington University. Waterston has the capacity to extrapolate—to start small and finish big. Accompanying his daughter on a jog, he found he liked running, and is now an accomplished marathon runner. During its first year of operation, his sequencing group produced just forty thousand base pairs of worm sequence, but within a few years it was cranking out enormous amounts, and Waterston was one of the earliest to urge an all-out human sequencing effort.

But even as those in the international HGP collaboration began to sequence model organisms, gearing up for the big one, the molecular biological equivalent of an earthquake shook the whole enterprise.

Craig Venter and TIGR had been doing well. Having milked his cDNA gene discovery strategy for several years, Venter became interested in sequencing whole genomes. In this, too, he was persuaded of the superiority of his own approach. The HGP had been carefully mapping the location of the different chunks of DNA on the chromosomes before actually sequencing them. That way, you already knew that chunk A was adjacent to chunk B and could look for overlaps between them when it came to knitting together the final sequence. Venter preferred a "whole genome shotgun" (WGS) approach, in which there was no initial mapping: you simply broke the genome up into random chunks, sequenced them all, fed all the sequences into a computer, and relied upon the computer to put them all in the right order on the basis of overlaps, without benefit of any prior positional information. Venter and his team at TIGR showed that this brute force method could indeed work, at least for simple genomes: in 1995 they published the genome sequence of a bacterium, *Haemophilus influenzae,* using this method.

It remained problematic, however, whether WGS would work for a large and complex genome like the human one. The problem is repeats—segments of the same sequence occurring in different places in the genome—which could in principle scupper a WGS sequencing attempt. These repeats might well mislead even the most sophisticated computer algorithm. If, for instance, a repeat occurs in chunks A and P, the computer could mistakenly situate A next to Q rather than in its proper position, next to B. For its part, the HGP itself had discussed this scenario when it considered using a WGS approach, and, based on careful calculations by Phil Green in Seattle, the consortium concluded that such an effort would likely be confounded by the human genome's massive amount of long-repeating sequences of junk DNA.

In January 1998, Mike Hunkapiller of ABI, maker of automated sequencing machines, invited Venter to check out his newest model, the PRISM 3700. Venter was impressed, but nothing could have prepared him for what was to follow. Hunkapiller suggested that Venter form a new company, funded by ABI's parent company, PerkinElmer, to sequence the human genome. Venter had no misgivings about forsaking TIGR—relations had long since soured with Haseltine at HGS. And so he wasted no time in founding the firm that was later to be called Celera Genomics. The company motto: "Speed matters. Discovery can't wait." The plan: to sequence the entire human genome by WGS using three hundred of Hunkapiller's machines and the single greatest concentration of computing power outside the Pentagon. The project would take two years and cost between $200 million and $500 million.

The news broke just before the leaders of what would come to be called the public (as opposed to private) Human Genome Project were meeting at Cold Spring Harbor Laboratory. To put it mildly, the news was not well received. The worldwide public project had already spent some $1.9 billion (of public money), and now, as the *New York Times* was spinning it, we might have nothing to show for the money except for the sequence of the mouse genome, while Venter waltzed off with the holy grail, the human genome. What was especially galling was Venter's flouting of what had come to be known as the Bermuda principles. In 1996, at an HGP conference in Bermuda—a meeting Venter attended—the HGP had agreed that sequence data should be released as soon as it was generated. The genome sequence, we all concurred, should be public property. Now a renegade, Venter had different ideas: he claimed he would defer releasing new sequence data for three months, selling licenses to pharmaceutical companies and any other parties seriously interested in buying a preview.

Fortuitously, the Wellcome Trust's Michael Morgan was able to give the public project a welcome boost just days after Venter's announcement by declaring that it would be doubling its support for the Sanger Centre, bringing the total up to around $350 million. Though the timing of the announcement made this look like a direct response to Venter's challenge, the increase in funding had in fact been in the works for quite some time. Shortly afterwards the U.S. Congress beefed up its own contribution to the public HGP's coffers. The race was on. In fact, from the outset there were always going to be at least two winners. Science only stood to benefit from two human genome sequences, one against which to check the other. (With over 3 billion base pairs involved, there was bound to be a typo or two.) Another winner would surely be ABI: they stood to sell a lot more PRISM sequencing machines, which most labs in the public project would now have to buy to keep up with Venter!

The acrimonious exchanges between the leaders of the private and public projects would become a fixture of newspaper science pages for the next couple of years. The back and forth got to a pitch that moved President Clinton to direct his science adviser, "Fix it . . . make these guys work together." But through it all, the sequencing moved ahead, and Venter did demonstrate that a WGS approach could work on a respectably sized genome when, in collaboration with the fruit fly wing of the public consortium, he announced the completion of an advanced draft of the *Drosophila* genome early in 2000. It, however, contains relatively little repetitive junk DNA, and Celera's success in assembling it in no way guaranteed WGS would work on the human genome.

No individual was more vital to meeting the Celera challenge than Eric Lander. It was he who envisioned an almost entirely automated sequencing process in which robots would take the place of technicians, and it was he who had the drive to make this vision a reality. Lander's résumé indeed shows he knew a thing or two about drive. A Brooklyn boy, he was a curve-busting math whiz at Stuyvesant High in Manhattan who went on to win first prize in the Westinghouse Talent Search; he then became valedictorian of his class at Princeton ('78) before earning his Ph.D. at Oxford on a Rhodes fellowship. A MacArthur "genius" award in 1987 seemed almost redundant. His mother, incidentally, has no idea how it all happened: "I'd love to say I'm responsible, but it's not true. . . . I'd have to say it was dumb luck."

Ultimately finding pure mathematics "an isolated, monastic kind of field," Lander, notably gregarious by the standards of his discipline, joined the jollier faculty of the Harvard Business School, but he soon found himself distracted and intrigued by the labors of his younger brother, a neuroscientist. Inspired, Lander taught himself biology by moonlighting in Harvard's and MIT's biology departments, all the while scarcely missing a beat on his day job at the B-school: "I pretty much picked up molecular biology on street corners," he says. "But around here, there are a lot of very good street corners." In 1989 he became a professor of biology on one of those street corners, MIT's Whitehead Institute.

Even among the so-called G5—the public effort's five major centers, which also included the Sanger Centre, Washington University's Genome Sequencing Center, Baylor College of Medicine, and the DOE in Walnut Creek—Lander's lab would be the largest single contributor of DNA sequences. His team at MIT would also be responsible for much of the enormous acceleration of productivity in the home stretch leading up to the release of the draft genome. On November 17, 1999, the public project celebrated its billionth base pair, with

Mass production meets DNA sequencing: MIT's Whitehead Institute.

the sequencing of a G. Just four months later, on March 9, 2000, a T was base number 2 billion. The G5 was cranking. Because Celera was using the public project's data, which were posted immediately on the Internet and were now pouring in thick and fast, Venter, perhaps finally breaking a sweat, halved the amount of sequencing he had originally projected Celera would do.

As the public/private race reached a climax in the media, behind the barricades the focus was increasingly shifting to the effort's mathematical brain trust, scientists hidden in back rooms among banks of computers. They were the ones who had to make sense of all those As, Ts, Gs, and Cs of crude sequence. They had two major tasks. First: To assemble a whole final sequence from the many, many discrete chunks on hand. Most parts had been sequenced numerous times, so there were several genomes' worth of sequence to sort out, all of which had to be distilled down into a single canonical genome sequence. Computationally, this was an enormous undertaking. Second: To figure out what was what in the final sequence, and above all where the genes were. Identifying the genome's components—distinguishing between one stretch of As, Ts, Gs, and Cs that encoded nothing but junk and another that encoded a protein—depended on extremely computer-intensive approaches.

At the heart of Celera's computer operations was Gene Myers, the computer scientist who had been the WGS approach's first and most forceful advocate. With James Weber of the Marshfield Medical Research Foundation in Wisconsin, he had proposed that the public effort adopt WGS long before Celera even came into existence. And so for Myers, the success of Celera's bid was a point of pride and vindication.

Anchored as it was by previously mapped genetic landmarks, the public project's job in assembling the sequence, though immense, seemed less daunting than the one confronting Myers in the landmark-free world of WGS. (In its final analysis, Celera used the public project's freely available map information.) In fact, in counting on these very landmarks, the public project had rather underestimated its own computational challenge, so, as Celera added computer muscle, the public project stayed focused on gearing up its sequencing operation. Only very late in the day did the leaders of the public project realize that, despite the map, they, too, like the proverbial father facing the parts of a new bike on Christmas Eve, had a major assembly problem on their hands. A date for completion (and assembly) of the "rough draft" had been fixed for the end of June. But at the beginning of May, the public project still had no working means of assembling all their sequences. Their deus ex machina took a strange form: a graduate student from UC Santa Cruz.

Jim Kent used a hundred PCs to assemble the rough draft of the genome for the public project.

His name was Jim Kent, and he looked like a member of the Grateful Dead. He had been programming computers since the beginning of the PC era, writing code for graphics and animations, but then decided on graduate school so he could be a part of bioinformatics, the new field dedicated to analyzing DNA and protein sequences. He realized that he was through with commercial programming when he received Microsoft's bulky twelve-CD-ROM package for developers of programs for Windows 95: "I was thinking to myself that the whole human genome could fit on one CD-ROM and didn't change every three months." Confident in May that he had a good way to crack the much-talked-about assembly problem, he induced his university to let him "borrow" 100 PCs recently bought for teaching purposes. He then embarked on a four-week programming marathon, icing his wrists at night to prevent them from seizing up as he churned out computer code by day. His deadline was June 26, when the completion of the rough draft was to be announced. The program finished, he set his 100 PCs to work, and, on June 22, his gang of PCs solved the public project's assembly problem. Myers at Celera cut it even closer, completing his assembly on the night of June 25.

Then came June 26, 2000. Bill Clinton at the White House and Tony Blair at 10 Downing Street simultaneously proclaimed the first draft of the HGP complete. The race was called as a tie, the honors to be shared equally. Happily the opposing parties managed to put the bad feelings behind them, for the morning at least. Clinton declared, "Today, we are learning the language in which God created life. With this profound new knowledge, humankind is on the verge of gaining immense, new power to heal." Grand words for a grand occasion. It was impossible not to feel some pride in an accomplishment that the press promptly compared to the first Apollo moon landing, even if the "official" date of the triumph was somewhat arbitrary. The sequencing was by no means over, and it would be more than six months before the scientific papers summarizing the genome were published. It has been suggested that the timing was dictated not by the HGP's timetable but by Clinton's and Blair's schedules.

Overlooked in the blaze of White House publicity was the fact that the object of celebration was but a *rough draft* of the human genome. Much work remained to be done. In fact, the sequences of only two of the smallest chromosomes, 21 and 22, were reasonably complete and had been published. And even these could not boast unbroken chromosome-tip-to-chromosome-tip sequences. As to the other chromosomes, some of their sequences were riddled with gaps. Since that big announcement, sights have been set on a new deadline of April 2003 for filling in the gaps and securing a full, accurate sequence. Some small regions, however, have proved literally unsequenceable, and in practice the goal has become to obtain an "essentially complete" sequence: at least 95 percent of the sequence finished with an error rate of less than 1 in 10,000 bases.

June 26, 2000: With a rough draft in hand, Craig Venter and Francis Collins temporarily set rivalries aside to bask in the presidential limelight.

Outside the White House: me with Eric Lander (Whitehead, MIT), Richard Gibbs (Baylor, Houston), Bob Waterston (St. Louis), Rick Wilson (St. Louis)

One of those responsible for coaxing the international herd of sequencing centers over the final hurdles was Rick Wilson, the bluff midwesterner who succeeded Bob Waterston as head of the Washington University center. Quality control is the name of the game, so each chromosome has been assigned a coordinator to oversee progress and ensure that his or her charge meets the project's overall specifications. Occasional glitches occur—for example, an errant piece of rice sequence crept mysteriously into one submission to the database—but screening procedures have proved effective in removing such contaminants. As I write this, the Human Genome Project is well on course to being "essentially complete" by the April 2003 deadline, which is also the fiftieth anniversary of the publication of the double helix.

The Human Genome Project is an extraordinary technological achievement. Had anyone suggested in 1953 that the entire human genome would be sequenced within fifty years, Crick and I would have laughed and bought them another drink. And such skepticism would have still seemed valid more than twenty years later when the first methods for sequencing DNA were finally devised. Those methods were, to be sure, a technical breakthrough but sequencing was a painfully slow business all the same—in those days it was a major undertaking to generate the sequence of even one small gene a few hundred base pairs in length. And now there we were, just another twenty-five years further on, celebrating the completion of some 3.1 billion base pairs of sequence. But we must also bear in mind that the genome is much more than a monument to our technological wizardry, astonishing though that may be: whatever its immediate political motivation, that White House celebration was perfectly justified in hailing the possibilities of a marvelous new weapon in our fight against disease and, even more, a whole new era in our understanding of how organisms are put together and how they operate, and of what it is that sets us apart biologically from other species—what, in other words, makes us human.

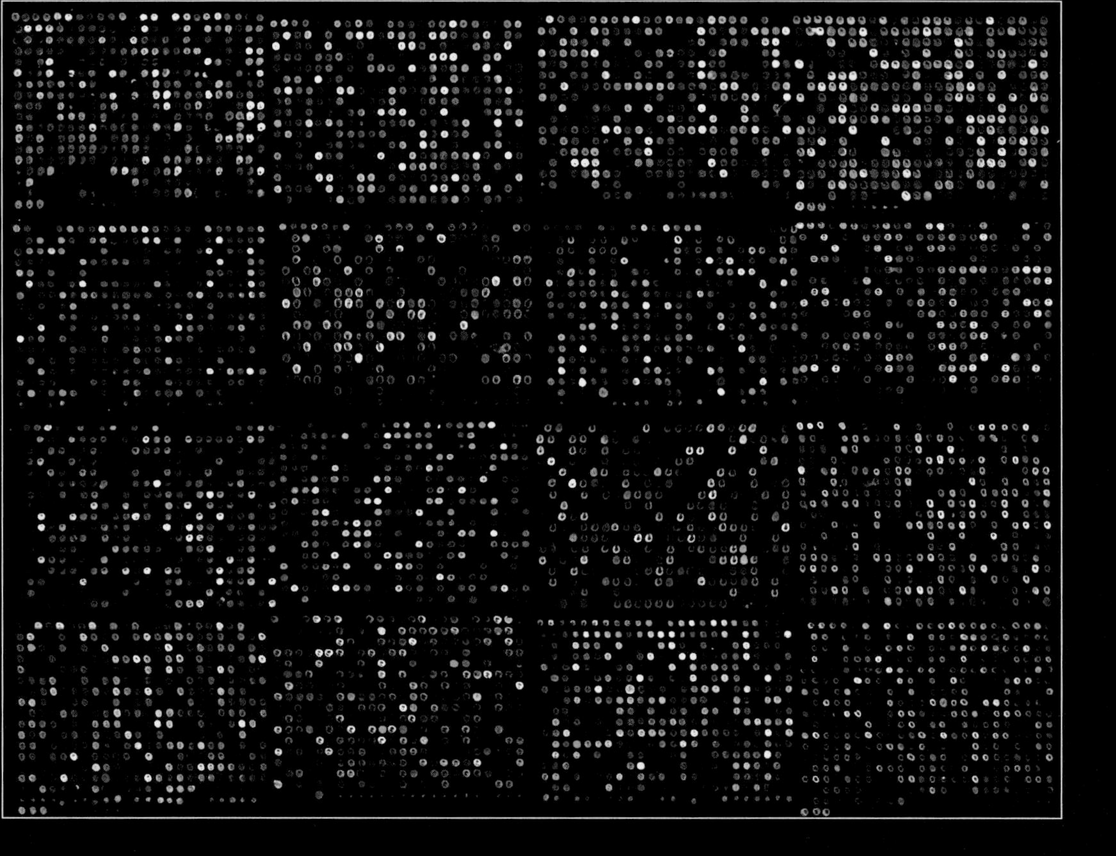

After the genome: microarray analysis of patterns of gene switching. In this case, each spot corresponds to one of 6,000 different genes in Plasmodium falciparum, *the cause of the severest form of malaria. In our search for a vaccine or a cure, we need to know what genes are active in different phases of the life cycle. Here a red spot indicates a gene active in one phase but not another, and a green spot the opposite. Genes that are active in both phases tend to show up yellow.*

CHAPTER EIGHT

READING GENOMES: EVOLUTION IN ACTION

I used to wish that when the human genome was finally completely sequenced it would turn out to contain 72,415 genes. My enthusiasm for this obscure number stemmed from the Human Genome Project's first big surprise. In December 1999, sandwiched between two major sequencing landmarks—billions one and two—came the first completed chromosome, number 22. Although a small one, constituting only 1.1 percent of the total genome, chromosome 22 was still 33.4 million base pairs long. This was our first glimpse of what the genome as a whole might look like; as one commentator for *Nature* wrote, it was like "seeing the surface or the landscape of a new planet for the first time." Most interesting was the density of genes along the chromosome. We had no reason to believe that chromosome 22 would not be representative of the entire genome, so we expected to find about 1.1 percent of all human genes in its sequence. That is to say, given the standard textbook estimate of about 100,000 human genes in total, we should have expected to see about 1,100 of them on chromosome 22. Almost exactly half that number were found: 545. Here was the first big hint that the human genome was not as gene-rich as we had supposed.

Suddenly the human gene count was a hot topic. At the Cold Spring Harbor Laboratory conference on the genome in May 2000, Ewan Birney, who was spearheading the Sanger Centre's computer analysis of the sequence, organized a contest he called Genesweep. It was a lottery based on estimating the correct gene count, which would finally be determined with the completion of the sequence in 2003; the winner would be the one who had come closest to the right answer. (That Birney should have become the HGP's unofficial bookie wasn't entirely surprising: numbers are his thing. After Eton, he took a year to tackle quantitative problems in biology while living in my house on Long Island—a far cry from trekking in the Himalayas or tending bar in Rio, just two of the more likely ways a young Briton might spend the "gap year" before university. Birney's CSHL work yielded two important research papers before he even set foot in Oxford.)

Originally Birney charged $1 per entry, but the price of admission to the pool increased with every published estimate that brought us closer to a final count. I was able to get in on the ground floor, putting $1 down on 72,415. My bet was a calculated attempt to reconcile the textbook figure, 100,000, and the new best guess of around 50,000, based on the chromosome 22 result. Birney announced the result in May 2003: 21,000 genes—many fewer than anyone had guessed. Lee Rowen, a Seattle-based expert in biological computing, won Genesweep with her bet of 25,947, still some 4000 off. Of course, I was wildly over the mark, so I'm now a dollar down on the genome.

Perhaps the only question to generate as much idle speculation as the gene count was that of whose genes we were sequencing. The information was in principle confidential, so money was not going to change hands on this one, but many wondered all the same. In the case of the public project, the DNA sample we sequenced had come from a number of randomly selected individuals from around Buffalo, New York, the same area where the processing work—isolating the DNA and inserting it into bacterial artificial chromosomes for mapping and sequencing—was taking place. Initially Celera claimed that its material too had been derived from six anonymous donors, a multicultural group, but in 2002 Craig Venter could not resist letting the world know that the main genome

Chromosome 2

sequenced was actually his own. Today that sequence is Venter's last remaining connection to the company. Concerned that sequencing genomes, though glamorous and newsworthy, was not proving viable from a business perspective, Celera reinvented itself as a drug company and bid farewell to its founder in 2002. As for Venter, he has established two new institutes, one to study the ethical issues raised by modern genetics, and the other to use the genomes of bacteria to find fresh sources of renewable energy.

With the whole rough draft in hand, it is now confirmed that there is nothing atypical about the gene density on chromosome 22. If anything, in fact, chromosome 22 with its 545 genes was for its size gene-rich rather than gene-poor. Only 236 genes have been definitively located on chromosome 21, which is about the same size. As we've seen, there are only an estimated 21,000 genes in total from the entire complement of 24 human chromosomes (22 + X + Y). And while we should stress that the final number can only rise as we make more discoveries, we are all but certain to end well below the 30,000 mark, never mind the 100,000 one.

As to how far below, only time will tell. Finding genes is not actually such a straightforward task: protein-coding regions are but strings of As, Ts, Gs, and Cs embedded among all the other As, Ts, Gs, and Cs of the genome—they do not stand out in any obvious way. And remember, only about 2 percent of the human genome actually codes for proteins; the rest, unflatteringly referred to as "junk," is made up of apparently functionless stretches of varying length, many of which occur repeatedly. And junk can even be found strewn within the genes themselves; studded with noncoding segments (introns), genes can sometimes

The genes on human chromosome 2: 255,000,000 base pairs long

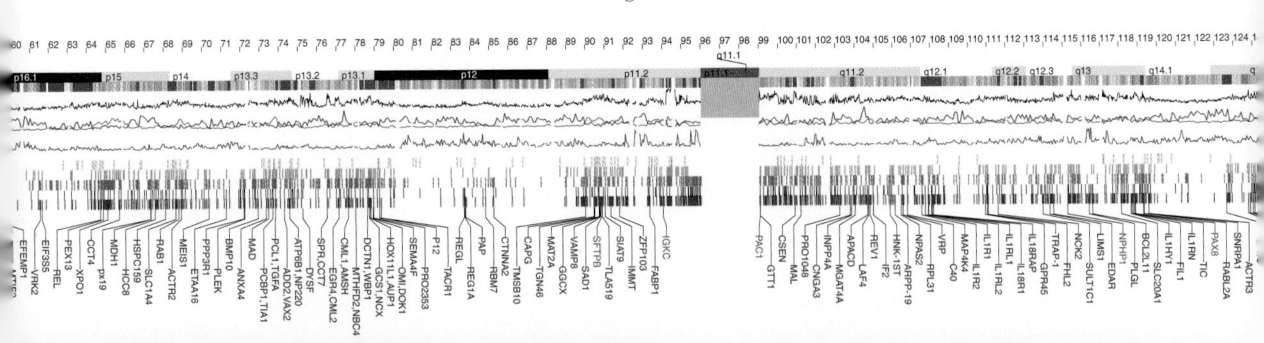

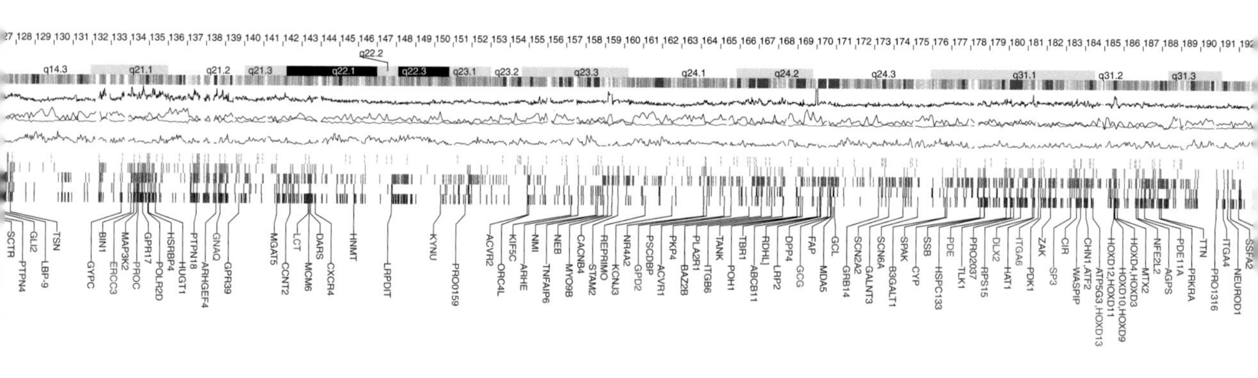

straddle enormous expanses of DNA, the coding parts like so many towns isolated between barren stretches of molecular highway. The longest human gene found so far, dystrophin (in which mutations cause muscular dystrophy), sprawls over some 2.4 million base pairs. Of these, a mere 11,055 (0.5 percent of the gene) encode the actual protein; the rest consists of the gene's seventy-nine introns (a typical human gene has eight). It is this awkward architecture of the genome that makes gene identification so difficult.

But human gene spotting has become less tricky now that the genome of the mouse is better known. The credit goes to evolution: in their functional parts

Comparison of mouse and human DNA for the same gene. Included is an intron—a noncoding region within the gene (shown in a box)—and parts of two exons—regions that code for the protein produced by the gene. The highlighted bases are where there has been no change over evolution between the two sequences. A dash implies the loss of a base in one species. The overall similarity of the mouse and human sequences suggests that natural selection has been tremendously effective in eliminating mutations. In the intron, where mutations are typically inconsequential, we see much more divergence than in the exons, where a change may impair the function of the protein.

```
Human  ATGGTTTGATGTCCTCCAGAAAGTGTCTACCCAGTTGAAGACAAACCTCACGAGTGTCACAAAGAACCGTGCAGATAAGG
Mouse  GTGGTTTGATGTACTCCAGAAAGTGTCTGCCCAATTGAAGACGAACCTAACAAGCGTCACAAAGAACCGTGCAGATAAGG

Human  TAAATGGTGCCGTTTGTGGCATGTGAACTCAGGCGTGTCAGTGCTAGAGAGGAAACTGGAGCTGAGACTTTCC-AGGTAT
Mouse  TGAATGGCAC----TGCAGCTAGAGATGACATGCG-GATATCACTGGGGTGGAAAC-AGAGCTCAGACTTTTCTAGATTA

Human  TTTGCTTGAAGCTTTTAGTTGAAGGCTTACTTATGGATTCTTTCTTTCTTTTTTTCTTTTTTATAGAATGCTATTCATAA
Mouse  GTTGCCAGAAGATTCTAATTGCAA--CTG----TGG------T--TTCTTTCACTTTTTCCTATAGAATGCTATTCATAA

Human  TCACATTCGTTTGTTTGAACCTCTTGTTATAAAAGCTTTAAAACAGTACACGACTACAACATGTGTGCAGTTACAGAAGC
Mouse  TCACATTAGGTTATTTGAGCCTCTTGTTATAAAAGCATTGAAGCAGTACACCACGACAACATCTGTACAATTGCAGAAGC

Human  AGGTTTTAGATTTGCTGGCGCAGCTGGTTCAGTTACGGGTTAATTACTGTCTTCTGGATTCAGATCAG
Mouse  AGGTTTTGGATTTGCTGGCACAGCTGGTTCAGCTACGGGTCAATTACTGTCTACTGGATTCAGACCAG
```

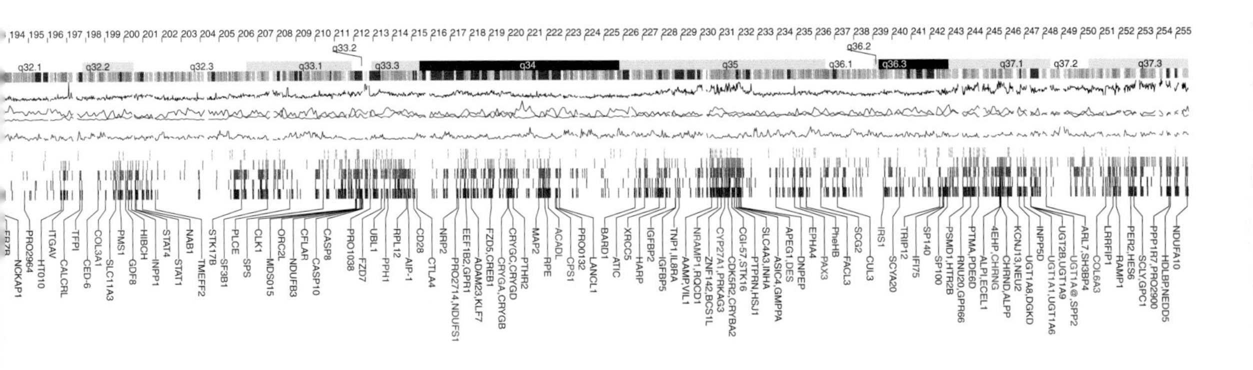

the human and mouse genomes, like the genomes of all mammals, are remarkably similar, having diverged relatively little over the eons intervening since the two species' common ancestor. The junk DNA regions, by contrast, have been evolution's wild frontier; without natural selection to keep mutation in check, as it does in coding segments, mutations aplenty have accumulated so that there is substantial genetic divergence between the two species in these regions. Looking for similarity in sequence between the human and mouse data is therefore an effective way of identifying functional areas, like genes.

Identifying human genes has also been facilitated by the completion of a rough draft of the puffer fish genome. Fugu, as it is better known to aficionados of Japanese cuisine, contains a potent neurotoxin; a competent chef removes the poison-containing organs, so dinner should produce only a little numbness in the mouth. But some eighty people die each year from poorly prepared fugu, and the Japanese imperial family is forbidden by law from enjoying this delicacy. More than a decade ago, Sydney Brenner developed a taste for the puffer, at least as an object of genetic inquiry. Its genome, just one-ninth the size of the human one, contains much less junk than ours: approximately one-third of it encodes proteins. Under Brenner's leadership, the fugu genome rough draft was completed for some $12 million, a genuine bargain by genome-sequencing standards. The gene count at present seems to fall between 32,000 and 40,000, in the same ballpark as humans. Interestingly, though, while fugu genes have roughly the same number of introns as human and mouse genes, the fugu introns are typically much shorter.

Even if we assume that plenty of genes remain to be discovered and generously increase our estimate of the human gene count from 21,000 to 25,000 we still get a somewhat exaggerated impression of our essential genetic complexity.

Over evolution, certain genes have spun off sets of related ones, resulting in groups of similar genes of like, but subtly different, function. These so-called gene families originate by accident when, in the course of producing egg or sperm cells, a chunk within a chromosome is inadvertently duplicated, so that there are now two copies of a particular gene on that chromosome. As long as one copy continues to function, the other is unchecked by natural selection, free to diverge in whatever direction evolution may choose as mutations accumulate. Occasionally the mutations will result in the gene acquiring a new function, usually one closely related to that of the original gene. In fact, many of our human genes consist of slight variations upon a relatively few genetic themes. Consider, for example, that 575 of our genes (nearly 2 percent of our total complement) are responsible for encoding different forms of protein kinase enzymes, chemical messengers that pass signals around the cell. Then, there are the 900 human genes underlying your nose's capacity to smell: the proteins encoded are odor receptors, each one recognizing a different smell molecule or class of molecules. Roughly these same 900 genes are present in the mouse as well. But here is the difference: the mouse, having adapted to a mainly nocturnal existence, has greater need of its sense of smell—natural selection has favored the keener sniffer and kept most of the 900 odor-detecting genes in service. In the human case, however, some 60 percent of these genes have been allowed to deteriorate over evolution. Presumably, as we became more dependent on sight, we have needed fewer smell receptors, so natural selection did not intervene when mutations caused many of our smell genes to be incapable of producing functioning proteins, making us relatively inept smellers compared with other warm-blooded creatures.

How does our gene number compare with that of other organisms?

COMMON NAME	SPECIES NAME	NUMBER OF GENES
Human	*Homo sapiens*	25,000
Mustard plant	*Arabidopsis thaliana*	27,000
Nematode worm	*Caenorhabditis elegans*	20,000
Fruit fly	*Drosophila melanogaster*	14,000
Baker's yeast	*Saccharomyces cerevisiae*	6,000
Gut bacterium	*Escherichia coli*	4,000

In terms of gene complement, then, we are only fractionally more complex than a weedy little plant. Even more sobering is the comparison with the nematode, a creature composed of only 959 cells (against our own estimated 100 trillion), of which some 302 are nerve cells that form the worm's decidedly simple brain (ours consists of 100 billion nerve cells)—orders of magnitude difference in structural complexity and yet we have not even double the worm's gene complement. How can we account for this embarrassing discrepancy? It's no cause for embarrassment at all: humans, it would appear, are simply able to do more with their genetic hardware.

In fact, I would propose there is a correlation between intelligence and low gene count. My guess is that being smart—having a decent nerve center like ours or even the fruit fly's—permits complex functioning with relatively few genes (if indeed "few" has any real meaning in relation to the number 35,000). Our brain gives us sensory and neuromotor capabilities far beyond those of the eyeless, inching nematode, and thus a greater range of behavioral response options. And the plant, being rooted, has fewer options still: it requires a full onboard set of genetic resources for dealing with every environmental contingency. A brainy species by contrast can respond to, say, a cold snap by using its nerve cells to seek out more favorable conditions (a warm cave will do).

Vertebrate complexity may also be enhanced by sophisticated genetic switches that are typically located near genes. With the genome sequencing accomplished, we can now analyze in detail these regions flanking genes. It is here that regulation occurs, with regulatory proteins binding to the DNA to turn the adjacent gene on or off. Vertebrate genes seem to be governed by a much more elaborate set of switching mechanisms than those of simpler organisms. It is this nimble and complicated coordination of genes that permits the complexities of vertebrate life. Moreover, a given gene may in addition yield many different proteins, either because different exons are coupled together to create slightly different proteins (a process known as "alternative splicing") or because biochemical changes are made to the proteins after they have been produced.

The unexpectedly low human gene count provoked several op-ed page ruminations on its significance. These tended toward a common theme. Stephen Jay Gould (whose recent premature death tragically silenced an impassioned voice), writing in the *New York Times,* hailed the low count as the death knell of reductionism, the reigning doctrine of virtually all biological inquiry. This doctrine holds that complex systems are built from the bottom up. Put another way: To understand events at complex levels of organization, we must first understand them at simpler levels and piece together these simpler dynamics. And so it follows that by understanding the workings of the genome, we will ultimately understand how organisms are assembled. Gould and others took the surprisingly small human gene count as evidence that such a bottom-up approach is not only unworkable but also invalid. In light of its unexpected genetic simplicity, the human organism, argued the antireductionists, was living proof that we cannot begin to understand ourselves in relation to a sum of smaller processes. To them, our low gene number implied that nurture, not nature, must be the primary determinant of who each one of us is. It was, in short, a declaration of independence from the tyranny supposedly exercised by our genes.

Like Gould, I well appreciate that nurture plays an important part in shaping each of us. His evaluation of nature's role, however, is utterly wrong: our low gene count by no means invalidates a reductionist approach to biological systems; nor does it justify any logical inference that we are *not* determined by our genes. A fertilized egg containing a chimp genome still inevitably produces a chimp, while a fertilized egg containing a human genome produces a human. No amount of exposure to classical music or violence on TV could make it otherwise. Yes, we have a long way to go in developing our understanding of just how the information in those two remarkably similar genomes is applied to the task of producing two apparently very different organisms, but the fact remains that the greatest part of what each individual organism will be is programmed ineluctably into its every cell, in the genome. In fact, I see our discovery of a low human gene count as good news for standard reductionist approaches to biology: it's much easier to sort through the effects of 25,000 genes than 100,000.

While humans may not have an enormous number of genes, we do have, as the sprawling dystrophin gene illustrates, a large, messy genome. Returning again to the worm comparison: while we have not even twice as many genes, our genome is thirty-three times larger. Why the discrepancy? Gene mappers describe the human genome as a desert spotted with occasional genetic oases—genes. Fifty percent of the genome is constituted of repetitive junklike sequences of no apparent function; a full 10 percent of our DNA consists of a million scattered copies of a single sequence, called *Alu*:

GGCCGGGCGCGGTGGCTCACGCCTGTAATCCCAGCACTTTGG
GAGGCCGAGGCGGGCGGATCACCTGAGGTCAGGAGTTCGAGA
CCAGCCTGGCCAACATGGTGAAACCCCGTCTCTACTAAAAATA
CAAAAATTAGCCGGGCGTGGTGGCGCGCCTGTAATCCCAG
CTACTCGGGAGGCTGAGGCAGGAGAATCGCTTGAACCCGGGA
GGCGGAGGTTGCAGTGAGCCGAGATCGCGCCACTGCACTCCA
GCCTGGGCGACAGAGCGAGACTCCGTCTCAAAAAA

Writing it out a million times would give a sense of the scale of the *Alu* presence in our DNA. In fact, levels of repetitive sequence are even higher than they would appear: sequences that would once have been instantly identifiable as repeats have, over many generations of mutation, diverged beyond recognition as members of a particular class of repetitive DNA. Imagine a set of three short repeats: ATTG ATTG ATTG. Over time mutation will change them, but if the period is short, we can still see where they came from: ACTG ATGG GTTG. Over a longer period, their original identity is completely lost in the welter of mutation: ACCT CGGG GTCG. Proportions of repetitive DNA are much lower in many other species: 11 percent of the mustard weed genome is repetitive, 7 percent of the nematode worm's, and just 3 percent of the fruit fly's. The large size of our genome is mostly due to its containing more junk than that of many other species.

These differences in the amounts of junk DNA explain a long-standing evolutionary conundrum. The basic expectation is that more complex organisms should have bigger genomes—they need to encode more information—than simple ones. There is indeed a correlation between genome size and an organism's level of complexity: the yeast genome is bigger than that of *E. coli* but smaller than ours. It is, however, only a weak correlation.

COMMON NAME	SPECIES NAME	APPROX. GENOME SIZE (MILLIONS OF BASE PAIRS)
Fruit Fly	*Drosophila melanogaster*	180
Fugu (puffer)	*Fugu rubripes*	400
Snake	*Boa constrictor*	2,100
Human	*Homo sapiens*	3,100
Locust	*Schistocerca gregaria*	9,300
Onion	*Allium cepa*	18,000
Newt	*Amphiuma means*	84,000
Lungfish	*Protopterus aethiopicus*	140,000
Fern	*Ophioglossum petiolatum*	160,000
Amoeba	*Amoeba dubia*	670,000

It is reasonable to suppose that natural selection operates to keep genome size as low as possible. After all, every time a cell divides, it must replicate all its DNA; the more it has to copy, the greater the room for error, and the more

Onions on top: the genome of one of his onions is six times larger than the onion vendor's.

energy and time the process requires. It is quite an undertaking for the amoeba (or newt, or lungfish). So what could have caused the amount of DNA in these species to get so out of hand? In cases of unusually large genomes, we can only infer that some other selective forces must have negated the selection-driven impulse to keep the genome slim. It could be, for instance, that large genomes are advantageous to species likely to be exposed to environmental extremes. Lungfish live at the interface of land and water, and they can survive protracted periods of drought by burying themselves in mud; it could be they need more genetic hardware than a species adapted to a single medium.

Two major evolutionary mechanisms account for this DNA excess: genome doubling, and the proliferation of particular sequences within a genome. Many species, particularly in the plant kingdom, are actually the product of a cross between two preexisting ones. The new species often simply combines the DNA complement from each of its parent species, yielding a double genome. Alternatively, through some kind of genetic accident, a genome may get doubled without input from another species. For example, one of the standbys of molecular biology, baker's yeast, has about 6,000 genes. But close inspection reveals that a large proportion of those genes are duplicates—baker's yeast often has two divergent copies of many of its genes. At some early stage in its evolutionary history, the yeast genome apparently got doubled. Initially the gene copies would have been identical, but, over time, they have diverged.

Barbara McClintock, discoverer of mobile genetic elements

An even richer source of excess DNA has arisen from the multiplication of genetic sequences capable of replicating and inserting themselves at more than one site in a given genome. These so-called mobile elements have been found to come in many varieties. But when their discovery was first announced by Barbara McClintock in 1950, the very idea of "jumping" genes was too far-fetched for most scientists accustomed to the simple logic of Mendel. McClintock, a superb corn geneticist, had already endured something of a bumpy career ride. When it became clear in 1941 that she would not be granted tenure at the University of Missouri, she came to Cold Spring Harbor Laboratory, where she would remain an active member of the staff until her death in 1992, at the age of ninety. McClintock once told a colleague, "Really trust what you see." This was exactly how she did her science: her revolutionary idea that some genetic elements could move around genomes followed simply from observable facts. She had been studying the genetics underlying the development of different-colored kernels in corn, and noticed that sometimes, part way through the development of an individual kernel, the color would switch. A single kernel might then turn out variegated, with both patches of the expected yellow cells and patches of purple ones. How to account for this sudden switch? McClintock inferred that a genetic element—a mobile element—had hopped into or out of the pigment gene.

Only with the advent of recombinant DNA technologies have we come to appreciate just how common mobile elements are; we now recognize them as major components of many, if not most, genomes, including our own. And some of the most common mobile elements, those that appear again and again in different sites in the same genome, have earned names reflecting their itinerant lifestyles: two fruit fly mobile elements, for example, are called "gypsy" and "hobo." And among those who study a simple plant called *Volvox* one mobile element is honored for its extraordinary capacity to jump around the genome: it is known as the "(Michael) Jordan element."

Mobile elements contain DNA sequences that code for enzymes that, through their capacity to cut and paste chromosomal DNA, work to ensure that copies of their particular element are inserted into new chromosomal sites. If a jump carries a mobile element into a junk sequence, the functioning of the organism is unaffected, and the only result is more junk DNA. But when the jump lands the mobile element in a vital gene, thereby disabling its function, then selection intervenes: the organism may die or otherwise be prevented from passing on the new jumped-in gene. Very rarely the movements of mobile elements may either create new genes or alter old ones in a way that benefits the host organism. Over the course of evolution, therefore, the effect of mobile elements seems mainly to have been the generation of novelty. And curiously, in recent human history, there is little evidence of active jumping: most of our junk DNA, it appears, was generated long ago. In contrast, the mouse genome contains many actively reinserting mobile elements, making for a much more dynamic genome. But this seems not to trouble the mouse species unduly; the intrinsically high reproductive potential of mice likely helps the species as a whole tolerate the genetic disasters attending frequent jumps into vitally functioning genetic regions.

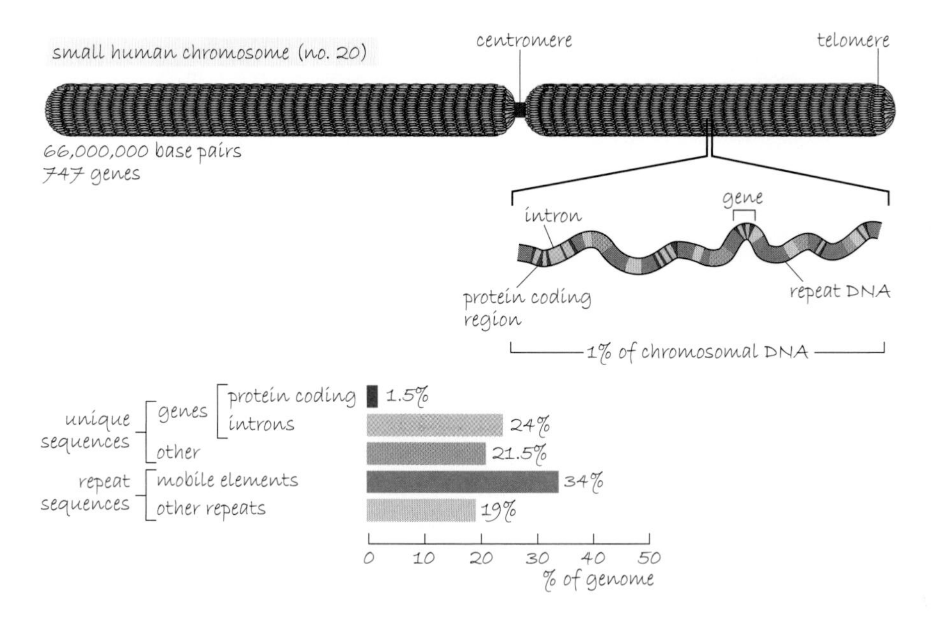

What our genome looks like: the major features of a small human chromosome, number 20

Having been used to establish many of the basic facts about how DNA functions, *E. coli*'s track record as a model organism was unparalleled. Not surprisingly, its genome therefore ranked high on the Human Genome Project's early "to do" list. It was Fred Blattner of the University of Wisconsin who was most eager to start sequencing *E. coli*. But his grant proposals went nowhere until the HGP got funded and he was awarded one of the first substantial sequencing grants. Were it not for his initial reluctance to adopt automated sequencing, his lab would have been the first to sequence a complete bacterial genome. But in 1991 his strategy for scaling up the operation was an

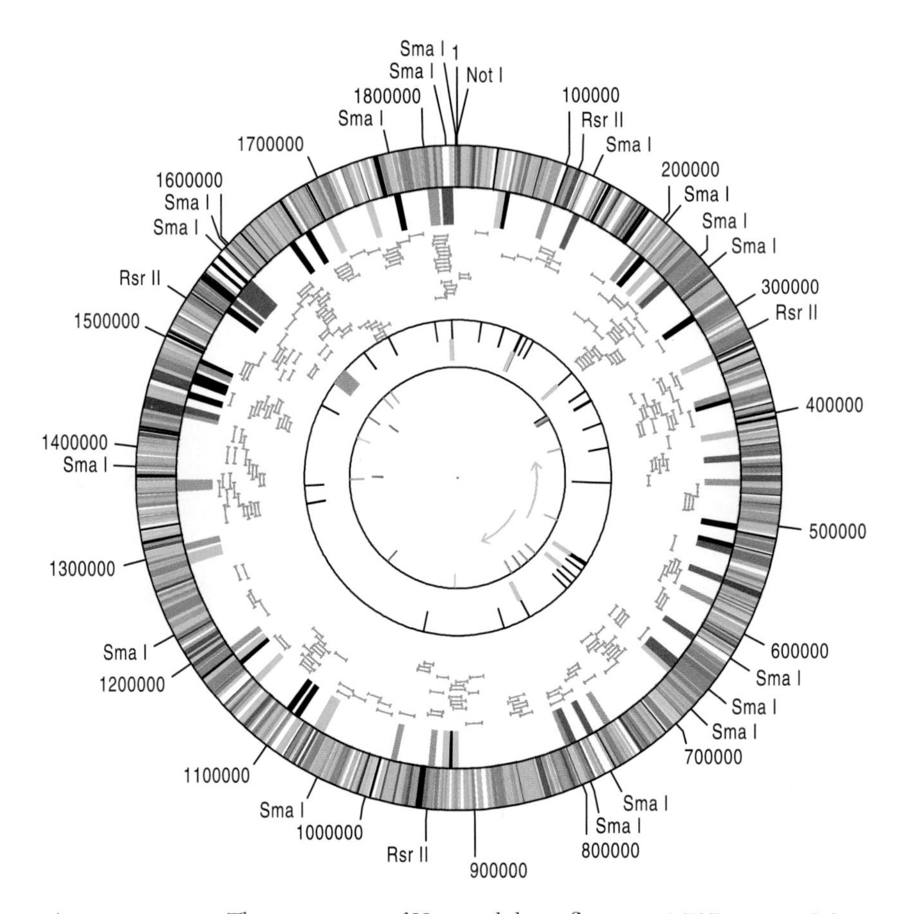

An entire genome. The genetic map of Haemophilus influenzae: *1,727 genes in 1.8 million base pairs.*

old-fashioned one: employ more undergraduates. Another latecomer to automation was Wally Gilbert, whom I had urged two years before to have a go at the smallest known bacterial genomes, those of the parasitic *Mycoplasma*—tiny bacteria that live within cells. Sadly, when a clever new manual sequencing strategy of his came to naught, his *Mycoplasma* project died with it. Blattner did, however, accept automation in time to establish in 1997 that the *E. coli* genome contains some 4,100 genes.

But the broader race to complete the first bacterial genome had been won two years before at The Institute for Genomic Research (TIGR) by a large team led by Hamilton Smith, Craig Venter, and his wife, Claire Fraser. And the bacterium they sequenced was *Haemophilus influenzae,* from which twenty years earlier Smith—a towering six-foot-six one-time math major who had gone on to medical school—had isolated the first useful DNA cutting (restriction) enzymes, a feat that won him the Nobel Prize in Physiology or Medicine in 1978. With *Haemophilus* DNA prepared by Smith, Venter and Fraser used a whole genome shotgun approach to sequence its 1.8 million base pairs. Just documenting the first "small" genome was enough to suggest the awesome size of the awaiting larger ones: if all the As, Ts, Gs, and Cs of the *Haemophilus* genome were printed on paper of this size, the resulting book would run some four thousand pages. Two pages on average would be needed for each of its 1,727 genes. Of these, only 55 percent have readily identifiable functions: for example, energy production involves at least 112 genes, and DNA replication, repair, and recombination requires a minimum of 87. We can tell from their sequences that the remaining 45 percent are functioning genes, but we simply can't at this stage be sure what it is they do.

By bacterial standards, the *Haemophilus* genome is pretty small. The size of a bacterial genome is related to the diversity of environments a particular species is likely to encounter. A species that leads a dull life in a single uniform setting—say, the gut of another creature—can well get by with a relatively small genome. One that hopes to see the world, however, and is apt to encounter more varied conditions, must be equipped to respond, and flexibility of response usually depends on having alternative sets of genes, each tailored to particular conditions, and ready at all times to be switched on.

Pseudomonas aeruginosa, a bacterium that can cause infections in humans (and poses a particular danger for cystic fibrosis [CF] patients), lives in many different environments. We saw in chapter 5 how a genetically doctored form of a related species became the first living organism to be patented; in that case, it was adapted to life in an oil slick, an environment notably different from the human lung. The *Pseudomonas aeruginosa* genome contains 6.4 million base pairs and 5,570 genes. About 7 percent of those genes encode transcription factors, proteins that switch genes on or off; a respectable proportion of its entire genetic complement is thus devoted to regulation. The *E. coli* "repressor" whose existence was predicted by Jacques Monod and François Jacob in the early sixties (see chapter 3) is just such a transcription factor. A rule of thumb then would go as follows: The greater the range of environments potentially encountered by a bacterial species, the larger its genome, and the greater the proportion of that genome dedicated to gene-switching.

TIGR did not stop at *Haemophilus*. In 1995, collaborating with Clyde Hutchison at the University of North Carolina, the institute sequenced the genome of *Mycoplasma genitalium* as part of what has been dubbed the "minimal genome project." *M. genitalium* (which, despite its ominous name, is a benign inhabitant of human plumbing) has the smallest known nonviral genome, some 580,000 base pairs. (Viruses have smaller genomes but, by co-opting the genomes of their hosts, can get away with not having the genetic wherewithal for many fundamental processes.) And that relatively short sequence was found to comprise 517 genes. So a question naturally arose: Is that the minimal gene complement necessary to sustain life? Subsequent research has set about knocking out *M. genitalium*'s genes to see which are absolutely vital and which are not. Currently it appears that the minimal genome contains no more than 350 genes and possibly as few as 260. Admittedly, this is a somewhat artificially defined "minimum" since the enfeebled bugs are supplied through their growth medium with every substance they could conceivably need. It's a bit like claiming kidneys are not necessary for life because patients can survive on dialysis machines.

Will we ever be able to construct a functioning minimal cell from scratch, by artificially combining its separate purified components? Considering there are more than a hundred *Mycoplasma genitalium* proteins whose functions remain a mystery, the achievement of such a goal seems for now a long way off. Even the five hundred proteins of *Mycoplasma,* some represented in the cell by a huge number of molecules, some by just a handful, constitute an enormously complex living system. I, for one, have enough difficulty following a movie like *Gosford Park* in which there are more than four or five major characters; the thought of blocking out the complexity of interactions among the vital players inside a living cell is nothing short of mind-blowing. For the living cell is no neat miniature machine; it is rather, as Sydney Brenner put it, "a snake pit of writhing molecules." Still, Craig Venter is confident that the era of the artificial cell is just around the corner, and he has wasted no time in assembling a panel of bioethicists to counsel him on whether to venture forth. They, like me, see no moral dilemma in trying to "create life" in this way. If such a feat were ever achieved, it would merely reaffirm what most of us in molecular biology have long known to be the truth: the essence of life is complicated chemistry and nothing more. Such a finding would have made headlines a century ago; today it's no big deal. Only the opposite conclusion—that there is more to the life of the cell than the sum of its basic components and processes—could generate deep excitement in today's scientific world.

DNA analysis has already changed the face of microbiology. Before DNA techniques were broadly applied, methods of identifying bacterial species were extremely limited in their powers of resolution: you could note the form of colonies growing in a petri dish, view the shape of individual cells through the microscope, or use such relatively crude biochemical assays as the Gram test, by which species can be sorted as either "negative" or "positive" depending on features of their cell wall. With DNA sequencing, microbiologists suddenly had an identification factor that was discernibly, definitively different in every species. Even species, like those inhabiting the ocean depths, that cannot be cultured in a laboratory because of the difficulty of mimicking their natural growing conditions are amenable to DNA analysis, providing a sample can be collected from the deep.

Now led by Claire Fraser, TIGR remains the leader of the bacterial genomics pack. In short order they have polished off the genomes of more than twenty different bacteria, including that of an ulcer-causing *Helicobacter,* a cholera-causing *Vibrio,* a meningitis-causing *Neisseria,* and a respiratory-disease-inducing *Chlamydia.* Their biggest competitor is a group at the Sanger Centre. The British contingent is led by Bart Barrell, who had the luck not to be in the United States, where his limited academic credentials would have barred him from top-gun status: he has no Ph.D., having come into science straight out of high school to work as Fred Sanger's assistant long before DNA sequencing became a reality. Before moving on to bacteria, Barrell made his name as an automation pioneer, having used several ABI sequencing machines to crank out some 40 percent of the baker's yeast genome of 14 million bases while the largely European yeast-sequencing consortium remained wedded to manual methods. Barrell's group later had the satisfaction of being the first to complete the sequence of *Mycobacterium tuberculosis,* the agent of the fearsome affliction once known as consumption.

In high school, Claire Fraser "had felt like an outcast because it wasn't cool to be a woman taking so many science courses." After studying at Rensselaer Polytechnic Institute, where she first became interested in microbes, she applied to medical school. Rather than accepting a place at prestigious Yale, she opted for SUNY Buffalo because her boyfriend was moving to Toronto. The director of admissions at Yale was nonplussed: "Well, young lady, I hope you know what you're doing." The Toronto connection would, alas, prove ephemeral; in 1981 Fraser married Venter, then a young assistant professor at SUNY Buffalo. "We went to a [scientific] meeting for our honeymoon," she recalled, "and wrote a grant proposal there."

Claire Fraser, TIGR's guide through the genomic jungle

The power of DNA analysis of microbes has been harnessed with great success in medical diagnostics: to treat an infection effectively, doctors must first identify the microbe causing it. Traditionally the identification has required culturing the bacteria from infected tissue—a process that is maddeningly slow, particularly in cases when time is of the essence. Using a fast, simple, and more accurate DNA test to recognize the microbe, doctors can start appropriate treatment that much sooner. And recently the same technology was pressed into service to deal with a national emergency: the hunt for the perpetrator of the anthrax outrage in the United States in the fall of 2001. By sequencing the anthrax bacteria from the first victim, TIGR investigators obtained a genetic fingerprint of the precise strain used. The hope is that this precise information on the source of the anthrax will lead eventually to the culprit.

As we learn more about microbial genomes, a striking pattern is emerging. As we have seen, vertebrate evolution is a story of progressive genetic economy: through a widening array of mechanisms for gene regulation, it has become possible to do more and more with the same genes. And even when new genes do appear, they tend to be merely variations on an existing genetic theme. Bacterial evolution, by contrast, is proving itself a saga of far more radical transformation, a dizzying process that favors the importation or generation of whole new genes, as opposed to merely tinkering with what already exists.

Indeed, recombinant technology owes its very existence to the extraordinary ability of bacteria to incorporate new pieces of DNA (usually plasmids). Not surprisingly, then, microbial evolution too bears the footprint of dramatic gene-importing events of the past. *E. coli,* normally a benign inhabitant of our intestines (and of petri dishes), has morphed through gene importation into a killer variant. The toxins produced by one strain that occasionally causes outbreaks of food poisoning (killing twenty-one people in Scotland in 1996–97) and headlines about "Killer Burgers" are attributable to massive genetic "borrowing" from other species.

Genetic material normally moves *vertically* down a lineage—from ancestor to descendant—so this importation of DNA from outside is known as "horizontal transfer." Comparison of the genome sequence of normal *E. coli* to that of the pathogenic strain has revealed a shared genetic "backbone," identifying both strains as members of a common species, but there are many "islands" of divergent DNA unique to the pathogen. Overall, the pathogen lacks 528 of the normal strain's genes and has instead a staggering 1,387 genes not present in the normal strain. In that 528-for-1,387 exchange lies the key to the transformation of one of nature's most innocuous products into a killer.

Other bacterial nasties also show similar evidence of wholesale horizontal transfer. *Vibrio cholerae,* the agent of cholera, is unusual for a bacterium in that it has two separate chromosomes. The larger one (about 3 million base pairs in length) appears to be the microbe's original equipment, containing most of the genes essential to the functioning of the cell. The smaller one (about 1 million base pairs in length) seems to be a mosaic, made up of bits and pieces of DNA imported from other species.

Complex organisms, especially large ones like humans, are by design fairly inviolable gatekeepers of their own internal biochemistry: in most cases, if we don't ingest or inhale a substance, it cannot alter us profoundly. And so the biochemical processes of all vertebrates have tended over time to remain very similar. Bacteria, on the other hand, are much more exposed to the chemical vagaries of the environment; a colony may find itself suddenly awash in a noxious chemical—say, a disinfectant like household bleach. Little wonder these highly vulnerable organisms have evolved a stunning variety of chemistries. Indeed bacterial evolution has been driven by chemical innovation, the invention of enzymes (or the retrofitting of old ones) to do new chemical tricks. One of the most fascinating and instructive instances of this evolutionary pattern occurs among bacteria whose secrets we have only recently begun to learn about, a group known collectively as the "extremophiles" because of its members' predilection for the most inhospitable environments.

Bacteria have been found in Yellowstone hot springs (*Pyrococcus furiosus* thrives in boiling water and freezes to death at temperatures below 70°C [158°F]) and in the superheated water of deep-sea vents (where the high pressure at depth prevents the water from boiling). They have been found living in environments as acidic as concentrated sulfuric acid and in acutely alkaline environments as well. *Thermophila acidophilum* is an all-around extremophile, withstanding, as its name suggests, both high temperatures and low pH. Some species have been discovered in rocks associated with oil deposits, converting oil and other organic material into sources of cellular energy, rather like so many tiny sophisticated automobiles. One of these species inhabits rocks a mile or more down and dies in the presence of oxygen; appropriately, it is named *Bacillus infernus.*

Perhaps the most remarkable microbes discovered in recent years are the ones that subvert what was once considered a key dogma of biological science—that all energy for living processes comes ultimately from the sun. Whereas even *Bacillus infernus* and oil-consuming bacteria found in sedimentary rocks are connected to the organic past—the sun shone eons ago on the plants and animals whose remains are today's fossil fuels—so-called lithoautotrophs are capable of extracting the nutrients they need from rocks created de novo by volcanoes. These rocks—granite is an example—bear no traces of organic material; they contain no vestige of the energy of sunny prehistoric days. Lithoautotrophs have to construct their own organic molecules out of these inorganic materials. They live, literally, on a diet of rock.

There has been no more persuasive indicator of our general ignorance of the microbial universe than our belated discovery of the bacterial genus *Prochlorococcus,* whose planktonic cells photosynthesize as they float in the open ocean.

As many as 200,000 may inhabit a single milliliter of seawater, making this arguably the most abundantly represented species on the planet. It is certainly responsible for a huge proportion of the ocean's contribution to the global food chain. And yet *Prochlorococcus* was unknown to us until 1988.

The extraordinary microbial universe around us reflects the phenomenal power of eons of natural selection. Indeed the history of life on our planet can be told mostly as a tale of bacteria; more complicated organisms, ourselves included, are embarrassingly late arrivals—a virtual afterthought. Life appears to have originated as bacteria some 3.5 billion years ago. The first eukaryotes—cells whose genes are enclosed within nuclei—arose around 800 million years later, but they remained single-celled for about a billion years after that. Only about half a billion years ago did the breakthroughs occur that would ultimately give rise to the likes of the earthworm, the fruit fly, and *Homo sapiens*. The predominance of bacteria is reflected in the DNA-based reconstruction of the tree of life first carried out by Carl Woese at the University of Illinois: the tree of life is a bacterial tree, with a few multicellular beings forming a late-growth twig. Now generally accepted, Woese's ideas were at first strenuously opposed within the biological establishment. Still some of the implications of the DNA-based approach to the tree of life have been difficult to take: they have shown, for instance, that animals are not, as was once supposed, closely related to plants; rather, the closest relatives of animals are fungi. Humans and mushrooms stem from the same evolutionary root.

The Human Genome Project has proved Darwin more right than Darwin himself would ever have dared dream. Molecular similarities stem ultimately from the way in which all organisms are related through common descent. A successful evolutionary "invention" (a mutation or set of mutations that is favored by natural selection) is passed down from one generation to the next. As the tree of life diversifies—existing lineages splitting to produce new ones (reptiles persist as such, but also bud off into both bird and mammal lineages)—that invention may eventually appear in a huge range of descendant species. Some 46 percent of the proteins we see in yeast, for example, also appear in humans. The yeast (fungal) lineage and the one that ultimately gave rise to humans probably split about 1 billion years ago. Since each has subsequently developed independently, free to follow its own evolutionary trajectory, there have been in effect 1 billion years of evolutionary activity since that yeast/human common ancestor; and yet, *through all that time,* that set of proteins that existed in the common ancestor has changed only minimally. Once evolution solves a particular problem—for example, designing an enzyme to catalyze a particular biochemical reaction—it tends to stick with that solution. We have seen how this kind of evolutionary inertia is responsible for the centrality of RNA in cellular processes: life started in an "RNA world," and the legacy remains with us to this day. And the inertia extends to the biochemical details: 43 percent of worm proteins, 61 percent of fruit fly proteins, and 75 percent of fugu proteins have marked sequence similarities to human proteins.

Comparing genomes has also revealed how proteins evolve. Protein molecules can typically be envisioned as collections of distinct *domains*—stretches of amino acid chains that have a particular function, or form a particular three-dimensional structure—and evolution seems to operate by shuffling domains, creating new permutations. Presumably most new permutations are as useless as they are random, doomed to be eliminated by natural selection; but in the rare instance that a new permutation proves beneficial, a new protein is born. Some 90 percent of the domains that have been identified in human proteins are also present in fruit fly and worm proteins. In effect, therefore, even a protein unique to humans is likely nothing more than a reshuffled version of one found in *Drosophila.*

There is no better demonstration of this fundamental biochemical similarity among organisms than so-called rescue experiments, the aim of which is to eliminate a particular protein in one species and then use the corresponding protein from another species to "rescue" the missing function. We have already seen this strategy implemented in the case of insulin. Because human and cow insulins are so similar, diabetics who fail to produce their own can be given the cow version as a substitute.

In an example evocative of B-movie science fiction, researchers have been able to induce fruit flies to grow eyes on their legs by manipulating a particular gene that specifies where an eye should go. That gene then induces the many genes involved in producing a complete eye to go to work in that designated location. The mouse's corresponding gene is so similar to the fruit fly's that it will perform the same function when situated—by the genetic engineer's sleight of hand—in a fruit fly whose gene has been eliminated. That this can be done is nothing less than remarkable. Fruit flies and mice have been separated by evolution for at least half a billion years, so—following the logic applied above to humans and yeast evolving simultaneously along independent lines—the gene has in fact been conserved over a billion years of evolution. This is all the more astonishing when we consider that fruit fly and mouse eyes have fundamentally different structures and optics. Presumably each lineage perfected an eye appropriate for its respective purposes, but the basic machinery for determining the location of that eye, needing no improvement, stayed the same.

The most humbling aspect of the Human Genome Project so far has been the realization that we know remarkably little about what the vast majority of human genes do. To use the hard-won information properly requires us to devise methods for studying the function of genes on a genomewide scale.

In the wake of the HGP, two new postgenomic fields have duly emerged, both of them burdened with unimaginative names incorporating the "-omic" of their ancestor: proteomics and transcriptomics. Proteomics is the study of the proteins encoded by genes. Transcriptomics is devoted to determining where and when genes are expressed—that is, which genes are transcriptionally active in a given cell. If the genome is ultimately to be understood in its more dynamic reality, not as a mere set of instructions for life's assembly but as the screenplay for life's movie—all the drama described in the precise order it is meant to occur—then proteomics and transcriptomics provide the keys to glimpsing the live action. The more we learn, the more we see of *Life, the Movie.*

We have long appreciated that a protein is a great deal more in biological terms than the linear string of amino acids that compose it. How the string folds up to produce a distinctive three-dimensional configuration is really the key to its function—what proteomics seeks to know. Structural analysis is still done using X-ray diffraction: the molecule is bombarded with X rays that bounce off its atoms and scatter in a pattern from which the three-dimensional shape may be inferred. In 1962, my one-time colleagues at the Cavendish Lab at Cam-

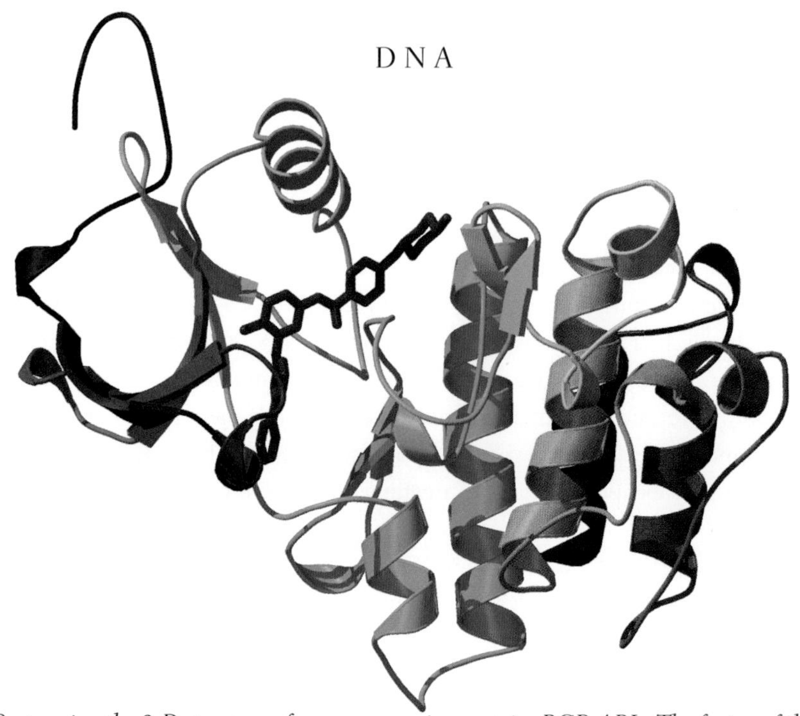

Proteomics: the 3-D structure of a cancer-causing protein, BCR-ABL. The fusion of the two genes caused by a chromosomal abnormality leads to the production of this protein, which stimulates cell proliferation and may cause a form of leukemia. Shown in purple is a small molecule drug, Gleevec, which inhibits BCR-ABL function (see chapter 5). It is with 3-D information like this that drugs will in the future be designed to target particular proteins. This representation of BCR-ABL's structure does not show the details of the atoms or individual amino acids, but nevertheless accurately reflects the protein's layout.

bridge University, John Kendrew and Max Perutz, received the Nobel Prize in Chemistry for their elucidation of the structures of, respectively, myoglobin (which stores oxygen in muscle) and hemoglobin (which transports oxygen in the bloodstream). Theirs was a monumental effort. The complexity of the X-ray diffraction images they had to interpret made me appreciate the relative simplicity of DNA!

Knowledge of a protein's three-dimensional structure greatly assists the work of medical chemists in their hunt for new drugs that work, as many do, by inhibiting protein functioning. In the ever more specialized and automated world of pharmaceutical research, several companies now offer to determine the structure of proteins as if they were production-line commodities. And the work is now immeasurably easier than it was in the day of Perutz and Kendrew: with more powerful X-ray sources, automated data recording, and faster computers driven by increasingly clever software, the time needed for solving a structure can be reduced from many years to a matter of weeks.

All too often, however, the three-dimensional structure itself provides no particular indication of that protein's function. Important clues may come instead from studying how the mystery protein interacts with other known ones. A simple way to identify such interactions involves spotting out samples of a set of known proteins on a microscope slide and then dousing them with the mystery protein, which has been previously treated so it will fluoresce under UV light. Where our test protein "sticks" to a particular spot on the slide's protein grid, it has become bound to the protein in that spot, causing it too to become fluorescent. Presumably, then, these two proteins are engineered to interact within the cell.

Ideally, to know life's screenplay, to "see" life's movie, we need to discover all the precise changes in protein composition that occur over the individual's development, from the moment of fertilization all the way through to adulthood. Though many proteins will be found to be active throughout the process, some will prove specific to a particular developmental stage, so in each growth phase we should expect to see different sets of proteins. Adult and fetal hemoglobins, for example, are subtly different. Similarly, each variety of tissue produces its own profile of proteins.

The most reliable way to sort out the various proteins from a given tissue sample is still the long-established method that uses two-dimensional gels to separate protein molecules on the basis of differences in their electrical charge and molecular weight. The several thousand protein spots thus differentiated can then be analyzed with a mass spectrometer, an instrument that can determine each one's amino acid sequence. Unfortunately, to apply proteomics like this to the vast number of proteins coded by an entire genome requires more funding than academic scientists typically have. For the most part, such expensive enterprises are left to the better-endowed researchers of large pharmaceutical companies. But because of the method's limitations, even their labs can't routinely find proteins that are present in very small amounts.

This type of high-throughput proteomics, with all its expensive hardware and industrial-scale automation of complicated procedures, is therefore not the way most scientists nowadays study gene function at the genome level. Instead, methods of transcriptomics have been adopted, because they are cheaper and easier to apply: the functioning of all genes in a genome can be tracked by measuring the relative amounts of their respective messenger RNA (mRNA) products. If you are interested in the genes being expressed in, say, a human liver cell, you isolate a sample of mRNAs from liver tissue. This represents a snapshot of the mRNA population in the liver cell: very active genes, those most heavily transcribed and that produce many mRNA molecules, will be more abundantly represented, whereas genes that are rarely transcribed will contribute only a few copies to the mRNA sample.

The key to transcriptomics is a surprisingly simple invention known as a DNA microarray. Imagine a microscope slide with a grid of 25,000 tiny dot-shaped wells etched onto it. Using precise micropipetting techniques, DNA sequences from just one gene are deposited in each well so that the grid contains every gene in the human genome. Critically, the location on the microscope slide of each gene's DNA is known. Affymetrix, a company near Stanford, has managed to miniaturize these arrays even further by etching them onto a sliver of silicon the size of a small computer chip, yielding a "DNA chip."

Using standard biochemical techniques, you can tag your liver mRNAs with a chemical marker so, like the proteins mentioned above, they will fluoresce obligingly under UV light. Then comes the step where the power and simplicity of the technique becomes wonderfully apparent: you simply dump your sample of mRNAs onto the microarray with its minuscule chessboard of 25,000 gene-filled wells. The very same base-pairing bonds that hold together the two strands of the double helix will compel each mRNA molecule to pair off with the gene from which it was derived. The complementarity is precise and foolproof: the mRNA from gene X will bond only to the very spot occupied by gene X on the microarray. The next step is merely to observe which spots have picked up the fluorescent mRNAs. One spot on the microarray may show no fluorescence, implying that there was no complementary mRNA in the sample—and thus, we may infer, no active transcription of that gene in the liver cell. On the other hand, a number of spots do fluoresce, some with particular intensity; this indicates that many mRNA molecules have bound to it. Conclusion: a very active gene. Thus, with a single simple experimental assay, you have identified every one of the genes active in the liver. And such molecular panoramas have been made possible thanks to the success of the Human Genome Project and the new mind-set it has ushered into biology: we no longer need be content to study bits and pieces—we can now see the whole picture in all its spectacular glory.

It is hardly surprising that Stanford's Pat Brown, one of the method's leading practitioners, sees DNA microarrays as "a new kind of microscope." Marveling at the technology's potential to reveal a whole new genetic universe, he has declared: "We're toddlers now just starting to discover our world."

Transcriptomics is more than just another brilliant technical innovation. It promises to take us to a new level in the hunt for the genes that cause illness: using microarray technology we can discover the chemical basis for particular afflictions by studying the differences between healthy and diseased tissue as a function of gene expression. The logic is simple. We carry out microarray gene expression analysis on both normal and cancerous tissue, and spot the difference between the two, the genes being expressed in one and not the other. Once we can identify which genes are malfunctioning—either over- or underexpressing themselves in the cancerous tissue, for instance—we may be able to establish a target that can be attacked with pinpoint molecular therapies as opposed to broadly toxic radio- and chemotherapies that destroy healthy as well as diseased cells.

And we can apply the same technologies to distinguish among different forms of the same disease. Standard microscopy has offered limited assistance in this task: cancers that look alike to the pathologist peering through the eyepiece can in fact be critically different at the molecular level. Lymphoma cells, for instance, come in varieties that are hard to tell apart visually, even with the highest powers of resolution, but the differences in their gene expression profiles are clear, and vitally important in devising the most effective treatment. Referring to the earlier tendency to assume all cancers of a particular tissue have the same root, Brown said, "It was like thinking a stomachache has only one cause. Recognizing the distinctions makes it possible for us to do a better job of treating these cancers."

At Cold Spring Harbor Laboratory, Michael Wigler is using the method in yet another way: rather than adding RNA to a microarray and looking for gene expression, he is adding DNA from cancer cells to create a profile of the genetic diversity present in tumors. Many cancers are caused by chromosomal rearrangements—such as might occur when segments of a chromosome are inadvertently duplicated, leading to an excess in the number of genes that code for growth-promoting proteins. Other cancers arise due to the loss of genes coding for proteins that repress cell growth. Applying Wigler's technique, clinicians biopsy cancerous and healthy tissues from the same person. DNA from the cancerous tissue is chemically tagged with a red dye while the DNA from the normal tissue is tagged green. DNA microarrays, containing all 35,000 of the known human genes, are exposed to a mixture of the two samples. Like mRNA in a standard microarray experiment, the labeled DNA molecules bind base pair to base pair to their complementary sequences in the array. Genes amplified in cancer cells are marked by red spots (because there are many more red-tagged molecules binding to that spot than green-tagged ones) while genes deleted in cancer cells show up as green spots on the microarray (because there is no red-tagged molecule to bind there). Such experiments have already greatly expanded the list of genes known to contribute to breast cancer.

Whenever we tackle a specific human disease, we realize the extent to which we are probing in the dark. We could move so much more quickly to the heart of the problem—know the exact nature of what is wrong and how we might fix it—if only we had a more detailed knowledge of how our genes express themselves when all is well. With a fully formed dynamic understanding of when and where each of our 25,000 genes functions during normal development from fertilized egg to functioning adult, we would have a basis of comparison by which to understand every affliction: what we need is the complete human "transcriptome." This is the next holy grail of genetics, the next big quest in need of superfunding. In the short term, a likelier, even more important objective will be to obtain the complete transcriptome for the mouse, whose advantage over humans is that we can both observe and intervene experimentally during the course of prenatal development. Even collecting all such relevant data from the mouse will require major investments of money and time. And, as proved by the experience of DNA sequencing, we will be well served to take the time to gain what expertise we can by completing transcriptomes for simpler model organisms before taking on the mouse, much less the human.

Microarray studies of gene expression during the yeast cell cycle have already

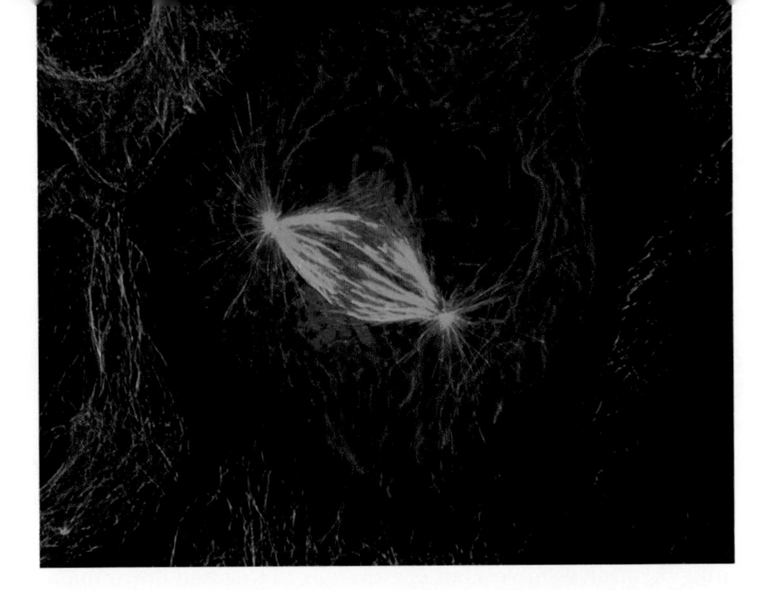

Cell division: a cell's chromosomes (blue) are duplicated and then lined up along a special "spindle" (green) prior to being assigned to each daughter cell. High-tech imaging techniques help bring to life the extraordinary chromosomal waltz underpinning life's ability to perpetuate itself.

revealed the staggering complexity inherent in the molecular dynamics of cell division alone. More than eight hundred genes are involved, each called into action at its precisely specified time in the cell cycle. Here too we may depend on evolution's reluctance to fix what ain't broke: a biological process, once successfully evolved, will likely continue to employ the same basic molecular actors for as long as life persists on earth. As far as we can tell, those same proteins that direct development through the course of the yeast cell cycle carry out similar roles in the human cell.

Ultimately the goal of all three "-omics" (gen-, prote-, and transcript-) is to create a full picture, detailed right down to the level of the individual molecule, of how living things are assembled and operate. As we have seen, in even the simplest cases, the complexity is bewildering, and, despite the spectacular progress of the last decade, there remain many daunting challenges. As they relate to complex organisms, the molecular underpinnings of development—that extraordinary egg-to-adult journey that is governed by a linear code strand composed of just four letters—are for now best understood in the case of the fruit fly.

The fly has, of course, been the focus of intensive genetic investigation ever since its adoption by T. H. Morgan, and through the ensuing years of continual innovation *Drosophila melanogaster* has remained a genetic gold mine. In the late seventies at the European Molecular Biology Laboratory in Heidelberg, Germany, Christiane "Janni" Nüsslein-Volhard and Eric Wieschaus undertook a spectacularly ambitious fruit fly project. They used chemicals to induce mutations and then looked for disruptions in the very early embryonic stages of the flies' progeny. Classically, the quarry of the fruit fly geneticist was mutations affecting adults, like the one Morgan found to produce white (rather than red) eyes. In focusing on embryos, Nüsslein-Volhard and Wieschaus were not only condemning themselves to years of eyestrain as they stared down microscopes in pursuit of those elusive mutants, they were also venturing into utterly uncharted territory. The payoff, however, was spectacular. Their analysis uncovered several suites of genes that lay out the fundamental body plan of the developing fly larva.

The more universal message of their work is that genetic information is hierarchically organized. Nüsslein-Volhard and Wieschaus noticed that some of their mutants showed very broad effects while others evinced more restricted ones; from this they inferred correctly that the broad-effect genes operate early in development—at the top of a switching hierarchy—while the restricted-effect genes operated later. What they had found was a cascade of transcription factors: genes switching on other genes that in turn switch on others still, and so on. Indeed, hierarchical gene-switching of this kind is the key to the construction of complex bodies. A gene producing the biological equivalent of a brick will, left to its own devices, produce a pile of bricks; with proper coordination, however, it can produce a wall, and ultimately a building.

Normal development depends on cells "knowing" where they are in a body. A cell in the tip of a fly's wing, after all, should develop along very different lines than one located in the region that will give rise to the fly's brain. The first piece of essential positional information is the simplest: How does the developing fruit fly embryo know which end is which? Where should the head go? Bicoid, a protein produced by a gene in the mother, is distributed in varying concentrations through the embryo. The effect is called a "concentration gradient": the protein levels are highest at the head end and fall off as you travel toward the rear. Thus the bicoid concentration gradient instructs all cells within the embryo as to where they fall on the head-to-tail axis. Fruit fly development is segmental, meaning that the body is organized into compartments, all of which have much in common but each of which has some features unique to it. In many respects, a head segment is organized just like one in the thorax (the middle part of the insect body), but the former has head-specific organs, like eyes, and the latter thorax-specific ones, like legs. Nüsslein-Volhard and Wieschaus found groups of genes that specify the identities of different segments. For instance, "pair-rule" genes encode transcription factors—genetic switches—expressed in alternating segments. Pair-rule gene mutants result in an embryo with developmental problems in every second segment.

In 1995, Nüsslein-Volhard and Wieschaus received the Nobel Prize in Physiology or Medicine for their pioneering work. Unlike most laureates, both have remained active lab scientists—not for them the retreat into a big diploma-festooned office. For Wieschaus, science is still irresistible: "Because embryos are beautiful and because cells do remarkable things, I still go into the lab every

Fruit fly faces. On the left is a normal individual, with a pair of feathery antennae protruding from its forehead. On the right is an antennapedia mutant, in which the antennae have been replaced by fully formed legs.

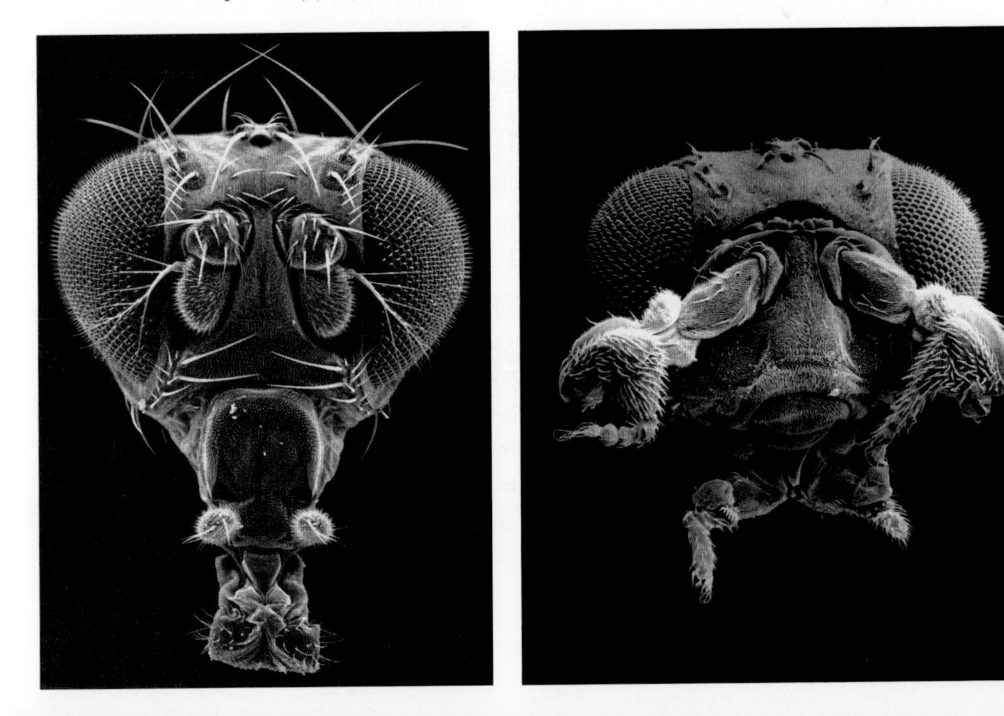

day with great enthusiasm." As a child in Birmingham, Alabama, he dreamed of becoming an artist. Short of money as a sophomore at the University of Notre Dame, however, he took on one of the smelliest and most menial jobs in all of science: making the "fly food" (a noxious gelatinous concoction consisting largely of molasses) for a research lab's experimental population of fruit flies. Most people who serve as chef to a few hundred thousand messy and unappreciative insects would likely develop a lifelong aversion to the critters. For Wieschaus the result was the opposite: a lifelong commitment to the fruit fly and the mysteries of its development.

Born into an artistic German family, Nüsslein-Volhard was one of those students who excels at everything that interests them but puts absolutely no effort into anything else. Her hard work in illuminating the fruit fly's developmental genetics would have been achievement enough to justify two careers, but in the wake of her Nobel she has redirected her formidable attention to the development of another species altogether, the zebra fish: new work that promises to unlock many of the secrets of vertebrate development. At the 2001 event marking the centenary of the Nobel Prize it struck me that she was the only woman scientist present among the throngs of gray-haired males. Indeed, she is one of only ten women ever to win a Nobel in science.

One of those no-longer-youthful men was Caltech's Ed Lewis, an old fruit fly hand who shared the prize with Nüsslein-Volhard and Wieschaus. Actually Lewis doesn't much fit the gray-hair stereotype: though in his eighties at that Stockholm event, when he wasn't obliged to wear tails he was often seen in running gear! He too had long been concerned with the genetic control of fruit fly development, but his special interest was "homeotic mutations." These produce a most bizarre result: one developing segment mistakenly acquires the identity of a neighboring segment. His long and painstaking dedication to the Hox genes, in which these mutations occur, exemplifies values vanishing in an era when fads too often set science's agenda.

Homeotic mutations—which we now know disrupt transcription factor-encoding genes (the genetic switches)—can have drastic effects. The "antennapedia" mutation results in the fly's growing legs where its antennae belong: a fully formed pair of legs protruding from its forehead. The "bithorax" mutation is almost as weird. Normally one of the segments making up the thorax produces the fly's pair of wings while the next thoracic segment toward the rear generates a pair of small stabilizing structures called "halteres." In a bithorax fly, the haltere segment mistakenly produces wings, so a fly that should have two wings in fact has four, the second pair just as perfectly formed as the first.

When they function properly, the genes regulating segment identity ensure that each body section acquires organs appropriate to its position: a head segment acquires antennae, and a thoracic segment acquires wings and legs. In the event of homeotic mutations, however, there is a confusion of segment identity. Thus, in the case of antennapedia, a head segment imagines itself a thoracic one and duly produces a leg rather than an antenna. Note, though, that while the leg is in the wrong place, it's still a perfectly good leg. Implication: The antennapedia positional gene switches on a whole suite of genes, typically those that produce an antenna, or, aberrantly, those that produce a leg; but the coordination within the suite is unhindered even when these genes are activated in the wrong place at the wrong time. Here again we see how genes high up in the developmental hierarchy control the fate of many, many genes farther down the line. As any librarian knows, hierarchical organization is an efficient way in which to store and retrieve information. With such a cascade arrangement, a surprisingly few genes can take you a long way.

Now that we are in the new era of comprehensiveness in biology ushered in by the once-unimaginable feat of the Human Genome Project, it may seem curious that we should find ourselves following the cutting edge of one of the next frontiers—that of developmental genetics—back into the realm of the fruit fly. But there is nowhere for us to go but back to the future, for even with the entire human genome in hand, the program and cues according to which its instructions are carried out remain a colossal mystery. Eventually we shall know the screenplay of human life as well as we know that of the fly. A comprehensive description of the patterns of human gene expression (the transcriptome) will be developed. A full inventory of the actions of all our proteins (the proteome) will be produced. And we will have a full and spectacularly complex picture of how each one of us is put together, and how each one of the multitudinous molecules we are made of figures in the functioning of you and me.

Matternes

CHAPTER NINE

. . .

OUT OF AFRICA: DNA AND THE HUMAN PAST

In August 1856 German quarry workers discovered part of a skeleton as they blasted their way into a limestone cave in the Neander Valley outside Dusseldorf. At first the remains appeared to be those of an extinct bear species whose bones often showed up in caves, but a local schoolteacher realized that the creature in fact belonged to a species much closer to our own. The exact identity of the owner of the bones, however, would prove a point of controversy. Particularly puzzling was the skull's thick brow ridge. One bizarre suggestion was that the bones belonged to an injured Cossack cavalryman who had crawled into the cave to die during the Napoleonic wars. Chronic pain from a preexisting condition, so the crackpot theory went, had produced a permanent furrow in the poor fellow's brow, deforming the bones of the skull to create the distinctive ridge. In 1863, in the midst of the debate about human origins provoked by the publication of Darwin's *Origin of Species* four years earlier, the original owner of the bones was given a name: *Homo neanderthalensis.* The bones belonged to a species distinct from, but similar to, *Homo sapiens.*

Though the German bones were the first to be officially designated Neanderthal, others found earlier in Belgium and Gibraltar were now recognized as being from members of the same species. More than a century later, many more specimens of *H. neanderthalensis* have been unearthed, and we now believe that Neanderthals settled throughout Europe, the Middle East, and parts of North Africa until about 30,000 years ago. French paleontologist Marcellin Boule is largely responsible for the popular image of Neanderthals as dim-

Barbecues, ancient and modern: an artist's reconstruction (top) of a Neanderthal encampment in southern Europe about 35,000 years ago, with the equivalent scene from the more recent past (bottom). Are we descended from Neanderthals? DNA evidence suggests not.

witted and hulking. But his reconstruction, which used material from a French site at La Chapelle-aux-Saints, was based on a single individual who turns out to have been elderly and arthritic. In fact, Neanderthal brains were slightly larger than ours (and of a different shape due to a flatter cranium) and evidence from burial sites suggests that Neanderthals were culturally sophisticated enough to engage in funeral rituals; they may then have even believed in an afterlife.

The biggest debate triggered by the discovery of Neanderthals, however, centered not on how smart they were but on how they might be related to us. Are we descended from them? Paleontology suggests that modern humans arrived in Europe at roughly the same time as the last of the Neanderthals disappeared. Did the two groups interbreed or were the Neanderthals simply eliminated? Because the events in question happened in the ancient past and the surviving evidence is fragmentary—little beyond the odd bone—debates like this can drag on and on, keeping academic paleontologists and anthropologists endlessly entertained. Is a particular bone specimen perhaps intermediate between the thick bones typical of Neanderthals and the lighter bones of modern humans? Such specimens may have belonged to a hybrid individual produced by interbreeding between the two groups—a missing link. But then again they might just as well have come from a full Neanderthal with atypically light bones, or, for that matter, a fully modern human with unusually thick ones.

To everyone's surprise, the debate has been resolved by DNA: 30,000-year-old DNA extracted in 1997 from the very bones that started it all in 1856. Having evolved in order to store information securely and transmit it from one generation to the next, DNA, no surprise, shows great chemical stability. It doesn't degrade spontaneously or react readily with other molecules. But it is not impervious to chemical damage. At the moment of death, the body's genetic materials, like all its other constituents, become susceptible to a horde of would-be degraders: reactive chemicals, and enzymes that break down the molecular fabric. These chemical reactions require the presence of water, so DNA may be preserved if a corpse dehydrates fast enough. But even under ideal preservation conditions, the molecule is likely to survive perhaps 50,000 years at the absolute maximum. To obtain a legible DNA sequence from 30,000-year-old Neanderthal remains, preserved imperfectly, was therefore a tall order at best.

But Svante Pääbo, a tall, laconic Swede at the University of Munich, decided to have a crack at the problem. If anyone could do it, he was the one. Pääbo had pioneered work on the retrieval of so-called ancient DNA; he had scored sequences from Egyptian mummies, frozen mammoths, and the 5,000-year-old "Ice Man" who melted out of an Alpine glacier in 1991. Despite this impressive résumé, though, the prospect of drilling into a precious Neanderthal relic to look for intact DNA, if indeed any was to be found inside, was daunting. As his archaeologist colleague Ralf Schmitz recalls, "It was like getting permission to cut into the Mona Lisa."

Matthias Krings, Pääbo's graduate student, took on the project. He was pessimistic at first, but favorable early analyses to assess the bones' state of preservation emboldened Krings to press ahead. His search for viable DNA was focused not in the cells' nuclei, as one might expect, but in the little bodies called mitochondria, which are scattered throughout the cell outside the nucleus and produce the cell's energy. Each mitochondrion contains a small loop of DNA, some 16,600 base pairs in length. And because there are from 500 to 1,000 mitochondria in every cell, but only two copies of the genome proper (in the nucleus), Krings knew that those decaying Neanderthal bones were much more likely to yield intact mitochondrial sequences than intact nuclear ones. Furthermore, since mitochondrial DNA (mtDNA) had long been a staple of studies of human evolution, Krings would have plenty of modern human sequences against which to make comparisons.

A major worry for Krings and Pääbo was contamination. In the past a number of claimed successes at sequencing ancient DNA had proved to be erroneous when the sequence turned out to be from a modern source that had contaminated the sample. Every day each of us sloughs off a vast number of dead skin cells, showering our DNA into the environment to wind up we know not where. The polymerase chain reaction (PCR), with which Krings expected to amplify the stretch of mtDNA he hoped to find, is so sensitive that it can act upon a single molecule, amplifying any DNA it might encounter regardless of whether the source is ancient or still kicking. What if the Neanderthal DNA was too degraded for PCR to work, but the reaction proceeded nevertheless, amplifying a DNA sequence from an invisible contaminating particle that had flaked off Krings himself? Krings might then have to explain how he and the Neanderthal happened to have the same mtDNA sequence—a result unlikely to please the young man's boss, and even less his parents. To insure against this possibility, Krings and Pääbo arranged for a separate laboratory, Mark Stoneking's at Pennsylvania State University, to replicate the study. Contamination might occur there, too, but probably not with DNA from Krings, a continent away. And if both labs obtained the same result from the sample, it would be reasonable to suppose they had found a bona fide Neanderthal sequence.

"I can't describe how exciting it was," says Krings of the moment he first glimpsed the sequencing results. "Something started to crawl up my spine." Although, as feared, some sequences showed evidence of contamination, in others he could see something wondrous: a collection of intriguing similarities to, and differences from, the modern human sequence. Piecing together segments, he was able to reconstruct a Neanderthal mtDNA stretch running 379 base pairs. But the results weren't yet in from Penn State. Those sequences, however, proved to be the same: the identical 379 base pairs. "That's when we opened the champagne," Krings recalls.

The Neanderthal sequence had more in common with modern human mtDNA sequences than with those of chimpanzees, telling us that Neanderthals were unquestionably part of the human evolutionary lineage. At the same time, however, there were dramatic differences between the Neanderthal sequences and all 986 available sequences of modern human mtDNA to which Krings compared his sample. And even the most similar of those 986 sequences still differed from the Neanderthal one by at least 20 base pairs (or 5 percent). Subsequently, mtDNA has been sequenced from two other Neanderthals, one found in southwest Russia, the other in Croatia. The sequences, as expected, are not identical to the original one—we would expect to see variation among Neanderthal individuals just as we would among modern humans—but they are similar. The sum of the genetic evidence leads us to conclude that while Neanderthals do have their place on the evolutionary tree of humans and their relatives, the Neanderthal branch is a long way from the modern human limb. If, when they encountered each other in Europe 30,000 years ago, Neanderthals and moderns had indeed interbred, Neanderthal mtDNA sequences would have entered the modern human gene pool. That we see no evidence of such Neanderthal input implies that modern humans eliminated the Neanderthals rather than interbreeding with them. But whether they achieved the lethal result by direct confrontation or by more subtle means is something the DNA can't tell us.

Studies of Neanderthal DNA have shown that we are genetically distinct from Neanderthals. But the overall lesson of molecular studies of human evolution has tended to run in the opposite direction, revealing just how astonishingly close we are genetically to the rest of the natural world. In fact, molecular data have often challenged (and overthrown) long-held assumptions about human origins.

The great chemist Linus Pauling was the father of modern molecular approaches to evolution. During the early 1960s, he and Emile Zuckerkandl compared the amino acid sequences of corresponding proteins from several species. These were the early days of protein sequencing, and their data were inevitably limited. Nevertheless, the pair noticed a striking pattern: the more closely related two species are in evolutionary terms, the more similar are the sequences of their corresponding proteins. For example, comparing one of the protein chains of hemoglobin molecules, Pauling and Zuckerkandl noted that over its total length of 141 amino acids, there is only one difference between the human version and the chimpanzee, but the difference between humans and horses is 18 amino acids. The molecular sequence data reflect the fact that horses have been evolutionarily separated from humans longer than chimpanzees. Unearthing evolutionary history buried in biological molecules is now common practice; at the time, however, the idea was novel and controversial.

Molecular approaches to studying evolution depend on the correlation of two variables: the length of time two species (or populations) have been separated and the extent of molecular divergence between them. The logic of this "molecular clock" is simple. To illustrate it, let us imagine some matchmaking between two pairs of identical twins, one of genetically identical females and one of identical males. Each female is wed to one of the males, and each couple is then placed on its own otherwise uninhabited island. From a genetic perspective, the populations of the two islands are at the outset indistinguishable. Now leave each couple and its descendants alone for a few million years. At the end of this period, mutations will have occurred in the population on one island that will not have occurred in the population on the other. And vice versa. Because mutations occur at a low rate and because individual genomes, being large, offer huge numbers of possible sites where mutations might occur, it is inconceivable that both populations will have acquired the same set of mutations. So when we sequence DNA from the descendants of each couple, we will find that many differences between the once-identical genomes have accumulated. We say that the populations have "diverged" genetically. The longer they have been separated, the more divergent they will be.

But how do we tell time, so to speak, by looking at this "molecular clock"? Put another way, how can we measure the genetic divergence between ourselves, say, and the rest of the natural world? In the late sixties, long before the advent of DNA sequencing, Allan Wilson, a whimsical New Zealander at UC Berkeley, together with his colleague Vince Sarich, set about applying the Pauling-Zuckerkandl logic to humans and their closest relatives. But at a time when protein sequencing was still a dauntingly cumbersome and laborious affair, Wilson and Sarich found an ingenious shortcut.

The strength of an immune reaction to a foreign protein reflects *how* foreign the protein is: if it is relatively similar to the body's own protein, then the immune reaction is relatively weak, but if it is very different the reaction is proportionately stronger. Wilson and Sarich compared reaction strengths by taking a protein from one species and measuring the immune responses it triggered in others. This gave them an index of the molecular divergence between two species, but to introduce a time dimension to this "molecular clock" they needed to calibrate it. Fossil evidence implied that New and Old World monkeys (the two major groups of monkeys) separated from their common ancestor around 30 million years ago—and so Wilson and Sarich set the immunological "distance" between New and Old World monkeys as equivalent to 30 million years' separation. Where did this put humans in relation to their closest evolutionary kin, chimpanzees and gorillas? In 1967 Wilson and Sarich published their estimate that the human lineage had separated from that of the great apes about 5 million years ago. Their claim provoked an uproar: in paleoanthropological circles conventional wisdom held that the divergence had occurred around 25 million years ago. Between humans and apes, the establishment insisted, there is clearly much more than 5 million years' worth of difference. It was, for many, cause enough to dismiss the Berkeley team's newfangled genetic method as untrustworthy, and, to declare that, anyway, geneticists should stick to their fruit flies, and leave humans to the anthropologists! Wilson and Sarich, however, weathered the storm. And subsequent research has shown that their dating of the human/great ape split was remarkably accurate.

When the time came to extend his analysis of the human/ape divide from proteins to DNA, Wilson entrusted the effort to his graduate student Mary-Claire King. The product, in 1975, was one of the outstanding scientific papers of the twentieth century. For a long time, though, such a triumphant outcome seemed unlikely, especially from King's perspective. Her work had not been going well, owing in part to the enormous distraction created at Berkeley by the anti–Vietnam War movement in the early 1970s. King considered going off to Washington, D.C., to work for Ralph Nader, but fortunately she sought Wilson's advice. "If everyone whose experiments failed stopped doing science," he wisely counseled, "there wouldn't be any science." King stuck with it.

Mary-Claire King

King and Wilson's comparison of the chimpanzee and human genomes combined a number of methods, including a clever technique called "DNA hybridization." When two complementary strands of DNA come together to form a double helix, they can be separated by heating the sample to 95°C—a phenomenon called "melting" in the molecular geneticist's jargon. What happens when the two strands are not perfectly complementary—when there are mutations in one of them? It turns out that two such strands will melt apart at a temperature lower than 95°C. How much lower will depend upon the degree of difference between the two strands: the greater the difference, the less the heat required to pry them apart. King and Wilson used this principle to compare human and chimpanzee DNAs. The closer the two were in sequence, the closer the double helix's melting point would be to the perfect-match standard of 95°C. The closeness observed was surprising indeed: King was able to infer that human and chimpanzee DNA differ in sequence by a mere 1 percent. In fact humans have more in common with chimpanzees than chimpanzees do with gorillas, the genomes of the latter two differing by about 3 percent.

So striking was the result that King and Wilson felt obliged to put forward an explanation for the apparent discrepancy between the rates of genetic evolution—slow—and of anatomical and behavioral evolution—fast. How could so little genetic change account for the substantial difference we see between the chimpanzee at the zoo and the species on the other side of the glass? They suggested that most of the important evolutionary changes had occurred in the pieces of DNA that control the switching on and off of genes. This way, a small genetic change could have a major effect by changing, say, the timing of the expression of a gene. In other words, nature can create two very different-looking creatures by orchestrating the same genes to work in different ways.

The next, and biggest, bombshell from Wilson's Berkeley lab came in 1987. Using patterns of DNA sequence variation, he and his colleague Rebecca Cann figured out the family tree for our entire species. It was one of the very few pieces of science ever to make the cover of *Newsweek*.

As Krings would in his analysis of Neanderthals a decade later, Cann and Wilson relied on mitochondrial DNA. There were several reasons for using mtDNA, but as usual the practical ones were most important. In the days before PCR technology had entered the research mainstream, getting enough DNA to probe a particular gene or region could be quite a headache. And Cann and Wilson's study called for analyzing not one but 147 samples. They therefore needed as much DNA as they could get their hands on. A human tissue sample is massively rich in mtDNA compared with the chromosomal DNA found in cell nuclei. Still, Cann and Wilson would need plenty of tissue if they were to have any hope of extracting even mtDNA in sufficient quantities. Their solution: placentas. Usually discarded by hospitals after babies are delivered, these are a rich source of mtDNA. All Cann and Wilson had to do was persuade 147 pregnant women to donate their babies' placentas to science—146, actually, because Mary-Claire King was more than willing to contribute her daughter's placenta. And they knew that to reconstruct the human family as completely as possible they would need tissue from the most genetically diverse range of donors they could assemble. Here America's melting-pot population offered a

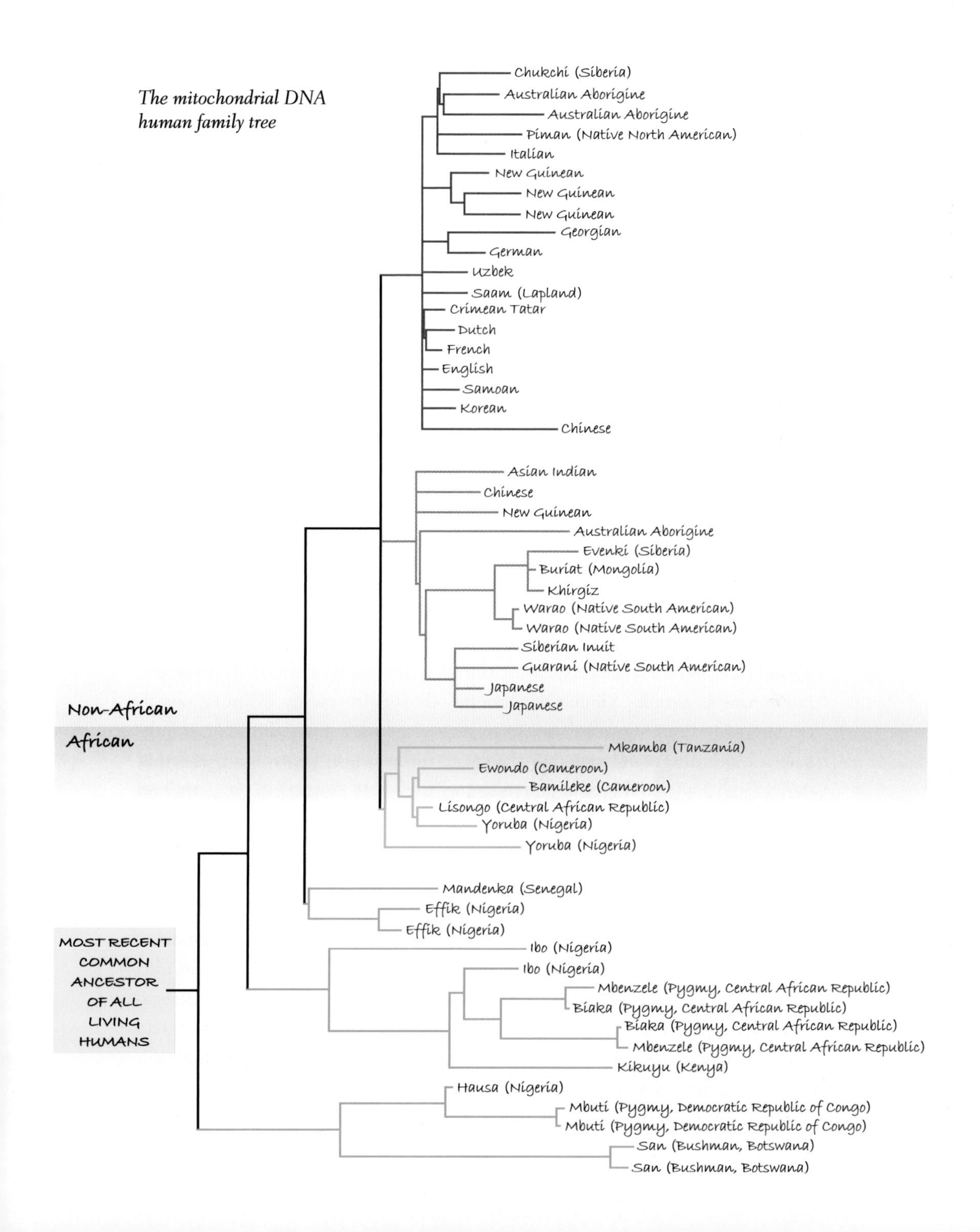

The mitochondrial DNA human family tree
Chukchi (Siberia)
Australian Aborigine
Australian Aborigine
Piman (Native North American)
Italian
New Guinean
New Guinean
New Guinean
Georgian
German
Uzbek
Saam (Lapland)
Crimean Tatar
Dutch
French
English
Samoan
Korean
Chinese
Asian Indian
Chinese
New Guinean
Australian Aborigine
Evenki (Siberia)
Buriat (Mongolia)
Khirgiz
Warao (Native South American)
Warao (Native South American)
Siberian Inuit
Guarani (Native South American)
Japanese
Japanese
Non-African
African
Mkamba (Tanzania)
Ewondo (Cameroon)
Bamileke (Cameroon)
Lisongo (Central African Republic)
Yoruba (Nigeria)
Yoruba (Nigeria)
Mandenka (Senegal)
Effik (Nigeria)
Effik (Nigeria)
Ibo (Nigeria)
Ibo (Nigeria)
Mbenzele (Pygmy, Central African Republic)
Biaka (Pygmy, Central African Republic)
Biaka (Pygmy, Central African Republic)
Mbenzele (Pygmy, Central African Republic)
Kikuyu (Kenya)
Hausa (Nigeria)
Mbuti (Pygmy, Democratic Republic of Congo)
Mbuti (Pygmy, Democratic Republic of Congo)
San (Bushman, Botswana)
San (Bushman, Botswana)
MOST RECENT COMMON ANCESTOR OF ALL LIVING HUMANS

distinct advantage: they would not have to travel to Africa to get hold of African DNA—the slave trade had brought African genes to our shores. But Cann and Wilson would have to depend on collaborators in New Guinea and Australia to find Aboriginal women (not much represented in the U.S. gene pool) who were willing to participate.

Your mtDNA is inherited from your mother. Your father's genetic contribution, contained in the head of a single sperm, did not include mitochondrial material. The sperm's DNA is injected into an egg cell that already contains mitochondria derived from the mother. Cann and Wilson would therefore be tracing the history of the human female line. Inherited from just one parent, mtDNA never gets an opportunity to undergo recombination, the process by which segments of chromosome arms are exchanged so that mutations are shuffled from one chromosome to another. The absence of recombination in mtDNA is a major advantage when we come to reconstruct the family tree based on similarity of DNA sequences. If two sequences have the same mutation, we know that they must be descended from a common ancestor (in whom that mutation originally arose). Were recombination occurring, however, one of the lineages could have acquired the mutation just recently through a recombinational shuffling event, so having a mutation in common would not necessarily indicate common ancestry. Now the logic for using mtDNA to make the family tree is simple. Similar sequences—those with plenty of mutations in common—indicate close relationship; sequences with many differences indicate a more distant relationship. In visual terms, close relatives—those that derive from a relatively recent common ancestor—will cluster close together on the family tree; distant relatives are more spread out, because their common ancestor is relatively far back.

Cann and Wilson found that the human family tree has two major branches, one comprising only various groups within Africa and the other consisting of some African groups plus everyone else. This implies that modern humans arose in Africa—that is where the ancestors common to all of us lived. This idea was hardly new. Noting that both our closest relatives, chimpanzees and gorillas, are native to Africa, Charles Darwin himself inferred that humans had evolved there too. The most striking, and controversial, aspect of Cann and Wilson's family tree is how far back it goes in time. By making a number of simple assumptions about the rate at which mutations accumulate through evolution, it is possible to calculate the age of the family tree—the time back to the great-great-great-great- . . . -grandmother of us all. Cann and Wilson came up with an estimate of about 150,000 years. Even the most distantly related currently living humans shared a common ancestor as recently as 150,000 years ago.

Like Sarich and Wilson's result two decades earlier, Cann and Wilson's was greeted by many in the anthropological community with outraged disbelief. One widely accepted view of human evolution held that our species was descended from individuals who left Africa about 2 million years ago before settling throughout the Old World. Such a model implied that the family tree should be about thirteen times deeper. Cann and Wilson's alternative, dubbed by the media "The Eve Hypothesis" or, less misleadingly, "Out of Africa," did not deny the more ancient migration, but rather implied that when modern humans arrived in Europe they displaced those populations of early hominids derived from the original exodus nearly 2 million years before. *Homo erectus,* the species that spread out from Africa 2 million years ago, migrated through the Old World and gave rise, about 700,000 years ago, to Neanderthals, who were thus in effect their European descendants. Then, no more than about 150,000 years ago, another group, *Homo sapiens* or modern humans—also descendants of *Homo erectus* but a group that had evolved without ever having left the mother continent—now chose to repeat the odyssey out of Africa made eons before by their *H. erectus* ancestors. We have seen how the Neanderthals failed to interbreed with the new arrivals in Europe, and the same seems to have been true whenever *H. sapiens* encountered *H. erectus.* Wherever they met, the former displaced the latter. And the disappearance of the last Neanderthal, around 29,000 years ago, represents the extinction of the last of the nonmodern descendants of *H. erectus.*

Mitochondrial Eve as cover girl

Cann, Wilson, and their colleagues had changed fundamentally the way we understand our human past.

Subsequent research has confirmed Cann and Wilson's conclusion. Much of the newer work has come out of the Stanford laboratory of Luigi Luca Cavalli-Sforza, who pioneered the application of genetic approaches to anthropological problems. Raised in a distinguished Milanese family, Cavalli-Sforza was fascinated with microscopes. And in 1938, he enrolled as a precocious sixteen-year-old in medical school at the University of Pavia. "It turned out to be a very lucky choice," he notes: the alternative would have been service in Mussolini's army. When I first met him in 1951, Cavalli-Sforza was still an up-and-coming bacterial geneticist. But a chance remark made by a graduate student would inspire a turn away from the genetics of bacteria toward the genetics of humans. The graduate student, who had trained to be a priest, mentioned that the Catholic Church had kept detailed records of marriages over the past three centuries. Realizing that in these records there lurked a wealth of research possibilities, Cavalli-Sforza began to apply himself more and more to human genetics, and he probably remains one of a very few human geneticists who can legitimately claim to have found their vocation via the Church.

Cavalli-Sforza understood that the most convincing confirmation of Cann and Wilson's assertions about human evolution would ideally come from genes only transmitted from father to son, i.e., some component of the human genome passed down through the male line. If one could arrive at their conclusions tracking the male lineage—taking a patrilineal route as opposed to the matrilineal path Cann and Wilson found through mtDNA analysis—one could be assured of a truly independent corroboration. The male-specific component of the genome is, of course, the Y chromosome. By definition, the possessor of a Y is male (the Y chromosome, remember, is inherited by men from their fathers, whose sperm cells can contain either an X or a Y; upon fusing with the egg cell, which always contains an X, the sperm thus determines our sex, XX combinations producing females and XY males). The Y chromosome, then, holds the key to the genetic history of men. In addition, because recombination occurs only between paired chromosomes, the use of the Y allows us to avoid that dreaded pitfall of evolutionary analysis, recombination: a Y is unique whenever it is present, and so there is never a matching Y with which it might trade material.

In a blockbuster paper published in 2000, Cavalli-Sforza's colleague Peter

Underhill did for the Y chromosome what Cann and Wilson had done for mtDNA. The findings were strikingly similar. Again the family tree was found to be rooted in Africa, and again it was shown to be remarkably shallow: not the ancient mighty oak imagined by anthropologists, but the shrub of Cann and Wilson's analysis, around 150,000 years old.

The existence of two independent data-sets yielding a similar picture of the human past is extremely compelling. When only one region, say mtDNA, is studied, the results, while suggestive, are still inconclusive; the pattern may simply reflect the peculiarities of the history of that particular region of DNA, rather than the impact of some major historical event on our species as a whole. Critically, the point at which a family tree converges—the most recent common ancestor of all the sequences in the study, that great-great- . . . -grandfather/mother of us all—is *not* necessarily associated with any particular event in human history. Though it *may* connote the origin of our species or some other historically significant demographic episode, it may just as likely signify something much more trivial from the point of view of human history—perhaps nothing more, say, than the effect of past natural selection on mtDNA. If, however, the same pattern of change can be observed in more than one region of the genome, the chances are that one has indeed found the genetic footprint of an important past event.

To better understand how natural selection can affect patterns of genetic variation (and the overall age of a family tree), imagine the following scenario: 150,000 years ago, the tribe of protohumans boasted a plethora of mtDNA sequences, just as our species does today, but then a beneficial mutation—one favored by natural selection—arose on one of those sequences. The mutation would increase in frequency until, after many generations, every member of the species would have it. Because there is no recombination in mitochondria, no exchange between mtDNAs, the selective process would affect the entire sequence in which a favored mutation first appeared, so every member of the species would end up with the same mtDNA sequence. So by the time that natural selection has finished its job and every individual possesses the favored mutation, there would be no mtDNA genetic variation in the species. Gradually over subsequent years, though, mutations would occur and variation would build up again, but all these new mtDNA sequences would ultimately be descended from that single sequence: the family tree's convergence point, the most recent common ancestor of all the sequences. The pattern would be exactly what Cann and Wilson found, but in this case the convergence point represents nothing more than an episode of evolution's fine-tuning of mtDNA.

This was the ambiguity that dogged Cann and Wilson's result: Was it produced by evolutionary tinkering, or by something much more significant in the overall scheme of human prehistory? But when Underhill observed a similar pattern for the Y chromosome, that ambiguity vanished. The coincidence suggested forcefully that at the moment in question (150,000 years ago), human populations did indeed undergo a radical genetic alteration, one capable of affecting mtDNAs and Y chromosomes simultaneously. The phenomenon involved, to which we shall turn in a moment, is called a "genetic bottleneck."

How can demographic factors affect a family tree? Any genealogy is the outcome of the waxing and waning of the lineages composing it: over time, some will thrive and others become extinct. Think of surnames. Assume that a thousand years ago on some remote island everyone had one of three surnames: Smith, Brown, and Watson. Assume, too, that small errors of transcription—"mutations"—occasionally occurred when the names of newborns were inscribed in the birth registers. The errors are infrequent and slight, so we can still tell which of the original names the altered forms derived from: "Browne" is clearly a mutation of "Brown." Now let us imagine that in the population today, a thousand years later, we find that everyone is called Brown, Browne, Bowne, Frown, or Broun. Smith and Watson have gone extinct while the Brown line has thrived (and diversified through mutation). What has happened? Pure chance has led to the loss of the Smith and Watson lines. Perhaps, for instance, several Mr. & Mrs. Smiths of one generation managed to produce mainly daughters. Assume (in accordance with tradition, though not the modern alternative convention) that surnames are transmitted along the male line; the bumper crop of daughters would thus have the effect of reducing the representation of Smiths in the next generation. Now say that the new generation of Smiths also overproduced daughters, and the demographic effect was heightened once again—well, you get the picture: eventually, the Smith name disappeared altogether. So did Watson.

This kind of random extinction is, in fact, statistically inevitable. Usually, however, it happens so slowly that its impact can be felt only over huge periods of time. Sometimes, though, a bottleneck—a period of very much reduced population size—will massively accelerate the process. With only three couples (six individuals) on the island at the beginning of its population history, it was reasonably likely that we would lose Smith and Watson within a single generation, the chances being fairly good that both the Smiths and the Watsons would have only daughters, or fail to procreate at all. In a large population, such abrupt disappearances of lines cannot occur; it is statistically inconceivable, given a population with many Smith couples, that they could all wind up producing only girls or simply fail to have children. Only over the course of many generations would the effects of the dwindling ranks gradually mount up. Indeed a real-life example of this hypothetical name-extinction process actually occurred in the South Pacific, when the six *Bounty* mutineers colonized Pitcairn Island with their thirteen Tahitian brides. Within seven generations, the number of surnames had dwindled to three.

When we look today at the surnames in our theoretical population, Brown, Browne, Bowne, Frown, or Broun, we can infer that they are all descended from just one of the three starting lineages, Brown. And so the implication of the human mtDNA and Y chromosome data should hardly surprise us: 150,000 years ago there were many different mtDNA sequences and many different Y chromosome sequences, but today's sequences are all descended from just one of each. All the others went extinct, most probably disappearing during some ancient bottleneck event—a population crash caused by plague, a change in climate, what have you. But whatever this cataclysmic event in our early history, one thing is clear: some time afterward, groups of our ancestors started to head out of Africa, beginning the epic saga of the human colonization of the planet.

Another interesting finding confirmed by both the mtDNA and Y chromosome data is the position on the human family tree of the San of southern Africa.* Theirs is the longest, and therefore the oldest, branch on the tree. This by no means implies that they are more "primitive" than the rest of us: every

*The San are also known as Bushmen (Sanqua, in Dutch), a derogatory term given to them by Dutch settlers in the late seventeenth century.

San hunters

human is at the same evolutionary and molecular remove from our closest relatives among the great apes. If we trace lineages back to the last common ancestor of both chimpanzees and humans, my lineage is about 5 million years old, and so is a San's. In fact, our two lineages are the same for most of those eons; only 150,000 years ago did the San lineage separate from other human lines.

It appears, from the genetic evidence, that after an initial migration into southern and eastern Africa, the San remained relatively isolated throughout history. This pattern is borne out by sociolinguistics when we consider the distribution of the San's unusual (at least to my ears) "click" languages. Their current distribution is extremely limited owing to the expansion of Bantu-speaking people from west central Africa starting about 1,500 years ago. The Bantu expansion displaced the San to marginal environments like the Kalahari Desert.

Given their relatively stable history, do the San provide a snapshot of what the ancestors of all modern humans were like? Possibly, but not necessarily—substantial change may well have occurred along the San lineage over the past 150,000 years. Even inferences from the San about our early ancestors' ways of living are questionable: the San's present lifestyle is an adaptation to the harsh desert environment to which they have been confined since the relatively recent arrival of the Bantu speakers. In 2000 I experienced the unique thrill of living for several days in a San community in the Kalahari. I was struck by their remarkable pragmatism, their efficient no-nonsense way of taking on all tasks before them, even those outside their normal experience, like fixing a flat tire. I found myself wishing that more of my colleagues were likewise adaptable. And if, in genetic terms, these people are as genetically "different" from me as any on the planet, I could not fail to be impressed by just how like-minded we were.

The genetic and cultural uniqueness of the San will disappear shortly. Young people in the Kalahari show little desire to continue the simple hunter-gatherer lifestyle of their nomadic parents. When, for instance, the group I visited staged a traditional "trance dance," the younger members were visibly embarrassed by their elders' antics. They will move away from their communities and marry into other groups.

In fact, history has already recorded a trend toward mixture between the San and other groups. Nelson Mandela's Xhosa tribe, for one, represents a biological mix of Bantu and San peoples, as the Xhosa language, though Bantu-based, reveals in its many typically San clicks. In our technologically accelerated day and age, it is unlikely that the genetic and cultural integrity of the San will survive much longer. It is, therefore, fortunate indeed that considerable efforts have been made over the past few decades to understand and document this unique people and their way of life. Philip Tobias of the University of Witwatersrand in Johannesburg both initiated these studies and, for many years, championed the San as an unofficial spokesman during the dark days of apartheid. And Trefor Jenkins, a voluble Welshman who arrived in South Africa after working as a doctor in Zambian copper-mining towns, has long spearheaded genetic studies of the San and other indigenous groups.

Sadly it currently remains beyond the reach of even the most sophisticated genetic methods to elucidate the origins of human culture. Archaeological evidence shows that our ancestors were up to much the same activities as other hominids, Neanderthals included, during the first phase of their evolution. Indeed, a cave site at Skhul in Israel offers proof that about 100,000 years ago populations of *Homo sapiens* and *Homo neanderthalensis* coexisted, neither apparently endangering the other. But, as we have seen, modern humans subsequently wiped out their heavy-browed cousins around 30,000 years ago. It therefore seems likely that in the intervening 70,000-year period modern humans, through technological and/or cultural advances, somehow acquired the edge.

Independent archaeological information supports this hypothesis. It would appear that, around 50,000 years ago, modern humans suddenly became *cul-*

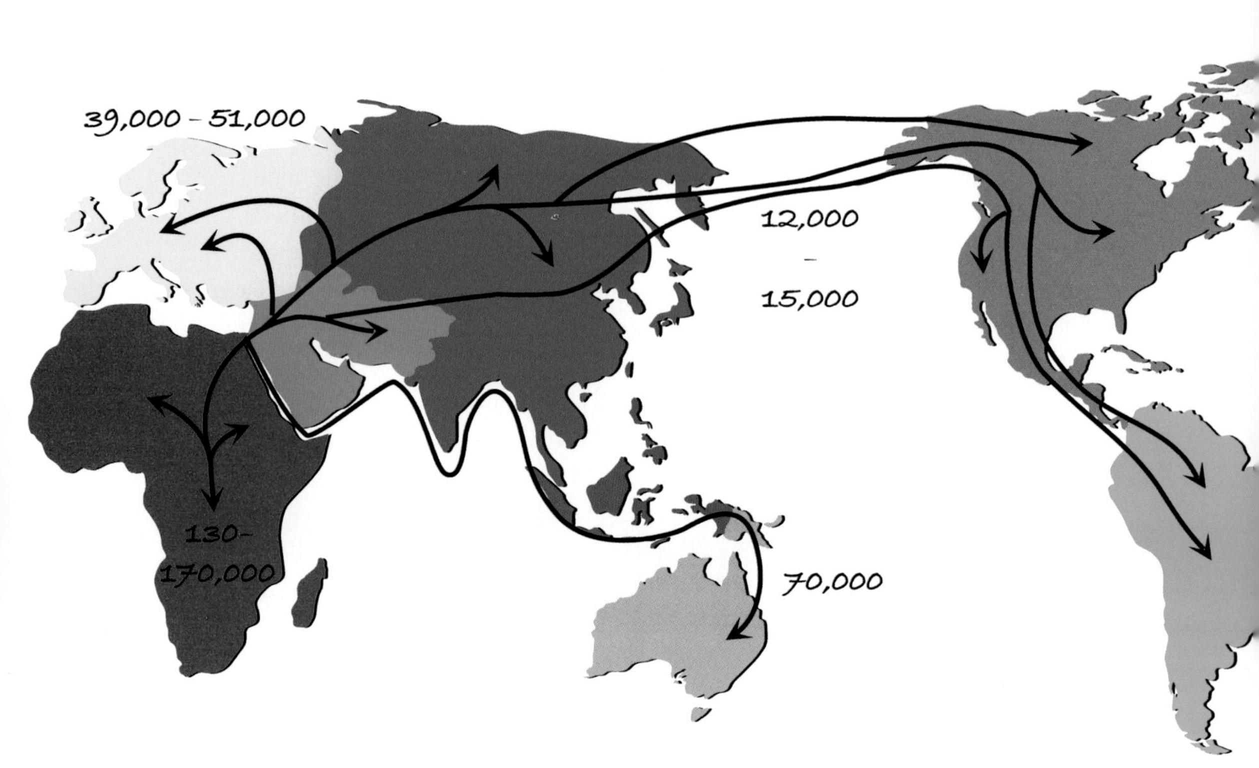

Out of Africa and beyond: Our species originated in Africa and spread out from there. Estimated colonization dates are based on mtDNA data.

turally modern: we see in the remains from this time the first indisputable ornaments, the first routine use of bone, ivory, and shell to produce familiar useful artifacts, and the first of many improvements in hunting and gathering technology. What happened? We shall probably never know. But one is tempted to speculate that it was the invention of language that made all of this—and all we have accomplished since—possible.

Prehistory by definition refers to the period prior to written records, and yet we find written in every individual's DNA sequences a record of our ancestors' respective journeys. The new science of molecular anthropology uses patterns of genetic variation among different groups to reconstruct this history of human colonization. Human "prehistory" has thus become accessible.

Studies of the distribution of genetic variation across the continents combined with archaeological information have revealed some details of our ancestors' global expansion. The journey along the fringes of Asia and through the archipelagoes of modern Indonesia to New Guinea and Australia was accomplished by about 60,000 years ago. Getting to Australia required crossing several substantial bodies of water, suggesting that our ancestors were already using boats at that early stage. Modern humans arrived in Europe around 40,000 years ago, and penetrated northern Asia, including Japan, some 10,000 years later.

Like so many other leaders in this field (including Rebecca Cann and Svante Pääbo), Michael Hammer, at the University of Arizona, received his training in Allan Wilson's Berkeley lab. And though Hammer's initial interest was mice, the publication of Cann and Wilson's mtDNA study diverted him from rodents to the human past. He was among the first to realize that information from the Y chromosome would provide the crucial test of Cann and Wilson's overall hypothesis. But the Y proved reluctant at first to yield its secrets. One study (done in Wally Gilbert's lab) sequenced the same chunk of DNA drawn from multiple individuals, only to find the sequence identical in every instance—a laborious effort that yielded zero information about genetic interrelations. Hammer persisted, however, and eventually he and others turned the Y chromosome into an anthropological gold mine, whose payoff culminated in Underhill's landmark paper.

A major vein in the Y chromosome mine has enriched our attempts to reconstruct the human colonization of the New World, a relatively late development. The identity of the oldest human settlement in the Americas remains contentious: a site in Clovis, New Mexico, is the traditional titleholder, dating back some 11,200 years; but fans of a site in Monte Verde, Chile, claim it to be at least 12,500 years old. It is also debated whether the first Amerindians crossed a land bridge across the Bering Strait during the last Ice Age or took a more southerly route in boats. What the genetic data make clear, however, is that the founding group was small: with only two major classes of Y chromosome sequences detected, there appear to have been just two distinct arrivals, each perhaps involving no more than a single family. Among Amerindians mtDNA variation is much more extensive than Y chromosome variation, suggesting that there were more women than men in each founding group. Probably the more common of the two Y chromosome sequences represents the first arrival; the descendant population would then already have been established before the arrival of the second group, which included the ancestors of today's Navajo and

Apache. The more common sequence also boasts another distinction: the presence (first noted in 2002) of a mutation that is rarely found elsewhere on the planet. Giving further evidence of its bearers' precedence as pioneers, this mutation is calculated to be about 15,000 years old, not much older than the earliest known archaeological sites.

Genetic analyses have permitted the reconstruction of more recent phases of prehistory as well. Hammer, for example, has shown that modern Japanese are a mix of the Jomon ancient hunter-gatherers, currently represented by Japan's aboriginal Ainu population, and relatively recent immigrants, the Yayoi, who arrived about 2,500 years ago from the Korean peninsula, bringing with them weaving, metalworking, and rice-based agriculture. In Europe, too, we see evidence of waves of migration, often associated with advances in agricultural technology. Groups like the Basques (who live in the mountainous Pyrenees on the French-Spanish border) and the Celts (who arrived later and are found throughout the northwest margin of Europe, from Brittany in France through Ireland and western Britain) are genetically distinct from the rest of Europe. One explanation is that each of these groups was displaced to relatively far-flung regions by more recent arrivals.

Bryan Sykes at Oxford has done much to reveal the complexity of the genetic map of modern Europe. Conventional wisdom had held that modern Europeans were largely derived from the Middle Eastern populations that invented agriculture in the Fertile Crescent, between the Mediterranean and the Persian Gulf. Sykes, however, has found that most European ancestry can be traced not to the Fertile Crescent but to older indigenous lines predating the incursions of Middle Easterners and to migrant groups from Central Eurasia. Such groups include the Celts and the Huns, who swept into Europe from the East around 500 B.C. and A.D. 400 respectively. And taking his analysis of mtDNA a step further, Sykes has argued as well that virtually all Europeans are descended from one of seven "daughters of Eve," his term for the surprisingly few major ancestral nodes in the European mtDNA family tree. A company he founded, called Oxford Ancestors, will, for a fee, sequence part of your mtDNA to determine from which of the seven "daughters" you are descended.

Another key to understanding the human past may rest with an observation fruitfully exploited by Cavalli-Sforza and others: patterns of genetic evolution often correlate with those of linguistic evolution. There are, of course, the obvious parallels between genes and words. Both are transmitted from one generation to the next; both undergo change, which in the case of language can be particularly fast, as any parent of a teenager knows. Likewise, American English is similar to but distinct from British English even though the two have been evolving separately for only a few hundred years. On the basis of the similarities and differences, then, the family tree of languages can be reconstructed in much the same way the genetic family tree can. But even more important, in many cases, as Darwin himself first predicted,* we can identify instructive correspondences between the two trees, such that what we learn about the one can deepen our understanding of the other. Both the Celts and the Basques offer dramatic cases in point: each people is genetically isolated from the rest of Europe, and each one's languages are correspondingly distinct from those of the rest of the continent. As for the New World, a controversial linguistic theory proposes that there are but three major language groups native to the Americas, and two of these correlate with the two early immigration events discerned in the Amerindian Y chromosome data. The third, by far the smallest, involves the isolated Inuit.

The availability of sex-specific genetic data—mtDNA for women, Y chromosomes for men—invites comparisons between male and female history. Mark Seielstad, a graduate student of Cavalli-Sforza's, chose to compare patterns of migration between the sexes. The logic is simple. Imagine a mutation that arises on a Y chromosome in Cape Town, South Africa. The speed with which it reaches, say, Cairo, is an index of rates of male migration. Similarly, the speed with which a Cape Town mutation in mtDNA reaches Cairo can be said to measure the rates of female migration.

For good or ill, history has been much more the chronicle of men, rather than women, on the move. Typically they were in search of plunder or empire: think of Alexander the Great's march from Macedonia into the northern reaches of India; of the Vikings and their sea-borne rampages from Scandinavia to Iceland

*In *The Origin of Species,* Darwin notes: "If we possessed a perfect pedigree of mankind, a genealogical arrangement of the races of man would afford the best classification of the various languages now spoken throughout the world."

and America beyond; or of Genghis Khan and his horsemen pouring across the steppe of Central Asia. But even without warfare as an excuse for travel, we still think of men as the more mobile members of human society. Men traditionally do the hunting, an activity that can often take them a long way from the hearth, whereas women in traditional hunter-gatherer societies stay close to home, gleaning food locally and raising the children. Therefore, Seielstad had reason to expect that men would be our species's genetic prime movers. The data proved him startlingly wrong. Women, on average, are *eight times more* mobile than men.

In fact, counterintuitive though it may be, the pattern can be simply explained. Almost universally, across all traditional societies, we humans engage in something anthropologists call "patrilocality": when individuals from two different villages get married, the woman moves to the man's village, and not vice versa. Imagine that a woman from village A has married a man from village B, and she moves to B. They have a daughter and a son. The daughter marries a man from village C and moves to C; the son marries a woman from village D and she moves to join him in B. Thus the male line stays put in B whereas the female line has moved, in two generations, from A to C via B. This process is carried out generation after generation, and as a result female migration proves extensive, but male migration does not. Men do indeed occasionally rush off to conquer distant lands, but these events are unimportant in the grand scheme of human migratory patterns: it's actually that step-by-step village-to-village migration of women that has shaped human history, at least on the genetic level.

Detailed regional studies of Y chromosome and mtDNA variation may also reveal something of the patterns of sexual relations and mating customs promoted in the course of colonization. In Iceland, for instance, which was uninhabited before the arrival of the Vikings, we find a marked asymmetry when we compare mtDNA and Y chromosomes. Most Ys are predictably Norse, but a large proportion of the mtDNA types are derived from Ireland. Apparently, the Norsemen colonizing Iceland took Irish women with them. Unfortunately, how the Irish women felt about this cannot be extracted from the mtDNA data.

A recent study of Y chromosome and mtDNA variation in Colombia shows a similar effect. In most segments of society, Colombian Y chromosomes are Spanish Y chromosomes, a direct biological legacy of the European conquest of the Spanish Main. In fact, approximately 94 percent of the Y chromosomes studied have a European origin. Interestingly, however, the mitochondrial pattern is quite different: modern Colombians have a range of Amerindian mtDNA types. The implication is clear: the invading Spaniards, who were men, took local women for their wives. The virtual absence of Amerindian Y chromosome types reveals the tragic story of colonial genocide: indigenous men were eliminated, while local women were sexually "assimilated" by the conquistadors.

Sometimes, however, enduring asymmetries are more a matter of cultural continuity than violent clash of cultures. The Parsees, a minority group in India, believe themselves to be descended from the Zoroastrians, an Indo-European Aryan people who fled religious persecution in Iran in the seventh century. Genetic analysis of modern Parsees indeed reveals that they have retained "Iranian" Y chromosomes, but their mtDNA tends to be of the "Indian" type. In this case the asymmetry is maintained by tradition. To be accepted as a true Zoroastrian Parsee, one has to have a Zoroastrian Parsee father. Thus membership in the Parsee community is paternally transmitted together with a Y chromosome. Here genetics confirms the hold of tradition.

Tradition has informed patterns of genetic variation among Jews as well. A recent study has shown that members of the priestly caste, the *kohanim* (and their descendants, usually identifiable today by the surname Cohen), have a Y chromosome distinctive enough to set them apart from all other groups. Even among the most obscure populations, those flung farthest by the Jewish Diaspora, such as South Africa's Lemba, the Cohen Y has been preserved—almost like a sacred religious text. Its source is thought to be Aaron, according to Scripture the founder of the *kohanim* caste and the brother of Moses. It is certainly not impossible that the *kohanim* Y chromosome sequence was indeed his and that it has been passed down intact, father to son, in every generation since. Such have been the rigors of tradition over the course of Jewish history.

Hammer and others have been able to use Y chromosomes to track the entire Diaspora with interesting results. The Ashkenazim, for example, who have lived in Europe for the past twelve hundred years (and now the United States and elsewhere), have nevertheless maintained the genetic indications of their Middle Eastern origins. In fact, molecular studies have made plain that the Jews, genetically at least, are virtually indistinguishable from all other Middle Eastern groups, including the Palestinians. So, too, is it written. Abraham, the great

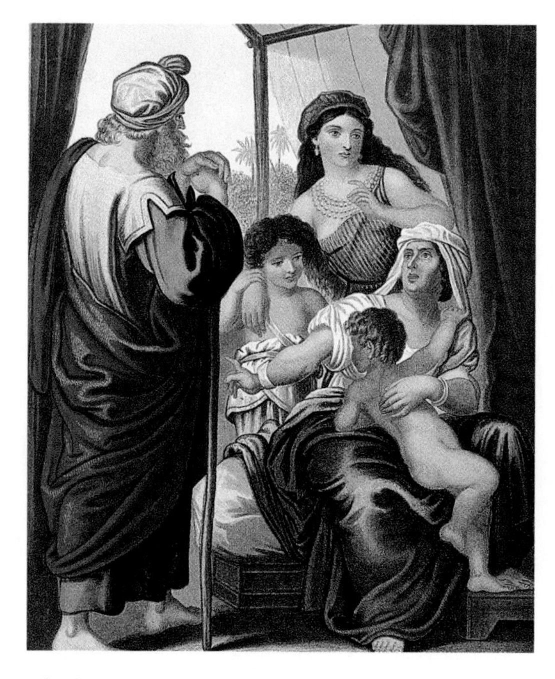

Abraham contemplates his complicated domestic arrangements.

patriarch, is said to have had two sons by different women: Isaac, from whom the Jews are descended, and Ishmael, forefather of the Arabs. That such a deadly enmity should have arisen between the descendants of one man is an irony that grows only more bitter when genes seem to verify tradition's narrative.

A simple stroll down a Manhattan street would suggest that ours is the most genetically variable species on the planet. In fact, though, the human genome is markedly less variable than those of most species for which we have genetic information. Only about 1 in every 1,000 human base pairs varies among individuals. Genetically, then, we are 99.9 percent alike, a minute degree of difference by the standards of other species. Fruit flies—even if they all look the same to us—have levels of variation some 10 times higher. Even Adélie penguins, those icons of sameness in their vast Antarctic colonies of indistinguishable individuals, are more than twice as variable as we are. Nor is this lack of variability found in our nearer relatives: chimpanzees are about 3 times as variable as we are, gorillas 2 times, and orangutans 3.5.

With the mtDNA and Y chromosome family results at hand, it is readily apparent why we humans are so alike. It's because our common ancestor was so recent; 150,000 years is a blink of an eye by evolutionary standards—insufficient time for substantial variation to arise through mutation.

Another counterintuitive finding about human variation, what little there may be, is that it does not correlate, for the most part, with race. Prior to Cann and Wilson's demonstration of humankind's surprisingly recent flight out of Africa, it was assumed that different groups had been isolated from one another on different continents for ages and ages, up to two million years. This would have permitted the accumulation of substantial genetic difference, in accordance with the Pauling-Zuckerkandl model, whereby the extent of genetic divergence between isolated populations is a function of the time over which they have been isolated. In light of Cann and Wilson's conclusion that we all share a much more recent common ancestor, it is clear that there has simply not been time enough for geographically separate populations to diverge significantly. Thus, though genetic differences, like skin color, are manifest across groups, race-specific genetic differences tend to be very limited. Most of our scant variation is actually spread rather uniformly across populations: one is as likely to find a particular genetic variant in an African population as in a European one. One is left to surmise that much of the genetic variation in our species arose in Africa *before* the out-of-Africa event, and so was already present in the groups that went forth to colonize the rest of the world.

As a final blow to any pride we may take in our own genetic variety: the Human Genome Project's conclusion that only about 2 percent of our DNA encodes genes would suggest that at least 98 percent of our variation falls in regions of the genome where it has no effect. And because natural selection very efficiently eliminates mutations that affect functionally important parts of the genome (such as genes), variation accumulates preferentially in noncoding (junk) regions. The difference between us is small; the difference it makes is even smaller.

Because of the short evolutionary timescales involved, most of the consistent differences we do see among groups are probably products of natural selection: skin color, for one.

Under their dense matted hair, the skin of our closest relatives, the chimpanzee, is largely unpigmented. (Chimpanzees, you might say, are white.) And presumably the common ancestor of chimpanzees and humans from which the human lineage spun off five million years ago was similar. And so we infer that the heavy skin pigmentation characteristic of Africans (and of the earliest modern humans, in Africa born) arose in the course of subsequent human evolution. With the loss of body hair, pigment became necessary to protect skin cells from the sun's damaging ultraviolet (UV) radiation. We now know at a molecular level how UV rays can cause skin cancer: they make the thymine bases of the double helix stick to one another, creating a kink, so to speak, in the DNA molecule. When that DNA replicates itself, this kink often promotes the insertion of a wrong base, producing a mutation. If, by chance, that mutation is in a gene that regulates patterns of cell growth, cancer may result. Melanin, the pigment produced by skin cells, reduces UV damage. As anyone with as hopelessly fair a complexion as mine knows too well, sunburn, though typically not lethal, can be a much more immediate health threat than skin cancer. Thus it is easy to imagine natural selection favoring the acquisition of dark skin in order to prevent not only cancer, but also the infections that can easily result from a severe sunburn.

Why did people living in higher latitudes lose melanin? The best explanation involves vitamin D_3 synthesis, a process carried out in the skin and requiring UV light. D_3 is essential for calcium uptake, which in turn is a critical ingredient of strong bones. (A deficiency of D_3 can result in rickets and osteoporosis.) It is possible that, as our ancestors moved out of Africa into highly seasonal environments, with less year-round UV radiation, natural selection favored pale-skinned variants because they, with less sun-blocking pigment in their skin, synthesized D_3 more efficiently with the limited UV available. The same logic may apply to the movements of our ancestors *within* Africa. The San, for instance, in South Africa, where UV intensities are similar to those of the Mediterranean, have a strikingly pale skin. But what about the Inuit peoples, who live in or close to the hardly sunny Arctic but are surprisingly dark? Their opportunities for producing the vitamin would appear to be further limited by the need to be fully clothed all the time in their climate. In fact, the selective pressure favoring lightness seems not to have asserted itself among them, and the reason appears to be that they have solved the D_3 problem in their own way: a diet with plenty of fish, a rich source of the essential nutrient.

Given what a powerful determinant, mostly for ill, skin color has been in human history and individual experience, it is surprising indeed how little we know about its underlying genetics. This deficit, however, may have less to do with the limitations of our science and more with the intrusion of politics into science; in an academic world tyrannized by political correctness, even to study the molecular basis of such a characteristic has been something of a taboo. What little we understand about it depends on old studies of mixed-race children, which established that several genes contribute to pigmentation. But our

Evolutionary response of body shape to climate: a heat-adapted Masai in Kenya and cold-adapted Inuit in Greenland

knowledge of other species and the similarity of basic biochemical processes among all mammals suggest a more complicated picture. We know, for instance, that many genes affect coat color in mice, and it is likely that these have direct human equivalents. So far, though, we have managed to identify only two genes involved in human pigmentation: the one that, when mutated, causes albinism, and the other, the "melanocortin receptor," associated with red hair and a pale (often freckled) complexion. The melanocortin receptor gene is variable among Europeans and Asians, but invariant among Africans, suggesting that there has been strong natural selection in Africa against mutations in the gene, i.e., against red-haired, fair-skinned individuals. Albinos, who lack pigment altogether, occasionally appear today in African populations (probably through de novo mutation) but their acute sensitivity to sunlight puts them at a severe disadvantage.

Another morphological trait likely determined by natural selection is body shape. In hot climates, where dissipating body heat is a priority, two basic types have evolved. The "Nilotic form," represented by the East African Masai, is tall and slender, maximizing the surface-area-to-volume ratio and thus facilitating heat loss. The Pygmy form, on the other hand, though still lightly built, is very short. In this case, a physically strenuous hunter-gather lifestyle has selected for small size to minimize the energy expended in movement—why lug a big body around to look for food? In high latitudes, by contrast, selection has favored body forms that promote heat retention: those with the lower ratios of surface area to volume. Neanderthals from Northern Europe were therefore heavily built, and so too on average are today's inhabitants of the same boreal climes. Some of the variation in athletic performance we see among groups is presumably attributable to these body-form differences. It should come as no surprise that in the high jump, for instance, a tall Nilotic body is better adapted than a short robust one.

If there is a trait whose distribution among human populations is hard to fathom, it is lactose intolerance. Mammalian milk, including the human variety, is rich in a sugar called lactose, and newborn mammals typically produce a special enzyme, lactase, to break it down in the intestine. Upon weaning, however, most mammals, including humans—at least, most Africans, Native Americans, and Asians—stop making lactase and so as adults cannot digest lactose. "Lactose intolerance" means that drinking a glass of milk can have unpleasant consequences, including diarrhea, gas, and abdominal bloating. Most Caucasians and the members of a few other groups, on the other hand, continue to produce lactase throughout their lives, and can therefore handle a lifelong dairy diet. The explanation has been advanced that lactose tolerance evolved in those groups historically most dependent upon dairy products, but the pattern of the trait is by no means fully convincing; there are, for example, groups of Central Asian animal herders—cheese for everyone—who are lactose intolerant. And despite belonging to an ethnic group that is typically lactose tolerant, I am intolerant. If natural selection had favored tolerance in a particular group, why would it leave its job undone? The most compelling evidence yet in support of the standard explanation is the presence of lactose tolerance in African groups traditionally associated with livestock. We may never fully understand the adaptive dimension of this trait, but molecular biologists working on a Finnish population have recently identified the mutation responsible for it. And so while we are by no means fighting a killer here, it is now possible, with a simple genetic test, to determine whether a newborn will grow up to face a choice between ice-cream deprivation and chronic gastric cramps.

More interesting than the relatively few differences we see among the races is what we all have in common—what it is that makes us so different from our closest relatives. As we have seen, our lineal split from the chimpanzee about 5 million years ago has barely given us enough time apart to evolve a 1 percent genetic difference. But in that 1 percent lie the critical mutations that make us the remarkable thinking, speaking creatures we are. It may be debated whether other species possess some limited form of consciousness, but clearly none of them has produced a Leonardo da Vinci or a Francis Crick.

The chromosomes of humans and chimpanzees are very similar. Chimpanzees, however, have 24 pairs whereas we have 23. It turns out that our chromosome 2 was produced by the fusion of two chimpanzee chromosomes. There are also differences in the human and chimpanzee versions of chromosomes 9 (bigger in humans) and 12 (bigger in chimpanzees) and several examples of inversions (or flips) within chromosomes that differ in humans and chimpanzees. Whether these chromosomal differences will prove significant is hard to say.

The relative merits are not much clearer at the biochemical level, where so far we know of only two differences between humans and chimpanzees. Difference 1: In both species a sugar molecule called sialic acid appears on the outside of every cell. But while the molecule is subtly modified in chimpanzees through the action of an enzyme, in humans, the gene encoding that enzyme is always mutated: no enzyme is produced, and human cell-surface sialic acid is unmodified. We have no clue at all as to whether this is significant. Difference 2: This one, discovered in 2002 by Svante Pääbo's group, is more suggestive: a difference in FOXP2, a gene known to be involved somehow in human language. (Because mutations in the human version have been found to cause linguistic impairment, FOXP2 has been misleadingly dubbed by the media as "the grammar gene.") Out of a chain running 715 amino acids, just two changes distinguish humans from chimpanzees and gorillas, whose FOXP2 proteins are identical. In fact, these amino acids are identical in *all* mammals tested except for humans. Moreover, statistical analysis of the pattern of DNA variation in and around the gene suggests that natural selection may have had a role in shaping the protein during human evolution. It is therefore tempting (but premature) to suggest that FOXP2 is the evolutionary equivalent of a smoking gun—a glimpse of a critical step in the origin of language.

Pääbo's lab has also pioneered a promising and original approach to identifying other genes that may encode the critical difference(s). Using DNA microarrays, which determine what genes are switched on in a particular tissue (see chapter 8), Pääbo has compared patterns of gene expression—which genes are switched on—in humans, chimpanzees, and macaque monkeys for three different tissues: white blood cells, liver, and brain. As would be expected on the basis of their close relationship, humans and chimpanzees fall out close to each other for both blood cells and liver. However, the pattern of gene expression in the brain tells a totally different story: the human brain is very different from those of the chimpanzee and macaque. Perhaps this is not entirely surprising: most of us would not need a laboratory full of equipment to figure out that human brains are distinct from chimpanzee brains. The research's significance lies instead in its ability to provide us with an inventory of the genes whose expression differs between human and chimpanzee brains. Even that will be only a start at best. It is unlikely that, even once we have a full catalog of the underlying mechanisms, we shall understand precisely *how* they set us apart. Our humanness is likely much more difficult to describe than even a precisely detailed list of controlled molecular events. But in our search for its genetic underpinning we are now at least beginning to assemble a list of suspects.

As I write this, the chimpanzee genome project is nearing completion. When it is done, the DNA making up the 1 percent difference that King and Wilson identified will be revealed. My guess is that they will be proved right: the critical differences will lie not in the genes themselves but in their regulation. Humans, I suspect, are simply great apes with a few unique—and special—genetic switches.

Molecular biology's grandest mission is surely to answer questions about ourselves and our origins as a species. But each human soul yearns to know its own story as well as that of its kind. DNA can provide a more individualized account of ancestry as well. In a sense, written in my DNA molecules is the history of my evolutionary lineage, a narrative that can be viewed at different levels. I can situate the sequence of my mtDNA into Cann and Wilson's human family tree, or I can look in greater detail at my known family's past. My Y chromosome and mtDNA will tell different stories—my mother's side, and my father's.

I was never interested in genealogy. But my family—like many, I suspect—had its own in-house archivist in the form of my aunt Betty, who spent a lifetime worrying about who was related to whom and how. It was she who found that the Watsons—of lowland Scots stock—first appeared in the United States in 1795 in Camden, New Jersey. And it was she who insisted that some paternal ancestor of mine designed Abe Lincoln's house in Springfield, Illinois. But I've always been more interested in my Irish side, my maternal grandmother's family. My mother's grandparents fled Ireland during the great potato famine of the 1840s, ending up in Indiana, where her grandfather, Michael Gleason, died in 1899, the year my mother was born. On his gravestone it says he had come from a town in Ireland called Glay.

On a visit to Ireland, I tried to find out more about my great-grandfather at the County Tipperary Records Office, whose quarters in Neneagh, twenty miles from Limerick, had formerly been a prison. My sleuthing was singularly unsuccessful. Finding no record at all of "Glay," I could only conclude that name as spelled on the tombstone of my probably illiterate ancestor was fanciful. Thus ended my only brush with genealogical research, until recently. Now that the framework of the human family tree has been laid out by Cann and others, I am keen to see where I fit in. Companies like Bryan Sykes's Oxford Ancestors represent the new face of genealogical research, with high-tech laboratories to replace dusty archives. With a sample of my DNA, Oxford Ancestors has conducted both mtDNA and Y chromosome analysis. Sadly the tests revealed nothing romantic, no exotic ancestry. I really am, as I feared, largely the product of generic Scots-Irish stock. I cannot even blame my more brutish attributes on ancient Viking incursions into my bloodline.

Barry Scheck (left) and Peter Neufeld (second from left) in action on their biggest case

CHAPTER TEN

GENETIC FINGERPRINTING: DNA'S DAY IN COURT

In 1998 Marvin Lamont Anderson, thirty-four years old, was released from the Virginia State Penitentiary. He'd been there for fifteen years, almost all his adult life, convicted of a horrific crime: the brutal rape of a young woman in July 1982. The prosecution had presented an unambiguous case: the victim recognized Anderson from a photograph; she picked him out in a lineup; and she identified him in court. Found guilty on all counts, he was given consecutive sentences totaling over two hundred years.

A clear-cut case. A better defense attorney, however, might have been more effective in countering the prosecution's efforts to stack the deck against the defendant. Anderson was picked up based exclusively on the (white) victim's report to the police that her (black) assailant had boasted of "having a white woman"; so far as the authorities knew, Anderson was the only local black man with a white girlfriend. Among the mug shots the victim looked at, only Anderson's was a color photograph. And of the men whose pictures she was shown, he alone was placed in the lineup. And although another man, John Otis Lincoln, was shown to have stolen, about thirty minutes before the attack took place, the bicycle used by the assailant, Anderson's attorney failed to call Lincoln as a witness.

Five years after Anderson's trial, Lincoln confessed under oath to the crime, but the trial judge declared him a liar and refused to act. Anderson meanwhile continued to protest his innocence and requested that DNA analysis be done on the physical evidence from the crime scene. But he was told that it had all been destroyed in accordance with standard procedure. It was then that Anderson contacted the lawyers of the Innocence Project, a group that had gained national attention using DNA analysis to establish definitive evidence of guilt or innocence in criminal proceedings. While the Innocence Project worked on Anderson's request, he was released on parole; assuming no violations, he would remain a parolee until 2088, easily the rest of his life.

In the end, Anderson's salvation was the sloppiness of the police technician who had performed the inconclusive blood group analysis on the crime scene material in 1982. She had failed to return the samples to the proper authorities for routine destruction, and so they still existed when Anderson asked for a reexamination. The director of the Virginia Department of Criminal Justice, however, refused the request, arguing it might establish an "unwelcome precedent." But under a new statute, the Innocence Project attorneys won a court order calling for the tests to be performed, and, in December 2001, the results proved categorically that Anderson could not have been the assailant. The DNA "fingerprint" matched Lincoln's. Lincoln has since been indicted and Anderson pardoned by Governor Mark Warner of Virginia.

DNA fingerprinting—the technique that rescued Marvin Anderson from an undeserved life sentence—was discovered by accident by a British geneticist, Alec Jeffreys. From the earliest days of the recombinant DNA revolution, Jeffreys had been interested in genetic differences among species. His research at Leicester University focused on the myoglobin gene, which produces a protein similar to hemoglobin, found mainly in muscle. It was in the course of this "molecular dissection" that Jeffreys found something very strange: a short piece of DNA that repeated over and over again. A similar phenomenon had been observed in 1980 by Ray White and Arlene Wyman, who, looking at a different gene, had shown that such repeats varied in number from individual to individual. Jeffreys determined that his repeats were junk DNA, not involved in coding for protein, but he was soon to discover that this particular junk could be put to good use.

Jeffreys found that this short stretch of repeating DNA existed not only in the myoglobin gene but was scattered throughout the genome. And although the stretches varied somewhat from one repetition to the next, all of them shared one short, virtually identical sequence of some fifteen nucleotides.

Alec Jeffreys, father of DNA fingerprinting

Jeffreys decided to apply this sequence as a "probe": using a purified sample of the sequence tagged with a radioactive molecule, he could hunt for the sequence genomewide. With DNA from the genome laid out on a special nylon sheet, the probe would stick down, by base-pairing, wherever it encountered its complementary sequence. By placing the nylon on a piece of X-ray film, Jeffreys could then record the pattern of radioactive spots. When he developed the film from the experiment, he was astonished by what he saw. The probe had detected many similar sequences across a range of DNA samples. But there was still so much variability from one sample to the next that even among ones taken from members of the same family you could tell the individuals apart. As he wrote in the resulting paper in *Nature* in 1985, the "profile provides an individual-specific DNA 'fingerprint.' "

Jeffreys's choice of the term "DNA fingerprint" was quite deliberate. This technology clearly had the power to identify an individual, just like traditional fingerprinting. Jeffreys and his staff obtained DNA samples from their own blood and subjected them to the same procedure. The images on X-ray film, as expected, made it possible to distinguish unambiguously between people. He realized the range of potential uses was extensive:

> In theory, we knew it could be used for forensic identification and for paternity testing. It could also be used to establish whether twins were identical—important information in transplantation operations. It could be applied to bone marrow grafts to see if they'd taken or not. We could also see that the technique [would work] on animals and birds. We could figure out how creatures are related to one another—if you want to understand the natural history of a species, this is basic information. We could also see it being applied to conservation biology. The list of applications seemed endless.

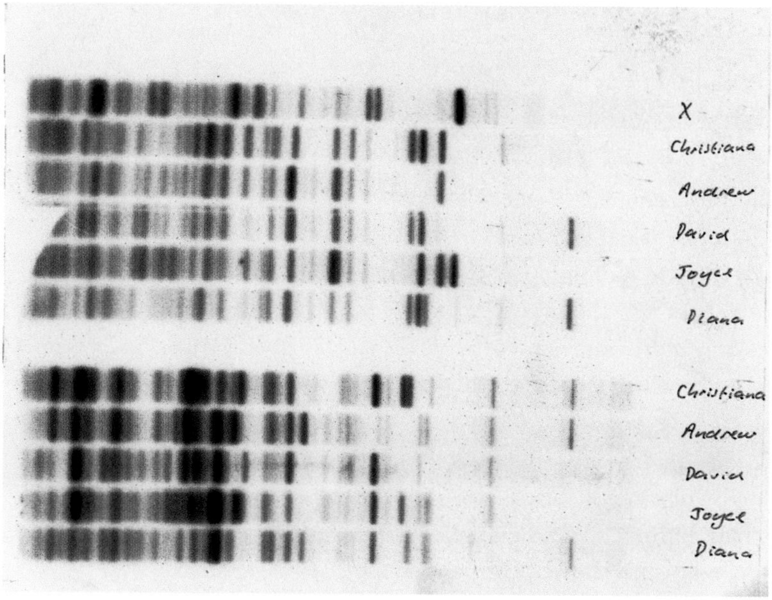

The first application of DNA fingerprinting: the gel used by Alec Jeffreys to determine Andrew Sarbah's true parentage

But the procedure's first practical application was stranger than any Jeffreys had anticipated.

In the summer of 1985, Christiana Sarbah was at her wits' end. Two years before, her son, Andrew, had returned to England after visiting his father in Ghana. But at Heathrow, British immigration authorities had refused to admit the boy, though he had been born in Britain and was a British subject. Denying that Sarbah was his mother, they alleged that Andrew was, in fact, the son of one of Sarbah's sisters and was trying to enter the country illegally on a forged passport. After reading a newspaper report about Jeffreys's work, a lawyer familiar with the case asked the geneticist for help. Could this new DNA test prove that Andrew was Mrs. Sarbah's son and not her nephew?

The analysis was complicated by the fact that neither the father nor Sarbah's sisters were available to give samples. Jeffreys prepared DNA from samples taken from the mother and three of her undisputed children. The analysis showed that Andrew had the same father as the other children, and that Sarbah was his mother. Or more specifically, that chances were less than 1 in 6 million that one of her sisters was his mother. The immigration authorities did not challenge Jeffreys's results but avoided formally admitting the error by simply dropping the case. Andrew was reunited with his mother. Jeffreys saw them afterward: "The look of relief on her face was pure magic!"

But would the technique work with blood, semen, and hair, the body tissues typically found at crime scenes? Jeffreys was quick to prove that it could indeed, and soon his DNA fingerprints would gain worldwide attention, revolutionizing forensic science.

On a Tuesday morning in November 1983, the body of a fifteen-year-old schoolgirl named Lynda Mann was found on the Black Pad, a footpath outside the village of Narborough, near Leicester in England. She had been sexually assaulted. Three years passed with no arrest in the case. Then, it happened again: on a Saturday in August 1986 the body of Dawn Ashworth, another fifteen-year-old, was found on Ten Pond Lane, another footpath in Narborough. The police were convinced that the same man had committed both murders and soon accused a seventeen-year-old kitchen assistant. But, while confessing to the Ashworth murder, the suspect denied involvement in the earlier case. So it was that the police consulted Alec Jeffreys to confirm that their suspect had killed both girls.

Jeffreys's fingerprint analysis contained both good and bad news for the authorities: Comparison of samples from the two victims showed that the same man had indeed carried out both murders, as the police believed. Unfortunately (for the police) the same test also proved that the kitchen worker in custody had not murdered either girl, a result confirmed by other experts the police called in. The suspect was released.

With their only lead now blown, and worries rising in the local community, the police took an extraordinary step. Confident that DNA fingerprinting would yet prove the key to success, they decided to request DNA samples from all adult males in and around Narborough. They set up stations to collect blood samples and were able to eliminate a great many candidates by the traditional, and cheaper, test for blood type. The remaining samples were sent for DNA fingerprinting. A good Hollywood version of the story would, of course, have Jeffreys identifying the true killer. And it did happen that way, but not without a further plot twist, also worthy of Tinseltown. The culprit initially eluded the genetic dragnet. When faced with providing the mandatory sample, Colin Pitchfork, pleading a terror of needles, persuaded a friend to furnish a sample in his stead. It was only later, when the friend was overheard telling of what he had done, that Pitchfork was picked up and thus gained the dubious honor of being the first criminal ever apprehended on the basis of DNA fingerprints.

The Narborough case showed law enforcement agencies worldwide that DNA fingerprinting was indeed the future of criminal prosecution. And it would not be long before such evidence was first adduced in an American legal proceeding.

Perhaps the British are, culturally, more accepting of authority, or perhaps recondite molecular mumbo jumbo was just more likely to rub Americans the wrong way, but in any case the introduction of DNA fingerprinting into the United States was highly controversial.

The law has always had difficulty assimilating the implications, if not the very idea, of scientific evidence. Even the most intelligent lawyers, judges, and juries have customarily found it difficult to understand at first. In one famous early instance of forensic courtroom drama, blood-typing had unequivocally ruled out Charlie Chaplin as the father of a child whose mother had slapped a paternity suit on the silent-screen legend. The jury nevertheless ruled in the mother's favor.

American courts had long applied the Frye test as their standard for admissibility of scientific evidence. Based on one of the first trials to introduce forensic proof, it tries to keep out unreliable evidence by requiring that the science on which it is based "must be sufficiently established to have gained general acceptance in the particular field in which it belongs." But being based on a poor understanding of what constitutes well-established science, the test proved an ineffective way of determining the credibility of "expert" testimony. It was not until 1993, in *Daubert vs. Merrell Dow Pharmaceuticals* that the Supreme Court ruled the Federal Rules of Evidence should be used: the judge in a trial should determine whether the proffered evidence is reliable (i.e., whether it can be trusted as scientifically valid).

Nowadays, with Court TV an established part of the television landscape, and with prime-time series focusing on forensic investigations a staple of the networks, it may be hard to appreciate how difficult it was for the American legal system to swallow DNA. Though everyone had been hearing about it since our landmark discovery in 1953, it still had about it an impenetrable scientific aura. Indeed, the field of genetics seemed only more arcane every time the popular media hailed a new advance. Perhaps worst of all was the fact that DNA-supported charges were presented not as dead certainties but as probabilities. And what probabilities they were! With figures like "1 in 50 billion" bandied about to establish the guilt or innocence of the accused, little wonder some questioned the value of lawyers, judges, juries, and expensive trials when a geneticist, wrapped in the authority of science, could settle a case.

But at all events, most trials depend on more than the comparison of two DNA samples. Meanwhile, the acceptance of the new methods progressed slowly but ineluctably. In some sense the cause of broader understanding and acceptance was aided by lawyers who made their name challenging the very cases that depended on DNA evidence. Skilled attorneys like Barry Scheck and Peter Neufeld became as knowledgeable as the experts they were cross-examining. Scheck—short, messy, and pugnacious—and Neufeld—tall, tidy, and pugnacious—gained attention searching for technical flaws in cases presented during the early days of genetic fingerprinting. The two first met in 1977 as colleagues at the office of the Bronx Legal Aid Society, a local center of legal advocacy for the indigent. After growing up in New York City, the son of a successful impresario who managed stars like Connie Francis, Scheck found his political calling when he went to college at Yale, taking part in the national student strike that followed the Kent State shootings in 1970. Ever suspicious of entrenched authority and the abuse of power, he volunteered to assist Bobby Seale's defense team during the Black Panther's trial in New Haven. Peter Neufeld grew up in suburban Long Island, where his mother still lives, not far from Cold Spring Harbor Laboratory. He was no less precocious in his leftward leanings, having been reprimanded in the eleventh grade for organizing antiwar protests.

It was little surprise when the two young bred-in-the-bone social progressives became crusading lawyers manning the barricades of legal aid in New York City—at a tumultuous moment in the life of the city, when rising crime rates made "justice for all" seem to some an ideal endangered in the pursuit of public safety. A decade later, Scheck would be professor at Cardozo School of Law, and Neufeld would be in private practice.

I first met Scheck and Neufeld at an historic conference on DNA fingerprinting held at Cold Spring Harbor Laboratory. The controversy was at its height in part because the forensic technology was being applied more and more broadly despite still being done with Jeffreys's as-yet-unrefined original technique, the arcane-sounding analysis of restriction fragment length polymorphisms, or RFLPs. Inevitably some results were difficult to interpret, and so DNA fingerprinting was being challenged on technical and legal grounds. The Cold Spring Harbor gathering was actually the first occasion on which molecular geneticists—including Alec Jeffreys—would confront the forensic specialists and lawyers now using DNA in the courtroom. The discussions were heated. The molecular geneticists accused the forensic scientists of sloppy laboratory techniques, of simply not doing the testing carefully enough. Indeed, in those days DNA fingerprinting in forensic laboratories was subject to little, if any, regulation or oversight. There were also challenges to the statistical assumptions, likewise unstandardized, used to calculate those imposing numbers suggesting virtual certainty. The geneticist Eric Lander spoke for more than a few concerned participants when he proclaimed bluntly: "The implementation [of DNA fingerprinting] has been far too hasty."

These practical problems were typified in a case Scheck and Neufeld were working on in New York. Joseph Castro was accused of murdering a pregnant woman and her two-year-old daughter. RFLP analysis, performed by a company called Lifecodes, had established that a bloodstain on his wristwatch was from the murdered mother. After a sustained examination of the DNA data, however, the expert witnesses of both the prosecution and defense jointly informed the judge in a pretrial hearing that, in their view, the DNA tests had not been done competently. The judge excluded the DNA evidence as inadmissible. The case never came to trial because Castro pleaded guilty to the murders in late 1989.

Despite the exclusion of the DNA evidence, the Castro case helped establish the legal standards for genetic forensics. These were the standards that would be applied in a much more prominent case Scheck and Neufeld were to take on, one that would make DNA fingerprinting a household term in America and indeed everywhere one could find a television: the trial of O. J. Simpson in 1994. The former sports icon was facing a possible death penalty if convicted of the heinous crimes he was charged with by the Los Angeles district attorney: the gory murder of Simpson's ex-wife, Nicole Brown Simpson, and her friend, Ronald Goldman. As part of the legal "dream team" assembled by the accused, Scheck and Neufeld would make critical contributions to Simpson's defense and acquittal. Forensic detectives had collected bloodstains from the crime scene at Nicole Brown Simpson's house, from O. J. Simpson's house, from an infamous glove and sock, and from Simpson's equally infamous white Bronco. The DNA evidence—forty-five blood specimens in all—contributed, according to the prosecution's case, a "mountain of evidence" pointing to Simpson's guilt. But Simpson had in his corner the most skillful mountaineers money could buy. The challenges from the defense came thick and fast, and as the whole world watched on TV, these counterclaims would bring some of the central controversies that had been simmering for years in forensic science up to a full-blown boil.

A decade before the Simpson trial, back in the days when prosecutors first began presenting DNA evidence, and only prosecutors commissioned the application of genetic technology, defense attorneys were quick to raise an obvious question: By what standard could one define a match between a DNA sample found at a crime scene and one derived from blood taken from the suspect? It was a particularly contentious issue when the technology still depended on RFLPs. In this method, the DNA fingerprint appears as a series of bands on an X-ray film. If bands produced by the crime scene DNA were not identical to those produced by the suspect's, just how much difference could be legitimately tolerated before one had to exclude the possibility of a match? Or how same does "the same" have to be? Technical competence came into question as well. Initially, when DNA fingerprinting was done in forensic laboratories without special expertise in handling and analyzing DNA, critical mistakes were not uncommon. Law enforcement agencies understood that if their powerful new weapon were to remain in commission, these questions would have to be answered. A new form of genetic marker—short tandem repeats (STRs)—replaced the RFLP method. The size of these STR genetic markers can be measured very accurately, doing away with the subjective assessment of RFLP bands on an X-ray film. The forensic science community itself dealt with the problem of variable technical competence by establishing a uniform code of procedures for doing DNA fingerprinting, as well as a system of accreditation.

Perhaps the toughest attacks, however, were launched against the numbers. While prosecutors were given to presenting DNA evidence in terms of dispassionate, seemingly incontrovertible statistics, sometimes, as defense lawyers began to argue, tendentious assumptions had been made in calculating the state's one-in-a-billion margins of certitude. If you have a DNA fingerprint from the crime scene, on what basis do you calculate the likelihood (or, more often, the unlikelihood) that it might belong to someone other than prime suspect A? Should you compare the DNA to that of a random cross-section of individuals? Or, if prime suspect A is, for instance, Caucasian, should your sample be compared only to DNA from other Caucasians (since genetic similarity tends to run higher among members of the same racial group than in a random cross-section of people)? The odds will vary depending on what one deems a reasonable assumption.

And an effort to defend a conclusion founded on the arcane principles of population genetics can backfire, confusing jurors or putting them to sleep. The sight of someone struggling manfully to put on a glove that simply doesn't fit is worth more—much more, experience tells us—than a mountain of statistics.

In fact, DNA fingerprinting evidence presented in the Simpson trial pointed to the accused. A blood drop collected close to Nicole Brown Simpson's body, as well as other drops found on the walkway at the crime scene, were shown with virtual certainty to be his. With an equal lack of doubt, the blood staining the glove retrieved from his home was determined to be a mixture of Simpson's and that of the two victims; the blood found on the socks and in the Bronco proved to match the blood of Simpson and that of his ex-wife.

No, finally, in the eyes of the jury, the undoing of the forensic case against Simpson had less to do with a failure to explain the arcana of population genetics than with the old charge of police incompetence. DNA is such a stable molecule that it can be extracted from semen stains several years old or from bloodstains scraped off sidewalks or from the steering wheel of an SUV. But it is also true that DNA can degrade, especially in moist conditions. Like any type of evidence, however, DNA is only as credible as the procedures for collecting,

HOW GENETIC FINGERPRINTING WORKS

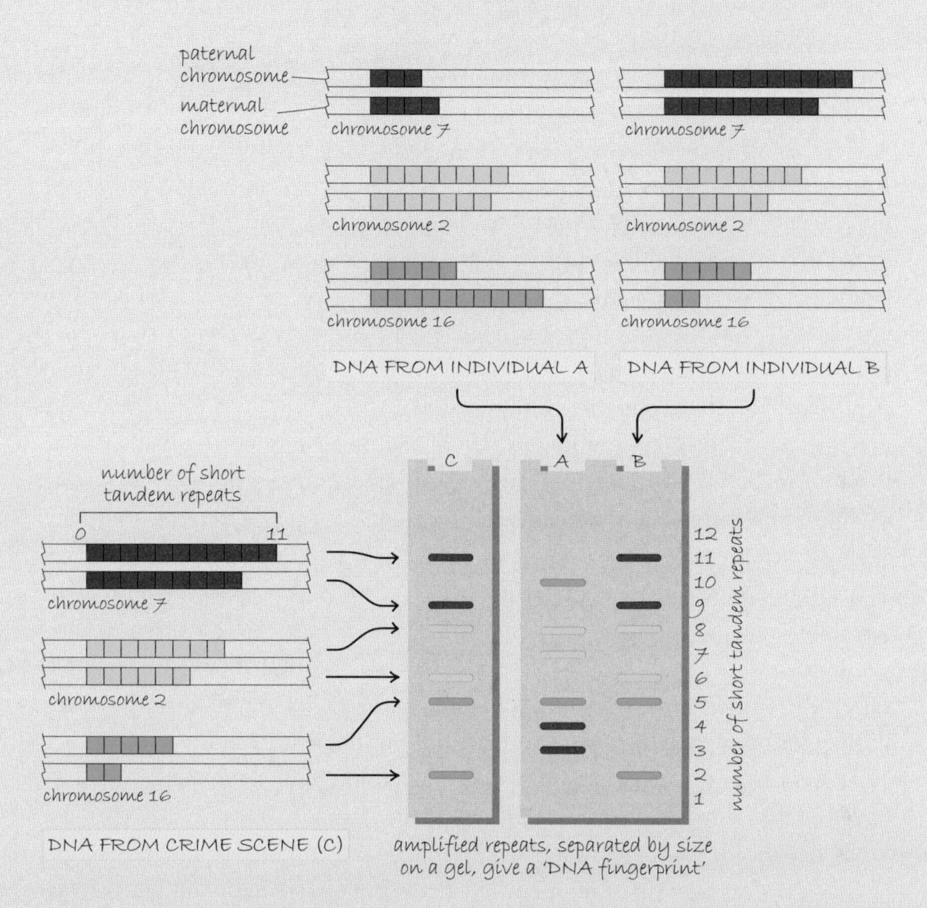

DNA fingerprinting using STRs. The DNA of two suspects is compared to DNA recovered from the crime scene. The fingerprint of B matches that of the crime scene DNA.

Today short tandem repeats (STRs) have replaced RFLPs as the keys to genetic identification. STRs, in which sequences of two to four bases recur as many as seventeen times, are the segments routinely amplified by PCR. For example, D7S820 is a region on chromosome 7 where the sequence AGAT can occur between 7 and 14 times. It happens that DNA polymerase, the enzyme that copies DNA, does a bad job of copying these repeating chunks of DNA—it tends to get the count wrong—so there is a high mutation rate in copy number of the AGAT sequence at D7S820. To put it another way, there is a great deal of variation in the number of AGAT copies among individual humans. With two copies of chromosome 7 (one from our father, the other from our mother), we typically have a different AGAT repeat count on each—say, 8 on one and 11 on the other. This is not to say, however, that an individual cannot be homozygous for a particular repeat count (e.g., 11 and 11). If we carry out DNA fingerprint analysis on a crime-scene blood sample and find it matches a suspect's fingerprint for D7S820 (say, 8 and 11 repeats), we have one indication of a match but not conclusive proof. After all, many others also have an 8/11 genotype for D7S820. It's therefore necessary to look at multiple regions; the more regions in which the crime scene DNA matches a suspect's, the greater the probability of a match, and the more remote the chances that the crime scene DNA could have come from anyone else. Under the FBI's system, a DNA fingerprint is produced from the analysis of twelve such regions, plus a marker that determines the sex of the individual from whom the DNA sample is derived.

sorting, and presenting it. Criminal trials always include the formality of establishing the "chain of evidence," verifying that what the police say was found in such-and-such a location did indeed start there before winding up in a Ziploc bag as Exhibit A. Keeping track of molecular evidence, as opposed to knives and guns, can be an especially demanding chore: scrapings from a sidewalk may be visually indistinguishable from scrapings from a gatepost, and the subsequently extracted DNA samples will doubtless look even more alike when placed in small plastic test tubes. Simpson's defense team was able to point to a number of instances when it seemed at least possible, if not probable, that samples had been confused or, even worse, contaminated.

There was, for example, the question of the bloodstain on the back gate of Nicole Brown Simpson's house. This was somehow missed in the early survey of the crime scene and not collected until three weeks after the murders. Forensic scientist Dennis Fung presented a photograph of the stain, but Barry Scheck countered it with another photograph taken the day after the murder, in which no stain appeared. "Where is it, Mr. Fung?" Scheck asked with a rhetorical flourish worthy of Perry Mason. There was no answer. The defense was able raise sufficient doubt in the minds of the jurors about the handling and sources of the DNA samples that the DNA evidence became irrelevant.

As we saw in the last chapter, sample contamination is one of the foremost banes of efforts to establish identity by genetic methods. Because it can yield a DNA fingerprint from even the tiniest sample, the polymerase chain reaction (PCR) is the modern forensic scientist's method of choice for amplifying particular segments of DNA. In the Simpson trial, for instance, crucial evidence included a single blood drop scraped from the sidewalk. But sufficient DNA for PCR can be extracted from cells in the saliva left on a cigarette butt. In fact, PCR can successfully amplify DNA from a single molecule, so if even the slightest trace of DNA from another source—someone handling the samples, for example—contaminates the evidence sample, the results are at best confused and at worst useless.

In the past decade, with the broadening application and acceptance of the DNA fingerprint as proof of identity, the law enforcement community had a flash of inspiration: Doesn't it make sense to DNA fingerprint, well, everyone—at least everyone who might be a criminal? Surely, the argument goes, the FBI should have a central database of DNA records, rather as it does for conventional fingerprints. Indeed, a number of states have passed laws requiring that DNA samples be taken from anyone convicted of a violent felony, like rape or murder. For example, in 1994 North Carolina passed legislation that authorizes taking blood samples from imprisoned felons, by force if necessary. And some of those states have since extended the mandate to cover all individuals who are arrested, whether they are ultimately found guilty of a crime or not.

The outcry from civil libertarians has been intense, and not without reason: DNA fingerprints are not like finger fingerprints. A DNA sample taken for fingerprinting purposes can, in principle, be used for a lot more than merely proving identity: it can tell you a lot about me—whether I carry mutations for disorders like cystic fibrosis, sickle-cell disease, or Tay-Sachs disease. Some time in the not so distant future, it may even tell you whether I carry the genetic variations predisposing me to schizophrenia or alcoholism—or traits even more likely to disturb the peace. Might the authorities, for instance, one day subject me to a more intensive scrutiny than would otherwise be the case simply because I have a mutation in the monoamine oxidase gene that reduces the activity of the enzyme? Some research suggests that this mutation may predispose me to antisocial behavior under certain circumstances. Could genetic profiling indeed become a new tool for preemptive action in law enforcement? Philip K. Dick's 1956 story (which inspired the 2002 movie) "The Minority Report" may not be such far-fetched science fiction as we like to imagine.

Whatever the outcome of the ongoing debate about who should be compelled to provide DNA samples and under what safeguards these ought to be maintained, the fact is that as I write there is a huge amount of DNA fingerprinting going on. In 1990, the FBI established its DNA database, CODIS (Combined DNA Index System), and by June 2002 it contained 1,013,746 DNA fingerprints. Of these, 977,895 are from convicted offenders and 35,851 are forensic crime scene samples for unsolved cases. Since its inception, CODIS has been used to make some 4,500 identifications that would not otherwise have been made.

One major justification for a national database is the potential for making "cold hits." Suppose investigators find some DNA—blood on a broken window, semen on underwear—at the crime scene and a fingerprint is made. Now suppose they have no leads by conventional investigative means, but when the fingerprint is entered into CODIS a match is found. That is what happened in St. Louis in 1996. The police were investigating the rapes of two young girls at opposite sides of the city, and although the two samples of semen revealed under RFLP fingerprint analysis that the same man had committed both crimes, a suspect could not be identified. Three years later, the samples were reanalyzed using STRs and the data compared with the entries in CODIS. In 2001, they found the rapist, Dominic Moore, whose DNA fingerprint was in CODIS because he had confessed to committing three other rapes in 1999.

The interval between a crime and a cold hit can be even more dramatic, and some malefactors have been shocked to face the molecular "j'accuse" of victims long buried. In Britain, fourteen-year-old Marion Crofts was raped and murdered in 1981, long before DNA fingerprinting was in use. Fortunately, some physical evidence was preserved, so it was possible to make a DNA fingerprint in 1999. The authorities and Crofts's bereaved family were disappointed again, this time in learning there was no match in the United Kingdom National DNA Database. In April 2001, however, when Tony Jasinskyj was arrested for assaulting his wife, a DNA sample was taken from him as a matter of routine procedure. When it was entered into the database, a match came up: Jasinskyj was found to be the unknown rapist of twenty years before.

In the United States, crimes like rape have customarily been subject to statutes of limitations in many states. In Wisconsin, for example, a warrant for the arrest of an alleged rapist cannot be issued more than six years after the crime has taken place. Although such statutes may seem devastatingly unfair to victims—after all, does the horror of a crime simply disappear after six years?—they have by tradition served the interests of due process. Eyewitness accounts in particular are notoriously unreliable, and all memories grow hazier over time; statutes of limitations are intended to prevent miscarriages of justice. But DNA is a witness of quite a different order. Samples stored properly remain stable for many years, and the DNA fingerprints themselves lose none of their authority to incriminate.

In 1997, Wisconsin's State Crime Laboratory established a DNA fingerprint registry and that same year the Milwaukee Police Department began reviewing all unsolved rape cases with physical evidence available for possible matching. They found fifty-three, and in six months they had scored eight cold hits against DNA fingerprints from felons already serving time. In one case, the identification was made so late the arrest warrant was issued only eight hours before the statute of limitations kicked in.

Among the cold cases, the State Police Department was also to establish evidence of a serial rapist—three separate assaults, three separate semen samples, the DNA fingerprints of all of them pointing to the same man. With the statute of limitations soon to take effect, Norm Gahn, an assistant district attorney, faced a dilemma. There was not enough time to identify the assailant in the database, but he could not draft a warrant without the suspect's name. Gahn hit on a clever strategy. The Wisconsin criminal code held that in the event a suspect's name was unknown, a valid warrant could be issued on the basis of "any description by which the person to be arrested can be identified with reasonable certainty." Surely, Gahn reasoned, any court would accept a DNA fingerprint as identifying someone by that standard. He made out the warrant: "State of Wisconsin vs. John Doe, unknown male with matching deoxyribonucleic acid (DNA) profile at genetic locations D1S7, D2S44, D5S110, D10S28, and D17S79." Despite Gahn's ingenuity, though, this John Doe still has not been caught.

Meanwhile the first challenge in court of a John Doe DNA warrant came in Sacramento, where one man, called the "Second Story Rapist," was believed to have committed three rapes over several years. Anne Marie Schubert, a local prosecutor, followed Gahn's lead in filing a John Doe DNA warrant just three days before the statute of limitations was to take effect. But she had to satisfy the requirements of her own jurisdiction, in particular the California law requiring that a warrant identify the suspect with "reasonable particularity"; toward this end she specified: "unknown male . . . with said genetic profile being unique, occurring in approximately 1 in 21 sextillion of the Caucasian population, 1 in 650 quadrillion of the African American population, 1 in 420 sextillion of the Hispanic population." Shortly after the warrant was issued, when John Doe's DNA fingerprint was entered into the state database, it turned out to match that of one Paul Eugene Robinson, who had been arrested in 1998 for violating parole. The warrant was amended with "Paul Eugene Robinson" in the place of John Doe and his STR markers, and Robinson was duly arrested. His attorney argued that the first warrant was invalid as it did not name Robinson. Fortunately, the judge upheld the validity of the warrant, remarking that "DNA appears to be the best identifier of a person that we have."

In the wake of the publicity stirred by these successful "John Doe DNA" warrants, many states have amended their rape statutes to permit an exception when DNA evidence is available.

The reach of DNA fingerprinting now even extends beyond the grave. In 1973, Sandra Newton, Pauline Floyd, and Geraldine Hughes, all teenagers, were raped and murdered in South Wales. Twenty-six years later, DNA fingerprints were prepared from samples saved from the crime scenes, but unfortunately the National DNA Database yielded no matches. So, rather than looking for an exact match, the forensic scientists looked for individuals whose DNA fingerprints indicated that they might be related to the murderer. They thus identified a hundred men, furnishing the police with a wealth of leads in light of which to reassess the masses of information that they had collected during the original investigation. Through a combination of state-of-the-art DNA forensics and good old-fashioned detective work they found a trail leading to one suspect, Joe Kappen. The only trouble was that Mr. Kappen had died of cancer in 1991—what was to be done?

In 1999 Kappen was exhumed and fingerprinted. And the fingerprints indeed matched those from DNA recovered from the three victims. Cancer may have exacted the ultimate price before the law could find him, but at least the girls' families had the long-postponed satisfaction of knowing his name.

DNA fingerprinting has solved mysteries involving bodies much more illustrious than Joe Kappen's. Take the extraordinary story of the Russian royal family, the Romanovs.

In July 1991, a small group of detectives, forensic experts, and police assembled in a muddy, rain-soaked clearing in the forest at Koptyaki, Siberia. Here, in July 1918, eleven bodies had been hurriedly buried. They were the remains of Tsar Nicholas II and Tsarina Alexandra; their son, Alexis, heir to the throne; their four daughters, Olga, Tatiana, Marie, and Anastasia; and four companions—all of whom had been brutally murdered a few days before, Anastasia still holding Jemmy, her pet King Charles spaniel, as she met her end in a hail of bullets. The killers initially tossed the bodies down a mine but, fearing discovery, recovered them the next day before finally burying them in that pit in the forest.

The grave had first been discovered in 1979 thanks to the detective work of Alexander Avdonin, a geologist obsessed with learning the fate of the tsar's family, and the filmmaker Geli Ryabov, who, having earned the privilege of making an official documentary of the Revolution, had gained access to relevant secret archives. In fact, it was a report written by the chief murderer for his bosses in Moscow that led Avdonin and Ryabov to the gravesite. They found three skulls and other bones. But as the chokehold of the Communist Party was then as tight as ever, they rightly realized they would do themselves no favors by drawing attention to the Bolsheviks' butchery of the royal family. They reburied the remains.

With the thawing of the political climate that culminated in the demise of the Soviet Union came the opportunity Avdonin and Ryabov had been waiting for. So it was that picks and shovels were again wielded in the forest clearing.

The exhumed remains—a total of more than one thousand pieces of skull and bone—were taken to a Moscow morgue, where the painstaking process of reassembling and identifying the skeletons began. There was an immediate surprise. The murdered were known to have numbered eleven, six females and five males, but the grave contained the bones of only nine bodies—five female and four male. It was clear from the skeletal remains that the missing bodies were those of Alexis (fourteen at his death) and Anastasia (who had been seventeen).

The claims of identification were viewed with some skepticism, especially as there had been disagreement between the Russian scientists and an American team that had come to assist. And so in September 1992, Dr. Pavel Ivanov brought nine bone samples to Peter Gill's laboratory at the British Forensic Science Service. Gill and his colleague David Werrett had been coauthors of

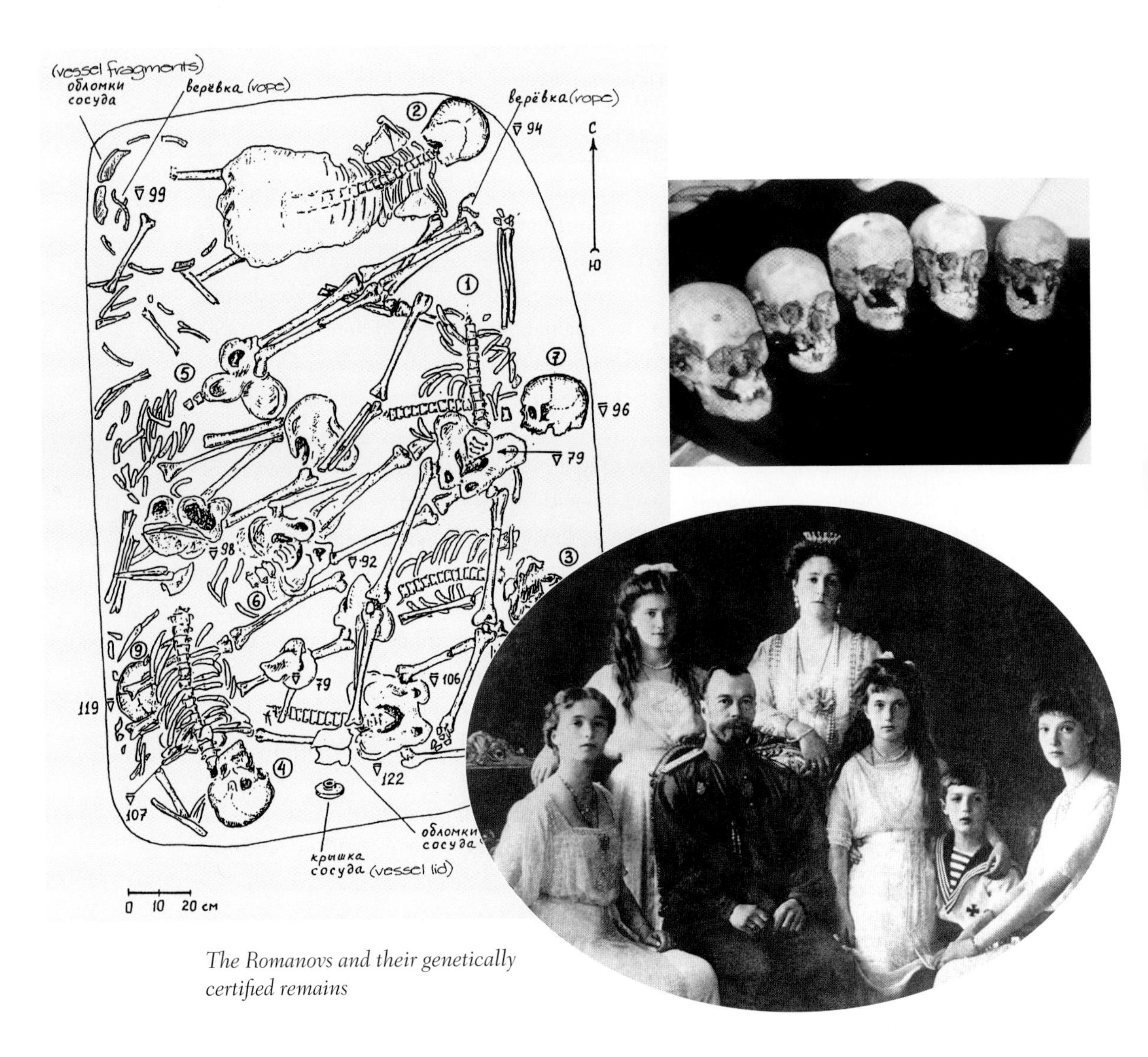

The Romanovs and their genetically certified remains

the first paper Alec Jeffreys published in this field and had since established the Forensic Science Service as the UK's premier laboratory for DNA fingerprinting.

Gill had developed a DNA fingerprinting method using mitochondrial DNA (mtDNA), which, as we saw in the analysis of Neanderthal mtDNA, has a special advantage in cases when DNA is old or difficult to obtain: it is far more abundant than the chromosomal DNA from the nucleus.

Gill and Ivanov's first task was the delicate job of extracting both nuclear and mtDNA from the bone samples. The analysis showed that five of the bodies were related and that three were female siblings. But were these the bones of the Romanovs? In the case of the Empress Alexandra at least, an answer could be found by comparing the mtDNA fingerprint from the bones thought to be hers with an mtDNA fingerprint from her grandnephew, Prince Phillip, the Duke of Edinburgh. The fingerprints matched.

It was rather more difficult to find a relative for the tsar. The body of the Grand Duke Georgij Romanov, his younger brother, dwelt in an exquisite marble sarcophagus deemed too precious to open. The tsar's nephew refused to help, still bitter over the British government's refusal to grant his family refuge at the onset of the Revolution. A bloodstained handkerchief was known to exist in Japan, one the tsar had used when he was attacked by a sword-wielding assassin in 1892. Gill and Ivanov secured a narrow strip of it but found that over the years the relic had been contaminated beyond usefulness with the DNA of others. It wasn't until two distant relatives were finally found that the mtDNA fingerprint was confirmed as the tsar's.

But the analysis had yet one more surprise in store: the mtDNA sequences from the presumed tsar and his modern relatives were similar but not identical. Specifically, at position 16,169, where the tsar's mtDNA had a C, that of the two relatives showed a T. And further testing revealed only further complications. The tsar's mitochondrial DNA was actually a mix of two types, both C and T. This unusual condition is called "heteroplasmy"—the coexistence within a single individual of more than one mtDNA type.

A few years later the worries of all but the most committed conspiracy theorists were finally put to rest. The Russian government finally agreed to crack the sarcophagus and provide Ivanov with a tissue sample from Georgij Romanov, the tsar's brother. The grand duke's mitochondria showed the very same heteroplasmy as those found in the bones from the pit. Those bones were without question the tsar's.

But what of the legendary Anastasia, whose skeleton was never recovered from the grave in the forest? There has been no lack of pretenders to the Romanov line, and among these none was more persistent than one Anna Anderson, who asserted for a lifetime that she was the lost grand duchess.

Pretender Anna Anderson in 1955, and Ingrid Bergman in the title role of Anastasia *(1956), the movie inspired by Anderson's claims*

She'd first made the claim as early as 1920 and went on to become the subject of many books as well as the film *Anastasia,* in which, played by Ingrid Bergman, she was indeed found to be the grand duchess. When Anderson died in 1984, her identity was still in dispute, but as the claims and counterclaims of her supporters and critics continued, the means for a resolution were at hand.

Anna Manahan (Anna Anderson's married name) had been cremated, making tissue retrieval from her remains impossible. But an alternative source of her DNA was discovered: in August 1970, she had undergone emergency abdominal surgery at the Martha Jefferson Hospital in Charlottesville. Tissue removed during the operation had been sent to a pathology laboratory where it was prepared for microscopy, and where, twenty-four years later, it was still filed away. After an appropriately Byzantine series of court cases over access to the specimen, Peter Gill traveled to Charlottesville in June 1994 and departed with a little preserved slice of Anna Manahan.

The results were crystal clear. Anna Anderson was related neither to Tsar Nicholas II nor to the Empress Alexandra. But in the wake of such a long odyssey, it is perhaps not surprising that some chose to ignore the DNA and believe what they would: the myth that Anna was Anastasia still lives on.

The fate of the Romanovs and Anna Anderson may be the stuff of fairy tales, remote from most of our lives, but DNA fingerprinting is ordinarily applied to grim realities painfully all too close. One of the most awful tasks facing investigators after a violent catastrophe like a plane crash is the identification of bodies. For various reasons—to permit the issuing of a death certificate, for instance—the law requires that it be done. And no one should underestimate the desperate emotional need of families to bury their loved ones with proper ceremony; for most of us, respect for the dead requires the recovery of their remains, however fragmented, and this task depends on positive identification.

In 1972, an American warplane believed to have been piloted by Michael Blassie was shot down during the Battle of An Loc in Vietnam. Remains were recovered from the crash site, but an inadequate forensic examination in 1978 based on blood type and analysis of the bones indicated that they were not Blassie's. The anonymous bones were labeled "X-26, Case 1853," and in a solemn ceremony attended by President Reagan, they were laid to rest in the Tomb of the Unknowns at Arlington National Cemetery. In 1994, CBS News picked up a story by Ted Sampley in the *U.S. Veteran Dispatch,* claiming that X-26 was Blassie. When the subsequent investigation by CBS uncovered evidence corroborating Sampley's claim, Blassie's family petitioned the Department of Defense to examine it. This time mtDNA fingerprints from the unknown's bones were found to match those of Blassie's mother and sister. Twenty years after his death, Blassie came back to St. Louis. Standing beside the gravestone, his mother was able to say, "My son is home. My son is finally home."

The Department of Defense has since established the Armed Forces Repository of Specimen Samples for the Identification of Remains. Blood samples are taken and DNA isolated from all new members of the military, both those on active duty and reservists. By March 2001, the repository contained more than 3 million samples.

DNA

I was on my way to my office when I heard that a plane had crashed into one of the World Trade Center towers. Like many others, I assumed initially it was an accident—anything else was unimaginable. But all too soon, when the second plane hit the other tower, it was apparent that a criminal act of the most ghastly kind had been perpetrated against thousands of innocent people. No one who watched that day is likely ever to forget the images of people leaning out of windows high on the towers, or falling to their deaths. And we were not shielded from the tragedy's immediate toll even on the tranquil campus of Cold Spring Harbor Laboratory, thirty miles from Manhattan: two of our staff lost sons that day.

The final loss of life has been reckoned at 2,792—an extraordinarily low number considering that as many as 50,000 may have been in the towers at the time of the attack. Nevertheless, given an event of such cataclysmic force, one can expect to find few bodies intact, much less alive. And so the search for survivors was transformed with a tragic inevitability into the hunt for remains; a million tons of mangled steel, pulverized concrete, and crushed glass were sifted for any human part they might yield. Some 20,000 were found and taken to twenty refrigerated semi trucks arrayed near the medical examiner's office. Since the beginning of this herculean forensic effort, many identifications have been made using dental records and conventional fingerprints, but as the easy cases are closed, increasingly the load shifts to DNA analysis. For comparison with all genetic traces from the site, relatives have supplied either samples of their own blood or items like the toothbrushes and hairbrushes of the dead, any possession that may have picked up even a few of its owner's cells from which DNA could be extracted. The task of carrying out the DNA fingerprinting has fallen to Myriad Genetics in Salt Lake City and Celera Genomics, both of which are accustomed to analyzing DNA on an enormous scale. But even with the very latest technology, this is a slow and painstaking process.

It is a common human desire to know one's forebears: who they were and where they came from. In the United States, a nation built by generation after generation of immigrants, the longing is especially intense. In recent years, the genealogical craze has been aided by the World Wide Web, which also supplies us with an informal measure of the phenomenon's dimensions: a Google search for "genealogy" yields over 10 million hits (a search for "DNA" gets you only 5 million). By comparing the fingerprints of individuals, DNA makes possible the highly specific sort of genealogical inquiry that Gill and Ivanov carried out to uncover, for instance, Anna Anderson's relationship to the Romanovs (none). But genealogies can also be constructed at a broader level, finding connections by comparing the DNA fingerprint of an individual with those of whole populations.

At Oxford, Brian Sykes used DNA analysis to delve into his own genetic history. Knowing that both surnames and Y chromosomes are transmitted down the male line, he surmised that all men born with the same surname should also have the same Y chromosome—the one belonging to the very first man to take that name. Of course, this linkage of Y chromosome and surname breaks down if a name should arise independently more than once, if men change the family name for one reason or another, or if many boys take the name of a man other than their biological father (a lad secretly sired by the milkman, for instance, would likely wind up with the surname of his mother's husband).

After contacting 269 men called Sykes, Professor Sykes managed to collect 48 samples for analysis. He found that about 50 percent of the Y chromosomes were indeed identical to his own "Sykes" chromosome; the rest bore evidence of conjugal lapses on the part of more than one Mrs. Sykes of generations gone by. Because the origin of the name is documented and can be dated to around seven hundred years ago, it is possible to work out the per-generation rate of infidelity. It averages out to a perfectly respectable 1 percent, suggesting that 99 percent of Sykes wives in every generation managed to resist extramarital temptation.

When Sykes set up a company to market genealogical DNA fingerprinting services, one of his first clients was the John Clough Society, whose members trace their ancestry back to a Briton of the same name who emigrated to Massachusetts in 1635. The society even knew that an ancestor of his, Richard, from the Welsh line of the family, had been knighted for his deeds on a crusade to the Holy Land. What they lacked, however, was any historical proof to link their families to those on the other side of the Atlantic. Sykes's company analyzed Y chromosome DNA from the Massachusetts Cloughs and from a direct male descendant of Sir Richard; the two were indeed identical—vindication for the Massachusetts branch. But not all the American Cloughs were as lucky; society members from Alabama and North Carolina were found to be unrelated not only to Sir Richard but to the Massachusetts Cloughs as well.

On *The Montel Williams Show,* or *Ricki Lake,* or *Jenny Jones,* you can see the young women and men looking nervous. The host opens an envelope, gives the couple a meaningful look, and then reads the card. The woman covers her face with her hands and bursts into tears, while the man leaps into the air, pumping his fist. Alternatively, the woman leaps to her feet, pointing triumphantly at the man who remains slumped, shoulders bowed, in his seat. In either case, we have just seen one of the more outlandish applications of DNA fingerprinting—the ultimate in infotainment.

Daytime television may make theater of the subject, but paternity testing is a serious business with a long tradition. Since the beginning of human history, much of one's life—its psychological, social, and legal realities—has depended on the identity of one's father. So, quite naturally, science has been drafted into the service of paternity testing ever since genetic techniques for distinguishing individuals were first developed. Until the advent of molecular genetics, blood itself was the most scientific clue to paternity. The patterns of inheritance were reliable and well understood, but with only a handful of blood groups to test for, the trait's power to discriminate was limited. Practically speaking, a test for blood type has limited power to exclude wrongly accused fathers, and it can never provide definitive affirmation of the right one. If our blood types are not compatible, I am assuredly not your father; but if they are, it's no certain proof

that I am—the same will be true of any number of men who have the same blood type I have. Using other markers in addition to the familiar ABO blood group markers improves the resolving power of this kind of test but it still cannot match the statistical muscle of STR-typing: an STR-based genetic fingerprint can establish proof positive of paternity. And in the era of PCR, it is convenient enough to use.

So convenient, in fact, that mail-order paternity testing companies do a thriving business. In some cities huge roadside billboards advertise a local paternity testing service with the none-too-subtle pitch line: "Who's the Daddy?" For a fee, these companies will mail you a DNA sampling kit that includes a swab to scrape some cells from the interior of the mouth. (Samples collected this way would not stand up in court. To be admissible, a DNA fingerprint must be based on a sample collected by a certified lab, which must verify the chain of evidence so as to prevent the sort of genetic switcheroo we saw in the Pitchfork case.) The tissue samples are sent by overnight courier to the testing laboratory, where the DNA is extracted.

The child's DNA fingerprint is compared with that of the mother; any STR repeats present in the child but not in the mother are presumed to have come from the father, whoever he may be. If the fingerprint of a supposed father lacks any of these repeats, he must be excluded. If none are missing, the number of repeats allows us to quantify the likelihood that a match is definitive by the so-called Paternity Index (PI). This measures the chances that some man other than the actual father could have contributed a particular STR, and it varies in relation to how common a given STR is in the population. The PIs for all STRs are multiplied together to give a Combined Paternity Index.

Most paternity tests are, of course, handled with the utmost discretion (unless you happen to be on a talk show), but one recent analysis drew many headlines owing to the great historical interest in the alleged father. It had long been suspected that Thomas Jefferson, third president of the United States and the principal author of the Declaration of Independence, was more than a founding father: he was thought to have had one or more children by his slave Sally Hemings. The first accusation was made in 1802, just twelve years after the birth of a boy, Tom, who later took the last name of one of his subsequent masters, Woodson. In addition a strong resemblance to Jefferson had been widely remarked in Hemings's last son, Eston. DNA was destined to set the record straight.

Jefferson had no legitimate male descendants so it is impossible to determine the markers on his Y chromosome. Instead, researchers took DNA samples from male descendants of Jefferson's paternal uncle, Field Jefferson (whose Y chromosome would have been identical to the president's), and compared them with samples from the male descendants of Tom and Eston. The results showed a distinct Jefferson fingerprint for the Y chromosome, but this DNA fingerprint was not present in the descendants of Tom Woodson. Jefferson's reputation had dodged that bullet. In Eston Hemings's descendants, however, the Jeffersonian Y chromosome signature came through loud and clear. But what the DNA cannot confirm beyond reasonable doubt is the source of that chromosome. We cannot say with certainty whether Eston's father was in fact Thomas Jefferson or some other male in the Jefferson lineage who might also have had access to Sally Hemings. Indeed, some suspicions have been cast on Isham Jefferson, the president's nephew.

Centuries of national reverence, then, are no protection against the harsh revealing light of DNA evidence. Nor, it seems, is any amount of celebrity or money. When the Brazilian model Luciana Morad claimed that Mick Jagger was the father of her son (whom she named Lucas Morad Jagger), the Rolling Stone denied it and demanded DNA testing. Perhaps Jagger was bluffing, hoping that the threat of a forensic denouement would weaken Ms. Morad's resolve and induce her to drop the case. But she did not. The tests were positive, and Jagger found himself legally obliged to contribute to the upbringing of his son. Boris Becker, too, submitted to a paternity test over a girl born to Russian model Angela Ermakova. The tabloids had a field day with stories that the tennis star believed himself the victim of a blackmail scheme contrived by the Russian mafia—the lurid details of how this plot was supposedly perpetrated are best left in the pages of the tabloids. Suffice it to say that when the DNA results were in, the swaggering Becker acknowledged his deed and pledged to support his daughter.

DNA fingerprinting to identify a child's biological relatives has been applied to causes rather more uplifting than those of Messrs. Jagger and Becker. In Argentina, between 1975 and 1983, 15,000 people were quietly eliminated for holding opinions unpopular with the ruling military junta. Many of the children of the "disappeared" were subsequently placed in orphanages or adopted illegally by military officers. Having lost their own children to the regime, the mothers of the disappeared then set about finding their children's children—to reclaim their grandchildren. Las Abuelas (grandmothers) drew attention to their nationwide quest by marching every Thursday in the central square in Buenos Aires. They continue their search to this day. Once a child has been located, genetic fingerprinting methods can be used to determine who is related. Since 1984, Mary-Claire King—whom we encountered earlier grappling with another set of relationships, that between humans and chimpanzees—has provided Las Abuelas with the genetic analysis needed to reunite families torn apart by eight nightmarish years of misrule.

DNA fingerprinting has come a long way since its first forensic applications. It is now a staple of our popular culture, a consumer good for the genealogically curious; a mousetrap in the ongoing spectacle of "gotcha" we play with celebrities and with those ordinary folk who wish only to be on television. But its most serious application remains in the resolution of legal questions involving life and death. The United States is the only nation in the Western world that still imposes the death penalty. Between 1976, when the Supreme Court reinstated capital punishment after a ten-year hiatus, and 2001, 749 convicts were put to death, and by the end of that period there were 3,593 prisoners on death row. It is against this background that we need to examine the work of the Innocence Project, and its founders, Barry Scheck and Peter Neufeld, some of the earliest and staunchest critics of DNA fingerprinting, at least as it was first practiced. Since the early days, Scheck, Neufeld, and other defense attorneys have come to realize that the forensic technology they opposed is actually a powerful tool for justice—more capable, in fact, of exculpating the innocent than of convicting the guilty. Proving innocence merely requires finding a single mismatch between a defendant's DNA fingerprint and that taken from the crime scene; proving guilt, on the other hand, requires demonstrating statistically that the chances of someone other than the accused having the specified fingerprint are negligible.

As of November 2002, the work of lawyers and students in Innocence Projects (there is now a whole network of them, based at law schools throughout the country) has led to the exoneration of 117 wrongfully convicted individuals. In Illinois, six of these mistaken convictions had resulted in death sentences, leading Governor George Ryan to take a remarkable and—given popular support for law-and-order palliatives like capital punishment—politically dangerous step of imposing an indefinite moratorium on executions in the state. In addition, Ryan appointed a special commission to review the handling of capital cases; published in April 2002, this commission's report listed among its strongest recommendations that provision be made to facilitate DNA testing of all defendants and convicts in the state's criminal justice system.

A father's lonely and ultimately forlorn crusade: The long-running campaign to clear James Hanratty's name ran aground on DNA evidence.

By no means has all DNA testing of those who insist on their innocence led to the overturning of convictions. James Hanratty was convicted of one of the most notorious murders in twentieth-century Britain. He accosted a young couple, shot the man fatally, and raped the woman before shooting her five times and leaving her for dead. Despite his insistence that he'd been miles away when the crime occurred, Hanratty was found guilty and sentenced to hang. In 1962, he became one of the last criminals to be executed in Britain.

Hanratty died proclaiming his innocence, and his family began a posthumous campaign to clear his name. Their efforts became a cause célèbre: they succeeded in compelling the authorities to have DNA extracted from the female victim's semen-stained underwear and from the handkerchief that had masked the assailant's face; both samples were then compared with DNA fingerprints from Hanratty's brother and mother. To their chagrin, it was determined that the crime scene DNA had indeed come from a member of the Hanratty family. Still unsatisfied, the Hanrattys had their black sheep's body exhumed in 2000 in order to retrieve tissue samples for DNA extraction. That more direct analysis showed it was unequivocally Hanratty's DNA on the underwear and the handkerchief. Finally, grasping for straws, the family argued, following the recently successful Simpson defense, that the sample sources had been handled improperly and become contaminated. But the Lord Chief Justice proved less distractable than the Simpson jury. He rejected this claim out of hand: "The DNA evidence establishes beyond doubt that James Hanratty was the murderer."

Usually the strongest objection to reopening a case comes from the district attorney, who is understandably reluctant to see a hard-won conviction subject to post-trial scrutiny. But sometimes such rigidity can be self-defeating, and if prosecutors have now learned that genetic evidence can nail a case, they should also recognize that DNA may also be the surest way to keep one shut. The example of Benjamin LaGuer illustrates the point. Sentenced in 1984 to forty years in prison for a rape in Worcester, Massachusetts, he never stopped protesting his innocence. Like Hanratty, he attracted a retinue of rich and famous sympathizers, who in 2001 arranged and paid for samples of DNA to be analyzed. The results must have surprised them all: LaGuer was the rapist. One can imagine that a man facing forty years behind bars rightly imagined he had nothing to lose in making such a demand. But ironically, it had taken two years to get the district attorney's office to agree to the DNA fingerprinting. As an editorial in the *St. Petersburg Times* sensibly remarked, "In hindsight, the prosecutor could have wasted less time arguing and gotten the pleasure of saying 'I told you so' much sooner had he consented early on to the DNA test."

Civil libertarians will always object to the broad application of DNA fingerprinting in society as a whole. But it is hard to argue with the social utility of applying the technology to those who, for whatever reason, pass through the criminal justice system; for the chances are, sadly, that those who pass through once will pass through again. Criminological data indicate that those convicted of minor crimes are likely to commit more serious offenses; 28 percent of homicides and 12 percent of sexual assaults in Florida have been linked to individuals previously convicted of burglary. And such patterns of recidivism can be detected among white-collar criminals as well: of twenty-two who had been convicted for forgery in Virginia, ten were linked through DNA fingerprinting to murders or sexual assaults. It would seem prudent to make the corporate bosses of Enron, ImClone, and Adelphia Communications provide DNA samples.

Efforts are under way to broaden DNA fingerprint databases. Recently, the British government has proposed allowing the police to keep DNA samples taken both from acquitted defendants and from those arrested but never charged. The same rule would permit the authorities to keep samples given voluntarily (when, for example, the police test everyone in a location, as they did in Narborough). These changes in collection rules will triple the number of entries in the police database within three years. In the United States, nineteen states now collect DNA samples from all felons, not just those involved in violent crime.

I think everyone should give a DNA sample. It is not that I am insensitive to the concerns about individual privacy or to the potential for inappropriate use of genetic information; as I have said earlier, in my role as the first director of the Human Genome Project, I set aside a substantial chunk of our funding to examine such questions in relation to clinically applied genetic information. But criminal justice is a different matter. Here by my calculation the potential for the greater social good far outweighs the risks of abuse. And since we must all surrender something for the benefit of living in a free society, the sacrifice of this particular form of anonymity does not seem an unreasonable price to pay, provided our laws see to a strict and judicious control over access to databases. Frankly, the remote possibility that Big Brother will one day be perusing my genetic fingerprint for some nefarious end worries me less than the thought that tomorrow a dangerous criminal may go free—perhaps only to do further evil—or an innocent individual may languish in prison for want of a simple DNA test.

But objections to DNA collection in general continue to be heard, and often from the most surprising and far-flung quarters. In both New York City and the Australian state of Tasmania, lawmakers have proposed that the entire police force be fingerprinted. The logic is simple: keep the police on file so their DNA can readily be excluded from any crime they might investigate. Remarkably, the measures were denounced by law enforcement bodies in both jurisdictions: those presumed to be the most law-upholding of citizens, those whose work only promises to be facilitated by the widespread availability of DNA fingerprinting, want no part of it where their own DNA is concerned. My suspicion is that there is something of the irrational at play here. As in the case of genetically modified foods, DNA has in the popular imagination a voodoo quality: there's something scary, mysterious about it. And a lack of understanding of genetic complexities leaves one susceptible to the worst anxieties and conspiracy theories. Once people understand the issues, I hope this hesitation in making the most of a new and powerful beneficial technology will vanish.

Barry Scheck and Peter Neufeld put it well in the preface to their book *Actual Innocence*: "DNA testing is to justice what the telescope is for the stars; not a lesson in biochemistry, not a display of the wonders of magnifying glass, but a way to see things as they really are." What could be wrong with that?

Nancy Wexler holding a child with early-onset Huntington disease, Lake Maracaibo, Venezuela

CHAPTER ELEVEN

GENE HUNTING: THE GENETICS OF HUMAN DISEASE

It was too early in the day for anyone, let alone an impeccably dressed middle-aged woman, to be drunk. But as she swayed unsteadily across the street, drunk is what she seemed, even to the cop on duty near the courthouse, who reprimanded her for creating a public spectacle. In fact, Leonore Wexler wasn't drunk at all. She was beginning to succumb to a ghastly fate that had already destroyed several close relatives before her eyes, a fate she had hoped would pass her by.

Not long thereafter, in 1968, Wexler's ex-husband, Milton, was to celebrate his sixtieth birthday in Los Angeles with their two daughters, Alice, 26, and Nancy, 23. But celebration, as it turned out, was not the order of the day. Milton told his daughters that their mother, 53, was suffering from Huntington disease (HD), a devastating neurological disorder that causes a progressive deterioration in brain function such that those afflicted gradually lose all knowledge of themselves and their loved ones. They also lose control of their arms and legs; at first walking is affected, as in Leonore's case, but as the decline continues patients also experience involuntary, jerky movements. There was no cure and no treatment to delay the relentless slide toward death.

Now Alice and Nancy could make sense of some disquieting facts about their mother's relatives as well as hints she herself had dropped that all was not right in the family. They knew that their uncles, Leonore's three brothers, had all died young; before his end, each had developed the same strange grimace, unsteady walk, and slurred speech. They knew that Leonore's father, their grandfather, Abraham Sabin, had also died young, though Leonore had carefully never mentioned he too had shown those symptoms. Huntington disease, it was becoming clear to them, ran in the family. It was Milton's grim task to answer their immediate question: What was the risk that Alice or Nancy might succumb? "Fifty-fifty," their father told them.

The disease that would afflict Abraham Sabin and his descendants was first identified by George Huntington. Born into a medical family, Huntington grew up in East Hampton, Long Island, where as a young boy he accompanied his father on his rounds. After qualifying as a physician at Columbia University, Huntington returned to the family practice on Long Island for a few years before moving to Pomeroy, Ohio. In 1872, he presented a paper at the Meigs and Mason Academy of Medicine in nearby Middleport entitled "On Chorea." Derived from the Greek word for dance, "chorea" was the name physicians had since the seventeenth century given to illnesses that produced jerky movements in their victims. Late in life Huntington would recount how he had come to be fascinated by the mysterious malady:

> Over 50 years ago, in riding with my father on his rounds I saw my first case of "that disorder," which was the way the natives always referred to the dreaded disease. I recall it as vividly as though it had occurred but yesterday. It made a most enduring impression upon my boyish mind, an impression which was the very first impulse to my choosing chorea as my virgin contribution to medical lore. Driving with my father through a wooded road leading from East Hampton to Amagansett, we suddenly came upon two women both bowing, twisting, grimacing. I stared in wonderment, almost in fear. What could it mean? My father paused to speak with them and we passed on. Then my Gamaliel-like* instruction began; my medical instruction had its inception. From this point on my interest in the disease has never wholly ceased.

Drawing on his own observations as well as the clinical notes of both his father and grandfather (the original manuscript has annotations penciled in by

*Gamaliel, a famous rabbi and teacher of St. Paul (Acts 22:3), believed in integrating book learning with everyday experience.

his father), the young physician's paper offered a masterful description of what became known as Huntington's chorea and is now called Huntington disease. The "chorea" movements, he explained, "gradually increase when muscles hitherto unaffected take on the spasmodic action, until every muscle in the body becomes affected." He noted the attendant mental deterioration: "As the disease progresses the mind becomes more or less impaired, in many amounting to insanity, while in others mind and body gradually fail until death relieves them of their suffering." And he recognized that the disorder was inherited: "When either or both the parents have shown manifestations of the disease, one or more of the offspring invariably suffer from the condition. It never skips a generation to again manifest itself in another. Once having yielded its claims, it never regains them."

Huntington correctly identified the key features of this kind of genetic disorder. He recognized that it affected both males and females and understood that it passed from generation to generation. Each child of a parent with Huntington disease has a 50-50 chance of inheriting it. By the luck of the draw, in some families everyone is affected; in others, none are. If a person does not inherit the abnormal gene from a parent, he or she cannot pass on the gene to the next generation. Today we know Huntington disease is caused by a mutation and since the gene is not preferentially expressed in one sex over the other (i.e., is not sex-linked), we have inferred that the affected gene is on neither the X nor Y sex chromosome. Let's call the normal version of the gene *H* and the mutant version *h*. We have two copies of each non-sex chromosome (called "autosomes") and so two copies of the Huntington gene. Individuals with the two copies of the normal gene (*HH*) are, predictably, disease free. But individuals with two (hh) or even one copy of the mutated gene (*Hh*) are bound to develop the disease. We call this pattern "autosomal dominant inheritance." ("Dominant" means that only one copy of a mutated gene is sufficient to cause disease—the abnormal gene dominates its normal partner.)

Since it is far likelier that a person will acquire one rather than two copies of the mutant form, most Huntington sufferers are *Hh*. Such individuals could pass on *H* or *h* to their children, yielding a 50 percent chance that a particular child would be affected, just as Milton Wexler told Alice and Nancy.

Back in 1968, not much was known about Huntington disease beyond these facts: it is heritable, and it makes its irreversible progress by killing nerve cells in specific areas of the brain. Milton Wexler resolved that he would take on the terror striking his family: he established the Hereditary Disease Foundation (HDF) to raise money and press for more government funding for Huntington disease research. His daughter Nancy was drawn in as well. While completing a doctorate in psychology at the University of Michigan—her thesis fittingly concerned the psychology of being at risk—she found herself increasingly involved in the affairs of the foundation. In the 1970s, when it became apparent that real progress would depend upon a better understanding of the genetics of the disease, Nancy Wexler began to reinvent herself as a geneticist.

On the shores of Lake Maracaibo, Venezuela, the burden of grinding poverty is compounded by a remarkably high incidence of Huntington disease. If Huntington were to divulge its genetic secrets anywhere, Lake Maracaibo seemed a likely place. In 1979, Wexler began to collect DNA samples and to record family histories with the goal of preparing a genealogy of all affected people. For the geneticist it was a great labor, but for Wexler, the daughter of a Huntington victim with the possibility of the disease in her own future, it was more than that. It involved seeing the familiar in such unfamiliar surroundings: people who lived in tin-roofed wooden huts on poles above the waters of the lake yet walked with that same drunken stagger that had overtaken her mother. Since her first trip to Lake Maracaibo in 1979, Wexler has returned annually to continue the work there. The people she works with have come to call her La Catira for her long blond hair. As Americo Negrette, her Venezuelan colleague and the scientist who first reported the occurrence of Huntington at Lake Maracaibo, describes it, she has made of them an extended family, greeting them each time "without theater, without simulation, without pose. With a tenderness that jumps from her eyes."

But tenderness could only mitigate the devastation Huntington disease had visited upon so many. The goal of Wexler's expeditions was ultimately to find the gene responsible for the disease. But how could her Maracaibo genealogies help to identify the culprit? The key lay in advances in human genetics.

If they were to home in on the Huntington gene, Wexler and others interested in genetic disease knew they would have to do for humans what Morgan and his students had started doing for fruit flies more than half a century earlier. As we have seen (in chapter 1), Morgan compared rates at which particular genetic markers—white (as opposed to red) eye color, say, and curly (as opposed to straight) wings—coincided in the offspring of crosses between parents showing various combinations of these traits; from these data he was able to determine how near each other on a chromosome were the genes governing those traits. But human genetics had lagged behind the fruit fly's for two major reasons. First was the impossibility—on moral and practical grounds—of doing the kind of experiments that were still the mainstay of genetic analysis: you can't simply breed two human beings you're interested in and then analyze the progeny two weeks later. Second, even if humans could be crossed at will, they were still lacking in genetic markers. Morgan was able to track a number of simple and obvious differences in appearance caused by specific mutations in individual genes. Humans unfortunately don't possess many easily analyzed traits that are inherited in this simple way; even the canonical example, eye color, turns out to be governed by several genes, not just one. Furthermore, with fruit flies, you can increase levels of genetic variation by subjecting individuals to X rays, or to other mutagenic agents: such options, happily, are not available in dealing with humans. Only with the advent of recombinant DNA did solutions to the two major obstacles present themselves.

In the age of DNA sequencing, genetic markers need no longer be visible, like white eyes in a fruit fly; a variation in the sequence itself will suffice, and you can track such a DNA marker through a family tree—that is, through a number of genetic crosses—simply by analyzing DNA from several generations. The revolution had begun the year before Wexler started her genealogical research. And, as with so many advances in science, a measure of serendipity was involved.

It had become an annual ritual: a small group of graduate students from the University of Utah would accompany their advisers to the Wasatch Mountain ski resort of Alta for an intensive workshop on their research (and, well, a little skiing on the side). Typically, a couple of big-shot scientists from other institutions would be invited, to cast a critical eye over the data presented by each nervous student. In 1978, the big shots included David Botstein from MIT and Ron Davis from Stanford.

David Botstein, it's been noted, "tends to think and talk excessively fast, and often at the same time." Ron Davis is quiet and retiring. That April in Utah, despite their contrasting styles, Botstein and Davis shared an epiphany. As they listened to Mark Skolnick's graduate students discuss genetic disorders traced in the very large pedigrees of Mormon families, Botstein's and Davis's eyes suddenly met as both registered simultaneously the same insight. Though both were experts on yeast, they saw a way to locate *human* genes! What they saw was that cutting-edge recombinant DNA techniques would allow them to apply to humans the very sort of genetic analysis first used by Morgan to study the fruit fly. In fact, DNA markers had already been used to map genes in a number of other species, but Botstein and Davis would be the first to develop the technique's potential in humans.

The technique, called "linkage analysis," determines the position of a gene in relation to the known positions of particular genetic landmarks. The principle is simple: it would be difficult for you, given no other information, to find Springfield on a map of the United States, but if I tell you that Springfield lies about halfway between New York and Boston—two landmarks labeled on the map—then your task is made very much easier. Linkage analysis aims to do this with genes: it establishes links between known genetic markers and unknown genes. It was a very successful method with the fruit fly, but, as we have seen, the dearth of known genetic markers in human beings prevented its application to human diseases—until Botstein and Davis recognized that advances in molecular biology had solved the problem.

The DNA markers that caught their eye were restriction fragment length polymorphisms (RFLPs). They occur when a DNA sequence cut by a particular restriction enzyme in one individual has changed in another so that it can no longer be cut by that enzyme. (Remember that restriction enzymes are sequence-specific: enzyme *Eco*R1 cuts only when it encounters GAATTC. That sequence occurs at a given location in the genome, but through mutation some individuals may have a variant form of that segment—say, GAAGTC. The enzyme will be able to cut only unchanged sequences, not the altered version.) These are naturally occurring differences in DNA sequence; they occur most

Genetic mapping of a disease gene. For convenience, only two generations and a few individuals are shown. If the analysis is to be statistically powerful, information is required from a large number of reated individuals.

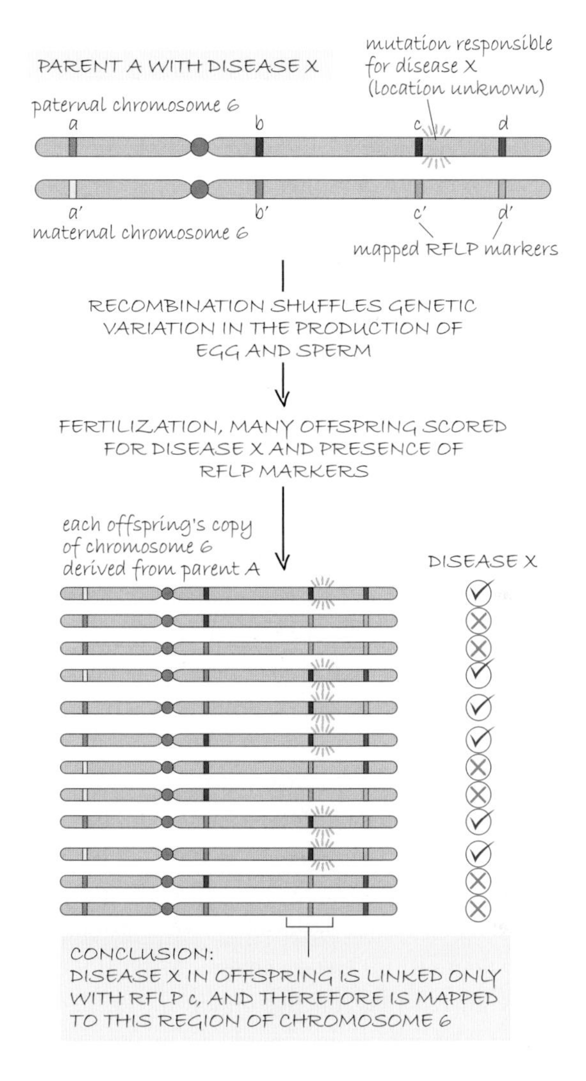

often in junk DNA, and so there is no functional effect. Still, literally millions of them are scattered through our genome.

In the months following the Alta meeting, Botstein, Davis, and Skolnick, together with Ray White, then at the University of Massachusetts, pursued the RFLP concept. In 1980, a landmark paper that grew out of this collaboration heralded the new age of molecular human genetics. They laid out a clear plan showing how RFLPs could be used, and they worked out the math concerning how many would be needed to ensure that every point in the human genome was within reasonable proximity of at least one RFLP marker—conditions that would in principle permit the mapping of the entire genome. It would be like having enough U.S. cities fixed on a map of North America to allow any unmarked place to be located with respectable accuracy using only information about how close it was to the labeled cities. But, for the genetic map, what was "reasonable" proximity? Botstein and his colleagues calculated that 150 RFLPs spread uniformly across the entire human genome would be enough. The most immediate benefit of the system was a new strategy for identifying genes that cause disease. Using families in which a disorder spanned several generations, they would take DNA samples from both affected and unaffected individuals. Then they would use recombinant methods to test RFLPs one after another, looking for ones that tracked the disease through the families.

In 1979, before the publication of the paper, White presented these ideas at a Cold Spring Harbor Laboratory conference. He noted that "among the more kosher molecular biologists, there was a lot of bitching and grumbling." What he was hearing was great skepticism as to whether the method would work at all; even those who thought it would couldn't agree on the best way to go about using it. These disagreements came into the open during a later meeting to discuss how RFLP linkage analysis could be used to find the gene involved in Huntington disease.

Nancy Wexler wanted her Lake Maracaibo genealogy to be considered immediately for linkage studies, but Botstein and White thought it was far too early to use RFLP linkage analysis to look for the Huntington gene or any other. They argued that much groundwork needed to be done first—the markers themselves had to be found and mapped—before the technique could be applied for such a specific purpose. In the end, Wexler's determination resulted in a parting of ways: while the Hereditary Disease Foundation pressed on with the hunt for the Huntington gene, Botstein and White pushed for a complete map of the human genome.

The latter goal required finding RFLP markers on every chromosome, and finding enough of them to ensure that at least one was close to every point in the genome. It was soon necessary to make an upward revision of the initial estimate of 150. But, undeterred, academic laboratories like White's began to isolate RFLPs, and soon commercial biotechnology was getting in on the action as well.

In 1983 Helen Donis-Keller, an experienced molecular biologist and in those days David Botstein's wife, established the Human Genetics Department at Collaborative Research, Inc., a Boston-area company. She aimed to produce an RFLP linkage map of the whole human genome, with sufficient markers to locate disease genes on any chromosome. The fruits of the effort were published four years later, in a paper aptly entitled "A Genetic Linkage Map of the Human Genome." The map included 403 loci—many more than Botstein's original estimate—and calculations showed that 95 percent of the genome was within reasonable proximity (or "linked") to a marker. It was a great day for genome mapping, but by 1987 rifts and rivalries were appearing once more among the researchers.

For one thing, there was resentment in academic quarters that Collaborative had incorporated freely available data from university labs while disclosing none of its own. (In this respect, Collaborative was pioneering the best-of-both-worlds strategy that Craig Venter and other would-be genome profiteers were soon to follow in the sequencing sweepstakes.) The French immunologist Jean Dausset, for instance, had been following a somewhat different course. His 1980 Nobel Prize in Physiology or Medicine attracted a generous benefactor, whose substantial gift allowed Dausset to pursue his own strategy for preparing a human linkage map. He realized the task would be much easier if all researchers worldwide were working with a standard set of pedigrees—DNA samples from the same families. So, Dausset created the Centre d'Etude Polymorphisme Humain (CEPH) in Paris to collect pedigrees optimal for genetic analysis: large families with three living generations from which to draw samples. The CEPH collection eventually contained DNA from sixty-one families, including many of the Mormons studied by Ray White, Nancy Wexler's Lake Maracaibo families, and Amish families catalogued by Victor McKusick of the Johns Hopkins Medical School. CEPH made DNA samples from all these families freely available to researchers, with the sole proviso that recipients give their analyses to CEPH for integration into the worldwide database. Collaborative Research took full and fair advantage of this resource.

By far the most serious criticism of the Collaborative map, however, was the patchiness of the distribution of its markers. Chromosome 7—linked to cystic fibrosis, one of Collaborative's targets—had 63 markers, but on chromosome 14 only 6 were identified. The distance between markers on the marker-poor chromosomes was very much greater than the average for the genome as a whole. Ray White was particularly upset by Collaborative's claims. He himself had found over 470 markers but had been publishing his data chromosome by chromosome as each was filled in with the required density of RFLPs. "We would never have dreamed of making such a publication with our data set, which is substantially larger than theirs, because we still have significant gaps," he remarked, rejecting Collaborative's grandiose claim. Whether the claims were grandiose or not, though, Collaborative's map had proved the feasibility of genome-wide mapping and was a significant advance.

But as we have noted, some, like Nancy Wexler, had seen another path opening up in the wake of the breakthrough 1980 RFLP paper. As efforts to produce a comprehensive map gathered steam, David Housman at MIT was gearing up for what David Botstein had declared to be mission impossible at this stage of the game: to discover the location of the Huntington disease gene. He placed this tall order in the hands of Jim Gusella, who had just completed his Ph.D. in Housman's lab. Now the mapping work would surge ahead on another front.

Botstein's initial pessimism stemmed from the lack of markers: RFLPs looked good on paper, but the work of actually collecting them had only just begun. Indeed, it would take years of effort on the part of White, Donis-Keller, and others for the number of known markers to creep up into the hundreds. Starting out in the dawn of the RFLP era, Gusella had his work cut out for him. By 1982, he had a total of only twelve markers, five he had found himself and seven supplied by others. Wexler meanwhile was back at Lake Maracaibo, trying to fine-tune her genealogy: working out who was married to whom, what children they had, who was whose cousin. Local custom was sometimes a hindrance: some names were quite common, and many individuals were known by more than one. The tree Wexler managed to construct for one family nevertheless wound up with seventeen thousand names on it! Periodically she and her colleagues would set aside a whole day just to collect blood; samples had to be dispatched to Boston all together lest the tropical heat of Lake Maracaibo accelerate the degradation of the DNA.

As for Gusella, he wasn't waiting for the Lake Maracaibo samples. I remember a meeting at Cold Spring Harbor in October 1982 at which he presented his earliest data. With a small Huntington-afflicted family from Iowa as his sample, he had tested just five of his twelve RFLPs, checking each to see whether it correlated with the disease. None did, and I couldn't help thinking that having set out to find a needle in a haystack, he was making rather much of having lifted out a few straws. Only with careful analysis of the whole haystack—the vast genome in its entirety—or, alternatively, with a lot of luck could anyone hope to find what Gusella was looking for. And so when he closed his talk by saying that the "localization of the HD gene is now just a matter of time," I said to myself, "Yes, a *very long* time."

But fortune favors the brave. Gusella returned to his laboratory and tried more RFLP markers. To his astonishment, the twelfth, called G8, seemed to show linkage with Huntington disease in the Iowa family. But the statistical correlation wasn't very strong. And so he eagerly awaited samples from Lake Maracaibo, testing them for G8 as soon as he received them. Now excitement was irrepressible: G8 indeed tracked with Huntington disease. By the summer of 1983, against all the odds, Gusella had discovered a linkage after trying only twelve RFLPs. But this was no ordinary stroke of luck: for the first time, the gene for a human disorder had been located on a chromosome without the helping hand of sex linkage and without any prior knowledge of the illness's biochemical basis. Suddenly a new scientific vista was opening up: it seemed we would finally be able to analyze rigorously all those genetic defects that have plagued our species for as long as it has existed. RFLPs had proved they were indeed an effective tool. And having traced the Huntington disease gene to a manageable portion of the human genome, it was surely just a matter of time before our powerful gene cloning techniques would lead to the isolation of the gene itself.

Huntington disease strikes its terrible blow in adulthood. But genetic disorders that strike in childhood have an added awfulness, afflicting those who have hardly had a chance to live. Following a diagnosis, it is often possible to predict with grim certainty the course of the child's life. Such is the case with Duchenne muscular dystrophy (DMD), a progressive muscle-wasting disease. DMD is a sex-linked disorder: the mutation responsible occurs in a gene carried on the X chromosome. Women may carry the mutation on one of their two X chromosomes, but they are usually protected by the presence of a normal version of the gene on their other X chromosome. It's highly unlikely a female will receive two defective copies since males carrying the mutation almost never survive to have children. If, however, the chromosome with the mutated gene is passed to a son, the boy will develop DMD because he has no other X chromosome to supply a normal copy of the gene. When he is about five years old, his parents will notice he has difficulty getting up from the floor or climbing stairs. By about ten he will need a wheelchair. He will probably die in his late teens or early twenties. DMD is not rare: it affects 1 in every 5,000 male children.

The hunt for genes involved in human disorders is a story dominated less by great research institutions and plucky entrepreneurs than by groups like the Hereditary Disease Foundation, organizations founded by those with firsthand experience of the devastation a particular genetic illness can bring. Led by people with something very precious at stake, these groups are by nature more willing to back risky or novel research, going where universities or biotech companies may fear to tread.

The Muscular Dystrophy Association of America and its counterparts in Europe had long supported laboratory research directed at understanding the basic biology of Duchenne muscular dystrophy. In the late seventies cytogeneticists (who study chromosomes microscopically) provided the first genetic clue. Among the very small number of girls who do develop DMD an abnormality was found on the short arm of one of their X chromosomes, at a location called Xp21. Could this be the location of the DMD gene?

Not long thereafter, Bob Williamson at St. Mary's Hospital Medical School in London initiated RFLP-based searches for both the gene causing cystic fibrosis and the one involved in DMD. His colleague Kay Davies hunted up RFLPs on the X chromosome and tested them for linkage to Duchenne. She was successful, and the clincher was their location: they were in the Xp21 region, just as would have been expected given those strange X chromosomes in the women with DMD.

While the gene hunters pushed ahead trying to isolate the genes involved in Huntington disease and DMD, a revolution of a quieter kind was taking place in the offices of clinical geneticists. From the first, Nancy Wexler and David Housman realized that RFLPs linked to a disease gene could be used not only to localize the gene itself but also as a diagnostic test to determine which members of a particular family were carrying the mutation. They could be used to test even an unborn child. Consider the case of a hypothetical family with DMD. At least one boy will be diagnosed—the "index case" that first reveals the presence of a DMD mutation in the family. His mother, the carrier of a mutated gene, also has one normal copy. Her sisters may also be carriers, so any sons they may have are at risk. Now suppose the mother becomes pregnant once again with a male fetus; the chances are 50-50 that the second son will be affected. But with RFLPs her physician can tell her what fate awaits that fetus if carried to term.

First, the affected son's X chromosome is analyzed to identify the particular RFLPs linked to the DMD gene in this family. Next, DNA is taken from the fetus, either a sample of the placenta or of the amniotic fluid, which contains fetal cells. If the fetus's RFLPs match those of the affected boy, then we can be pretty certain that the unborn fetus will also be affected. Why only pretty certain? As we saw in chapter 1, when egg cells are produced, the chromosome pairs undergo recombination, exchanging DNA: the two copies of chromosome 1 trade with each other, as do the two copies of chromosome 2, the two copies of the X chromosome, and so on. If this swap should occur at a point on the X chromosome between the RFLP markers and the DMD gene, the RFLPs we have found to be associated with the normal version could possibly wind up associated with the mutated (DMD) copy. Experience taught us that with the first RFLPs for DMD this happens about 5 percent of the time, and so RFLP-based diagnosis has only a 95 percent chance of being accurate. This degree of imprecision is an unavoidable consequence of recombination. So while such diagnosis represented a tremendous advance, absolute certainty depended on identifying the gene itself, not merely the markers associated with it.

The key to isolating the DMD gene was a young boy named Bruce Bryer, whose X chromosome was missing a very large piece from the short arm. The piece was so large that Bruce suffered from three other genetic disorders in addition to DMD. In 1985 Lou Kunkel at Harvard Medical School reasoned he could use Bruce's DNA to "fish out" a normal gene from the DNA of an unaffected boy. Bruce's case was special because the disease was caused not by a defective copy of the gene but by its complete absence. Kunkel realized that all of Bruce's DNA should be present in a normal boy's, but whatever sequences the latter had and Bruce lacked would hold the key. Using recombinant meth-

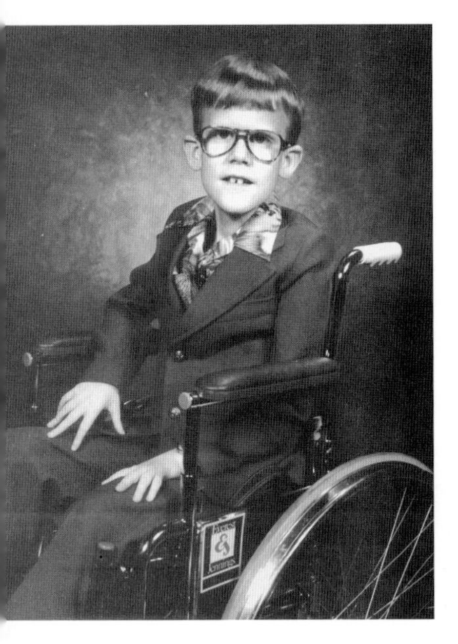

Bruce Bryer, whose X chromosome deletion led to the identification of the DMD gene. He managed to lead a remarkably normal life and was an accomplished organist by the time he died after a car accident at the age of seventeen.

ods, Kunkel subtracted Bryer's DNA from the normal DNA and kept the difference—the DNA that should contain the DMD gene. The subtraction didn't work perfectly, but it did work well enough that he could find the DNA pieces he wanted by using genetic markers associated with the Xp21 region.

Tony Monaco, a graduate student with Kunkel, took on the job of determining which, if any, of these Xp21 pieces of DNA might constitute part of the DMD gene itself. The only way to do this was to test each piece against DNA from several unrelated patients with DMD. Monaco hit the jackpot with the eighth try: a sequence called pERT87 was found to be absent in five of his DMD boys. This meant almost certainly that pERT87 was very close to the gene and perhaps even a part of it. Monaco began to isolate other sequences close to pERT87, and these too proved to be missing in the DNA of DMD patients. By 1987, Kunkel's group had isolated the complete gene. Now it could be given a proper name: dystrophin. Even with the genome sequence completed, it still holds the record for largest gene in the human genome, owing mainly to its many large introns.

Immediately the new knowledge was applied to produce foolproof prenatal diagnosis for DMD. And soon scientists discovered that a range of different mutations could impair dystrophin and cause the disease. But it remained unclear what the gene actually did. Would its function give us clues to developing effective therapies for DMD?

The first step was to locate the protein produced by the gene in muscle cells. Eric Hoffman in Lou Kunkel's laboratory found that the dystrophin protein was typically located in muscle cells just below the membrane that encloses the muscle fiber. Further studies have revealed dystrophin's critical role in connecting proteins that make up the muscle cell's interior architecture to a set of molecules that span the cell membrane and interact with other proteins outside the cell. The linking of the interior molecules to those in the membrane somehow secures the cell membrane when muscles contract and relax. Without dystrophin, the membrane suffers damage and the muscle dies cell by cell. Given our new and detailed knowledge of dystrophin and its function, it may seem remarkable that there is still no cure for DMD. This is the central frustration inherent in the current state of the art: genetics has made it possible to identify and understand disease, without yet permitting us in most cases to right the genetic wrong.

Kunkel's approach typifies the modern mapping-based approach to dissecting a disorder. Though now common practice, when Kunkel applied it the method was far enough beyond the bounds of research orthodoxy that the Muscular Dystrophy Association was taking something of a gamble in supporting four years of his efforts—a gamble that paid off handsomely. In the old days you tried to use biochemical analyses of a disease's symptoms to identify the disease gene; these days, following Kunkel, you map the gene, and then interpret the symptoms in the light of the gene's function.

One of the strongest arguments in favor of the mapping approach was that the work produced was useful even before the gene had finally been identified. The hunt for the Huntington disease and Duchenne muscular dystrophy genes yielded genetic markers that could be applied in diagnosis before the genes themselves were found. So it was too with one of the most prevalent genetic disorders, cystic fibrosis (CF). But the hunt for the CF gene would prove particularly notable for two reasons: it marked the first time that a company became involved in mapping a human disease gene, and the first instance of brutal competition among the scientists involved in such an endeavor.

In cystic fibrosis patients, a thick mucus accumulates in the lungs, making it difficult to breathe. The cells lining the tubes of the lungs can't clear out the mucus in which bacteria thrive, producing pulmonary infections. Before antibiotics those afflicted had a life expectancy of just ten years; today survival rates are substantially better. CF is also one of the most common genetic disorders, with about 1 in 2,500 individuals of northern European descent affected. It follows a recessive pattern of inheritance: you need two mutant versions of the gene to be affected. But since as many as 1 in 25 people of northern European descent carry a single mutant version (though are themselves protected because they have one normal copy), there is a relatively high risk that two carriers will get together and both pass it on to their children. It therefore became a medical priority to devise a diagnostic test as soon as this became a realistic goal.

Born in Shanghai, and raised and educated in Hong Kong, Lap-Chee Tsui came to the United States as a graduate student in 1974. Tsui learned his molecular genetics doing research on viruses before moving in 1981 to Manuel Buchwald's laboratory in Toronto to work on cystic fibrosis. Tsui is a quiet, pleasant man who is nevertheless intense and passionate about his goals. Planning to track down the gene via RFLP linkage analysis, he spent the first couple of years finding CF families before starting the painstaking process of testing their DNA with every RFLP he could lay his hands on. But the luck that had smiled upon Jim Gusella in his pursuit of the Huntington disease gene did not favor Tsui: after about a year all he had managed to do was eliminate a lot of RFLPs. He needed more, and was thrilled when Collaborative Research offered to share its RFLP markers with him.

Tsui's Toronto group was not alone in pursuing the CF gene: Bob Williamson in London, who had worked on DMD, also took up the hunt, as did Ray White, now in Utah, attracted by access to the very extensive pedigrees assembled by the Mormon church. These records, the Ancestral File, permit present-day members of the church to make provision for deceased forebears, who lived their lives outside the fold or died before the church was founded in 1830. The aim is to unite families for eternity. Seldom have the needs of religion and genetics been so happily aligned.

But it was the Toronto group that would notch the first success, when it found in 1985 a linkage between one of Collaborative Research's RFLPs and the CF gene. At the time, the location of that RFLP was unknown, but, seeing its potential as a golden egg, Collaborative Research quickly set about locating it. They soon determined that it was on chromosome 7 but did not immediately inform Tsui, their collaborator. Nor did they mention the chromosomal location when they announced the discovery in the November 22 issue of the prestigious journal *Science.* Clearly they were trying to preserve their monopoly on the new information, but secrecy and science often don't mix well: word soon spread on the grapevine that 7 was the place to be.

As Collaborative kept quiet, Williamson and White were just days away from the same discovery. Their own two papers, published in *Nature, Science*'s British rival, both mentioned that the key RFLPs were on 7. Tsui was incensed: he was about to lose his claim to the linkage discovery thanks to his partners' shenanigans—in science there are no prizes for coming in second—but Helen Donis-Keller persuaded *Nature* to accept a paper from the Toronto-Collaborative team announcing the location. So it was that three papers appeared in the November 28 issue of *Nature,* along with an editorial explaining how it had all come about.

Lap-Chee Tsui, gene tracker

The Toronto-Collaborative partnership did not survive the clash of academic and commercial cultures. Collaborative Research would find that the academic world had become wary of collaborating with them, a situation hardly helped by the crass and not very sturdy claim made by Orrie Friedman, Collaborative's CEO, that "we own chromosome 7." Fortunately, this soap opera saw its final episode in December 1985, when all the research groups agreed to pool their resources in order to test 211 families for linkage to chromosome 7 RFLPs. The results were spectacular. The RFLPs were very close to the gene, within 1 million base pairs—which made them useful in diagnosis, one of the major goals of the CF research.

The next step promised to be even more difficult. Learning that New York is halfway between Washington, D.C., and Boston is better than merely knowing it is somewhere in the United States. But when one must set out on foot from Washington to Boston, looking yard by yard for a sign that reads "Welcome to New York," the clue may seem not so helpful after all. One million base pairs might be close by the standards of linkage analysis, but it is a very long way by the standards of gene cloners who analyze regions one base pair at a time. To go the distance from the two RFLPs nearest the CF gene, Tsui teamed up with Francis Collins, who was then at the University of Michigan and would later succeed me as director of the Human Genome Project.

Collins had developed "jumping" techniques to facilitate the cloning of a gene between a pair of known RFLPs, but he was under no more illusion than

Tsui about the magnitude of the problems facing them. After two years of work, they managed to localize the CF gene to a 280,000-base-pair segment of DNA, within which they found the sequence of a gene known to play an important role in human sweat glands, which are dysfunctional in cystic fibrosis patients. It seemed the complete CF gene might finally have been corralled.

The only way to be sure they had got it right was to sequence the cDNA and search for the disease-causing mutations. Given a region 6,500 base pairs long, this was quite a challenge in 1989, and it had to be done twice: once using DNA from a CF patient and once with DNA from a healthy individual. The result, however, was clear-cut: the patient's DNA was missing a stretch of three base pairs, resulting in the absence of just one amino acid in the protein. This one mutation accounts for about 70 percent of CF cases, but over a thousand others found in the CF gene also cause the disease. This multiplicity of harmful variants has greatly complicated the task of DNA-based diagnosis.

Let us now return to Nancy Wexler, David Housman, Jim Gusella, and their colleagues, whom we left back in 1983 at the triumphant moment when a particular RFLP, G8, had been linked to the gene for Huntington disease. If it seemed up until then that they had enjoyed more than their collective share of good luck in locating the HD gene with astonishing speed, the gods were soon to redress the imbalance. Finding the gene had taken a mere three years; isolating it for detailed analysis would take ten years and an international team of 150 scientists. In this case, the region where the gene had been localized was 4 million base pairs long. The Huntington disease geneticists worked hard to narrow that window, but genetic mapping gets more difficult as the genetic distance gets smaller, and finally these efforts were rewarded only with ambiguous data. Imagine the foot journey from Washington to Boston, in search of New York. Now imagine arriving at an intersection in Philadelphia to find a signpost indicating New York in both directions.

Giving up on the contradictory linkage analysis, the Huntington gene hunters devised an alternative strategy, focusing on the region that was most similar among Huntington disease patients. This approach eventually reduced the region to only 500,000 base pairs, and the time had come to turn to gene-cloning techniques. The first results were disappointing: they found three genes in the right-hand half of the region, but none showed any abnormalities in patients with Huntington. Undaunted, they explored the left-hand side and found a single gene, with the prosaic name IT15. Finally, after ten years and many losing lottery tickets, luck had begun to smile on them once more. The gene contained a short sequence, CAG, that repeated over and over again, like the short tandem repeats (STRs) used in DNA fingerprinting. It turned out that unaffected people have fewer than thirty-five CAGs, while people with more than forty will develop Huntington as adults; in the rare instance of more than sixty, a severe form of Huntington develops before the age of twenty. CAG is the genetic code for the amino acid glutamine, so each of the CAG repeats adds an extra glutamine to the protein. In the case of Huntington sufferers, the protein coded by the HD gene—the rather difficult-to-say huntingtin—contains extra glutamines. This difference likely affects the behavior of the protein in brain cells, probably by causing molecules to stick together in gluey lumps within the cell, somehow causing its death.

It had been a tremendous effort by all the laboratories in the Hereditary Disease Foundation's team, and in recognition that it was truly a collaboration, the only name appearing as the author of the article was that of the Huntington Disease Collaborative Research Group. The same strange type of mutation—repeats of the three-base-pair sequence—had already been implicated in three other disorders, remarkably all of them also neurological diseases. We now know of fourteen of these "trinucleotide repeat disorders," but we still are no closer to understanding why brain cells are so susceptible to this kind of mutation.

It may be depressing to know that despite the substantial time it has taken to hunt down their respective genes, these disorders—Huntington, Duchenne, and cystic fibrosis—are, by the standards of geneticists, "simple." They are caused by mutations in a single gene and not much affected by environment. If you have the three-base-pair deletion in both your cystic fibrosis genes or more than forty CAG repeats in one of your Huntington disease genes, you will develop those disorders no matter where you live or what you eat or drink.

There is a large number of single-gene disorders—the current genetic disease database lists several thousand—but the majority are extremely rare, each occurring in just a few families.

Much more common are "complex" or "polygenic" disorders, which include many of our most common ills: asthma, schizophrenia, depression, congenital heart disease, hypertension, diabetes, and cancer. These are caused by the interaction of several—perhaps many—genes, each of which alone has only a small effect and perhaps no detectable effect at all. And typically in polygenic disorders, there is a further complication: these sets of interacting genes may create a predisposition to a particular disease, but whether you actually develop a case of it depends on environmental factors. Suppose that you have a set of gene variants that predisposes you to alcoholism. Whether or not you actually become an alcoholic depends on your exposure to the environmental trigger, alcohol. Your fate may be quite different growing up in a dry county in Texas as compared with Manhattan. The same principle holds for asthma; in a "good" summer, when the pollen and spore counts are low, you may develop no symptoms despite being genetically disposed to the disease.

The complex interplay of genes and environment is nowhere more evident than in cancer. Cancer is fundamentally a genetic disorder caused by mutations in several genes. Each mutation alters one more element in the cell's behavior until it acquires all the characteristics of a fully malignant cell. Cancer mutations arise in two ways. Some are inherited. We have all heard the phrase "it runs in the family," and while some traits described this way—Catholicism for one—are not necessarily heritable, some kinds of cancer are. Still, the disease is so lamentably common that it is not so unusual to have two or even three cases in one family even without a hereditary component. (Geneticists studying "cancer families" therefore apply very strict criteria in deciding whether a cancer is inherited.) Plenty of cancer mutations also arise in the normal course of living. DNA can become damaged owing to errors the enzymes make in the course of duplicating or repairing the genetic molecule, or as a consequence of the side effects of the normal chemical reactions within the cell. And many cancers arise thanks to our own foolishness. Ultraviolet rays in sunlight are potent mutagenic agents to which sun worshippers willingly expose themselves, and cigarettes are a very efficient way to deliver carcinogens straight into your lungs, where they cause lung cancer. Other environmental factors, for instance asbestos in the workplace, have also been shown to promote cancer. The point is that DNA can get damaged quite naturally, but it is up to us to minimize the damage through informed social and personal choices.

In 1974, Mary-Claire King (of human/chimpanzee and Las Abuelas fame) moved to UC San Francisco to work in a laboratory studying breast cancer, where she decided to commit herself to the hunt for a breast cancer gene. At the time, the RFLP linkage approach was still six years away, but King knew that there would be clues in pedigrees, so she set about collecting families. She looked for families in which members had developed breast cancer at an early age, and in which there was also ovarian cancer, reasoning the odds favored a hereditary culprit in such cases. The only genetic markers available to her were protein markers, and after a few years she published her first breast cancer paper, describing unsuccessful tests for linkage with cell surface proteins. This was followed by other papers showing similarly negative results. Naysayers were equally negative: breast cancer is too heavily affected by the environment to permit genetic analysis, they said, referring predictably to needles and haystacks. Undeterred, King continued to refine her data-set, and by 1988, with an analysis of 1,579 families, she thought she had good evidence for a breast cancer gene in these high-risk families.

The medical world was astonished when in 1990 she reported that she had found an RFLP on chromosome 17 linked to breast cancer in a subset of 23 of her families, involving a total of 146 cases of breast cancer over three generations. She checked factors that might have confounded the analysis—perhaps these women had been exposed to more X rays, or they differed from others in their pregnancy histories—but her data held up. There was a gene at chromosome location 17q21 that when mutated greatly increased a woman's risk. King's paper set off a race to isolate the gene itself, called BRCA1 (for Breast Cancer 1), and an ongoing controversy about the commercial exploitation of genes.

Isolating the BRCA1 gene would inevitably be a big event. Even if it was important only in a small subset of high-risk families (i.e., it would only be responsible for a small proportion of all breast cancers), the insights that might come from knowing what the gene did would be cause enough for excitement. King teamed up with Francis Collins, whose gene-hunting credentials were impeccable, but the pair had tough competition. Mark Skolnick, the Utah population geneticist involved in the RFLP linkage breakthrough, formed a company, Myriad Genetics, with Wally Gilbert, whose entrepreneurial spirit had survived his uneasy tenure at the helm of Biogen. Myriad's business plan was to use the power of the Mormon family pedigrees to map and clone genes, and BRCA1 came within their crosshairs very soon. In 1994, a consortium of geneticists from Myriad, the University of Utah, NIH, McGill University, and Eli Lilly beat the rest of the world, announcing what they rather coyly called a "strong candidate" for the BRCA1 gene. They had found it. Everyone involved filed for a patent (although Myriad initially saw to the exclusion of the NIH scientists). In 1997, Myriad's application was approved.

At the moment BRCA1 was being cloned, a different consortium of geneticists, including scientists from Myriad and the Institute for Cancer Research in England, reported they had located a second breast cancer gene, BRCA2, on human chromosome 13. Once again a race began, and within a year the English group claimed success in isolating BRCA2. They knew they had bagged their quarry once they had determined about two-thirds of the gene's DNA sequence and shown it to be defective in six different families. Not to be outdone, Myriad formed yet another consortium, this one comprising institutes in Canada and France; soon they would publish the complete sequence of BRCA2, a very large gene. Of course both Myriad and the Institute for Cancer Research filed patent claims.

It was clear that these were going to be commercially important genes. Mutations in them have very serious consequences for women. The risk of a woman developing breast cancer by age seventy because of a mutated copy of either BRCA1 or BRCA2 can be as high as 80 percent. And it has been established that the same mutations also raise the risk of ovarian cancer to as high as 45 percent. Women in whose families these mutations run need to be informed as early as possible whether they are carrying a defective variant of either gene. There are difficult but potentially life-saving choices to be made: an elective bilateral mastectomy in high-risk women reduces cancer incidence by 90 percent. At the same time, genetic screening can identify individuals in these families who have normal genes; this affords them the comfort of knowing they are not at increased risk.

It sounds like a worthy thing to have brought to market: a genetic test for a very serious disease, a means to help women make informed decisions about their health. Why, then, is Myriad frequently portrayed as exemplifying all that is wrong when commerce is married to science? Myriad now has nine U.S. patents covering BRCA1 and BRCA2, and in 2001 it was granted one in the European Union, one in New Zealand, four in Canada, and two in Australia. In effect, the company now enjoys a global monopoly on these genes and worldwide control over how they are used. It is entirely reasonable that Myriad should make money from testing for BRCA1 and BRCA2 mutations—the company provides a valuable service and has invested a great deal of money to develop the test. But how much money should the company reasonably be making? Today each test costs more than $2,700. At the same time, Myriad restricts academic researchers from using the BRCA gene sequences to develop alternative tests. And information about BRCA mutations gleaned from DNA sequencing among patients enrolled in academic research projects is withheld even from the patients themselves; to do otherwise would be a diagnostic clinical use, infringing the Myriad patents.

Myriad has lately made some concessions. A government deal now permits scientists doing NIH-funded research to use the test at a cut rate of $1,400. But critics have viewed this as a token gesture and remain particularly vocal in Canada and Europe. The European Parliament has passed a resolution expressing "dismay" at the actions of the European Patent Office and instructing the Parliament's staff to prepare challenges to Myriad's patents on both genes. Myriad's French partners in the BRCA2 sequencing—the Institut Curie and the Institut Gustave-Roussy—were particularly incensed about Myriad's BRCA2 patent and have filed a joint complaint to the European Patent Office. In any case, Myriad's monopoly may not be good for patients. The company's test fails to detect all possible cancer-causing changes affecting the gene, so people who test negative for the screened mutations may nevertheless still be at risk. Now when you're tested Myriad has you sign a waiver to the effect that a negative result does not necessarily indicate a clean bill of genetic health. Developing a more comprehensive test is difficult for technical reasons, but more breast cancer labs around the world would likely be trying it right now if not for Myriad's research-stifling patent.

Over the past dozen or so years, linkage analysis has also zeroed in on several other important cancer genes, including ones involved in neurofibromatosis ("Elephant Man disease," not commonly understood to be a form of cancer), colorectal cancer, and prostate cancer. But while effective, the gene-by-gene approach is slow and painstaking, with each study being dependent upon finding appropriate families for analysis. It is here that the Human Genome Project will prove its tremendous value. The DNA and protein microarrays we looked at in chapter 8 will furnish cancer-gene hunters with a powerful high-caliber weapon. When I first became interested in cancer research in the 1960s, we knew so little about the underlying genetics and relied on such primitive tools that I turned to viruses that cause cancer in animals. It was my hope that by studying such viruses—which, having very few genes, were manageable even then—we might glean some insight into human cancer. Nowadays, cancer research is no longer confined to viruses; the tens of thousands of genes in actual human tumors are within our powers to map and clone. An enormous wealth of knowledge awaits us as we discover, in greater and greater detail, all the minute biochemical deviations that contribute to turning a normal cell into a cancerous one.

Most linkage analysis studies depend on tracking one's genetic quarry through as large a pedigree as possible. But there is another strategy that looks at small populations with a high incidence of a disorder. And you can't get much smaller than the population of Tristan da Cunha.

A volcanic island rising steeply and inhospitably out of the sea, Tristan da Cunha is a speck of land—just forty square miles—in the middle of the South Atlantic, among the most remote places on the planet. The first permanent settlement was a British garrison established there in 1816 to prevent the French from using the island as a base from which to spring Napoleon from his exile on St. Helena, an island 1,200 miles to the north. Subsequent population growth was sporadic—a few settlers here, a few survivors of shipwrecks there—and as of the unofficial census of 1993, the total was only 301. That year a team from

Arguably the remotest inhabited spot on the planet: Tristan da Cunha viewed from an uninhabited island nearby

the University of Toronto went to the island to follow up on medical studies done on the islanders in 1961, when the entire population was evacuated to England because the island's dormant volcano had become temporarily active. The most surprising finding had been that about half of the evacuees had a history of asthma.

When the Toronto Genetics of Asthma Program examined 282 of the inhabitants in 1993, they found that 161 (57 percent) showed some symptoms of asthma. The Canadians prepared a genealogy of all the local families; it wasn't overly difficult since all the islanders are descendants of fifteen early settlers and so their lineages are closely interrelated. Asthma was apparently introduced into the island by two women who settled there in 1827. A population like this is a boon for gene hunters: the island is essentially an extended family, and so the genes causing any observable disorder are likely to be the same throughout the population—a best-case scenario for linkage analysis. In a larger, more mixed population, some people's asthma may be caused by one set of genes while that of others is caused by another set. Such heterogeneity is what makes nailing the genetic determinants of complex diseases so difficult.

The Toronto team collected blood samples and prepared DNA, but needed major funding to complete the study. It was then that they heard from Sequana, a company founded to hunt down disease genes. Sequana financed the study, immediately provoking charges that the company was exploiting the islanders, who did not perhaps fully appreciate their role in Sequana's business strategy.

Canadian activists calling themselves the Rural Advancement Foundation International claimed that Sequana was "committing an act of biopiracy . . . violating the fundamental human rights of the people from whom the DNA samples are taken." Sequana claimed—in a move that was guaranteed to provoke more accusations of "biopiracy"— to have found two genes conferring susceptibility to asthma but refused to reveal them publicly until the European patent application was filed. The genes are located on chromosome 11, and subsequent studies of heterogeneous mainland populations have confirmed a role for chromosome 11 in asthma. It would thus seem that the genetic factors underlying the high incidence of asthma on Tristan da Cunha are not relevant solely to the isolated inhabitants of South Atlantic islands.

The storm over Sequana's "biopiracy" was nothing compared to the hurricane that was to envelop Kari Stefansson and his company, deCODE Genetics, a few years later. Recognizing that it was tedious and inefficient to look for a different Tristan da Cunha–like micropopulation for every disorder, Stefansson reasoned that what he needed was an isolated island but one with a much larger population, among whose members one could look for a number of disease genes at once. It just so happened that Kari Stefansson was born on such an island.

Iceland is about the size of Kentucky but its population of only 272,512 is one-fifteenth that of the Bluegrass State. The island was settled in the ninth and tenth centuries by the Vikings, who brought with them women kidnapped from Ireland during the voyage. Iceland offers several advantages for the enterprising gene hunter. First, the population is very homogeneous, derived almost entirely from the original settlers; there has been very little immigration since Viking days. Second, there are detailed genealogical records going back many generations; many Icelanders can trace their ancestry back five hundred years. This handy resource is supplemented by a detailed register of births begun in 1840 at the University of Iceland. Third, Iceland has had a nationalized health care service since 1914, so the entire nation's medical records are uniformly ordered and readily accessible—at least in principle.

A Harvard neurologist, Stefansson was interested in genetically complex disorders such as multiple sclerosis and Alzheimer disease. Recognizing his own people as a nearly perfect population for genetic research, he devised a project to link the genealogical and medical records to create a database for gene-hunting. Despite the project's worthy purpose, local privacy statutes stood in the way until the Althingi (the Icelandic Parliament, founded in A.D. 930) passed the Law on a Health Sector Database in 1998. The legislation authorized "the creation and operation of a centralized database of non-personally identifiable health data with the aim of increasing knowledge in order to improve health and health services."

In 2000, deCODE was awarded a twelve-year license to build and run the Icelandic Healthcare Database at its own expense in exchange for an annual fee payable to the national government. The genealogical component of the database contains information in the public domain, but access to the medical records database is more restrictive, operating on a basis of "presumed consent"—information about people's health is entered into the database unless they take the initiative to opt out. The genotype component is most restrictive, relying on informed consent—individuals must actively agree to give tissue samples for DNA extraction. Here's the hot spot: although deCODE has a system in place to protect donor privacy, critics argue it is inadequate. Since a person's DNA must be correlated with genealogical and medical records, samples are not taken anonymously; instead the source's identity is encrypted. In theory such encryption could be broken. Word might get around, especially in such a small population, that one family or another is carrying "bad" genes, opening up the possibility of gene-based discrimination. The deCODE project crystallized in microcosm many of the issues concerning genetic privacy that had been discussed rather more hypothetically elsewhere. Nevertheless, despite the controversy, most Icelanders were in favor of the company, viewing it as a means of combining a noble mission—fighting genetic disease—with the happy prospect of serious money swelling the country's small economy. Caught up in the excitement, many Icelanders invested heavily in the company, snapping up shares for as much as sixty-five dollars long before deCODE was officially floated on NASDAQ. But the economic downturn has not been kind to biotech in general, or to deCODE in particular. At the time of writing, shares are worth around two dollars. Many Icelanders have been left to rue those heady *buy!-buy!-buy!* days. Still, deCODE has undeniably brought money into Iceland through its lucrative collaborations with both Hoffmann–La Roche and with Merck. But with the government of Iceland willing to provide a $200 million loan guarantee and the company forced to lay off a substantial proportion of its workforce, the financial reality is that deCODE may yet prove to be less of an economic boon to the country than had been hoped.

The real test of deCODE, however, lies not in the vagaries of the stock market, but in the science it produces. Here unfortunately the commercial imperative of non- or delayed disclosure makes evaluation difficult. The company, we learn via press releases, is now carrying out linkage analysis on forty-six disorders, including asthma, depression, cancer, osteoporosis, and hypertension. It has found linkage markers for twenty-three of these disorders, and it has isolated genes contributing to peripheral blood disease, stroke, and schizophrenia. But with few of the details published in scientific journals, it is too often difficult to distinguish science from hype. Still, deCODE has certainly shown itself capable of making useful contributions: in June 2002, deCODE researchers published a new map of the human genome with significantly higher resolution than that of the old CEPH map. In addition, the long-anticipated publication of the company's research on schizophrenia hints at a scientifically—and commercially—productive future for the company.

Whatever the years to come hold for deCODE, it's clear that its overall approach—a three-way marriage of medical, genealogical, and genetic records for a well-defined population—has great potential. The company is, therefore, not alone in subjecting the population of a whole country to genetic scrutiny. Finland, for example, has a population of 6 million and notable incidence of some thirty-five genetic disorders, some unique, others simply more common there than in other European countries. The Finns have duly attracted considerable interest among human geneticists. Other countries, too, are leaping on the genetic database bandwagon. In April 2002, Britain launched its "Biobank" program, and the government of Estonia is promoting a comparable nationwide effort as well. Large-scale population studies like these will ultimately help us track down even the most elusive of genes.

Human genetics has a long history, starting with our early ancestors' curiosity about how certain characteristics were passed down through generations. But for virtually all of that history, the scientific foundation of the inquiry was weak at best. Charles Davenport's attempts to bolster his eugenics program by searching for the genetic basis of what he called "feeblemindedness" scarcely rate as science. As a measure of just how slow the field was to develop, it's worth noting that for a long time the accepted figure for a fundamental genetic parameter defining our species was wrong. It was not until 1956, three years after the discovery of the double helix, that the number of human chromosomes was determined correctly to be forty-six, and not forty-eight, as had been supposed without question since 1935. But the flood of knowledge let loose in the twenty years since RFLP linkage studies began has, with fantastic speed, created a fertile field from formerly barren ground. Having completed the sequencing of the human genome, we will likely soon find the genes underlying virtually all important genetic diseases. The question then becomes: What do we do with them?

David Vetter, whose inherited immune system disorder made him susceptible to the slightest infection, was raised in a sterile world and became the original "bubble boy."

CHAPTER TWELVE

DEFYING DISEASE: TREATING AND PREVENTING GENETIC DISORDERS

From the moment he was born, David Vetter never felt the direct touch of another human being. David suffered from an inherited condition called severe combined immunodeficiency disorder (SCID). The failure of his body to develop B and T cells, both crucial elements in the immune response to disease, left him susceptible to the slightest infection.

David's parents knew before he was born that he might have SCID: their firstborn son had died of it. The Vetters and the doctors were ready. They decided early on that if the infant should prove to have SCID, he would be isolated in a germ-free environment until a treatment could be developed—surely it would not be long given the rate of progress in medicine. David was delivered by cesarean section in September 1971 and immediately placed in a sterile incubator. All contact with him was made using latex gloves built into the little chamber. As he grew older he was transferred into larger and larger sterile environments, plastic "bubbles," but one constant remained: the gloves. They would continue to be his only way of feeling anyone or anything in the outside world.

The hoped-for cure proved elusive. David remained in his bubble, where he drew national attention. NASA tried to help him with a Mobile Biologistical Isolation System, essentially a space suit permitting the boy the freedom to venture beyond the bubble. But a space suit is really only a bubble of a different kind.

Advances in transplantation methods looked promising, and in October 1983, a month after his twelfth birthday, David received a bone marrow transplant from his older sister. Unfortunately, her marrow proved to contain a virus that caused a pernicious lymphoma to develop in David's defenseless system. By February 1984, he had to forsake his bubble and be placed in intensive care. He died soon after, but at least in those final days he was able to experience at last the warmth of human touch.

We can be thankful that SCID is rare, but genetic disorders are surprisingly common among children. In fact, about 2 percent of all babies are born with some kind of serious genetic abnormality. It's estimated that genes are directly responsible for one-tenth of admissions to children's hospitals, and indirectly implicated in about half. David Vetter's case is sadly representative of where our knowledge stands regarding most genetic diseases: we can understand what's wrong and we can diagnose them, but there is relatively little we can do to treat, much less cure, them.

It is interesting to chart the image of SCID in popular culture. In the seventies the condition inspired a made-for-television tearjerker called *The Boy in the Plastic Bubble.* By the nineties, the Bubble Boy had become a figure of fun in the sitcom *Seinfeld.* And in 2001, Disney released a tasteless film reimagining as a series of goofy adventures the life of a boy confined in a bubble on account of an unnamed but unmistakable condition.* This enduring powerlessness of science in the face of such a horrific illness must to some degree account for this trajectory from sentimentalism to farce. But the same powerlessness makes the diseases only harder to bear for the afflicted and their families. Especially with diseases that cause a progressive and inexorable decline, a diagnosis is virtually a death sentence. In the absence of treatment, some would prefer not to know their awful fate, particularly if they have witnessed its ravages in loved ones. In the previous chapter we met Nancy Wexler; with a 50 percent chance of developing Huntington disease, the scourge that had claimed her mother and uncles, Wexler worked so long and so hard at Lake Maracaibo and in genetics laboratories in the United States to track down the genetic culprit. But even though her extraordinary crusade resulted in isolating the gene and identifying the lethal mutations, a cure is still nowhere in sight. And although she has done so much to make a diagnostic genetic test available, Wexler herself has said that she will not be tested—at least, not until a viable treatment seems near. She

*The Disney ending: The movie's bubble inhabitant turns out to be healthy after all.

would prefer to live with a huge uncertainty than discover the truth in a 50-50 gamble: the odds are even that she will face a mental and physical decline leaving her a shell of the dynamic woman she is now.

Sometimes it is almost more unbearable to care for a sufferer than to become one. Carol Carr of Hampton, Georgia, watched her husband, Hoyt, develop Huntington disease in his thirties. His sister Roslyn died of it, and his brother George committed suicide soon after being diagnosed. Carol quit her job and became Hoyt's full-time nurse for the next twenty years, as he fell apart. They already had three sons by the time Hoyt was diagnosed, and when he died in 1995, Carol was already nursing her two eldest, Randy and Andy, as she had her husband—feeding and bathing them, giving them their medicine, helping them to the bathroom. Soon James, her youngest, was also developing symptoms. In despair, Carol reluctantly placed Randy and Andy in a nursing home, where, on June 8, 2002, she shot them both dead. The *New York Times* reported James's opinion that Huntington had killed his brothers long before his heartbroken mother ever pulled the trigger.

Not all genetic diseases are tragedies of medical helplessness. Perhaps the best example to the contrary is the disorder responsible for that strange fine-print warning appearing on some food products, especially soft drinks: "Contains Phenylalanine." Phenylalanine is an amino acid—a common component of proteins—that cannot be processed by people with a genetic disorder called phenylketonuria (PKU).

The story starts in Norway in 1934. A young mother was determined to find out what was wrong with her two children, ages four and seven, both of whom had seemed perfectly normal at birth. The elder was not fully toilet trained and was barely capable of speaking a few words, let alone forming a complete sentence. The case came to the attention of Asbjørn Følling, a biochemist and physician. After conducting a battery of tests, Følling found a biochemical abnormality he linked to their condition: they had too much phenylalanine in their urine. But he also learned that theirs was no isolated case: he discovered thirty-four others in twenty-two families across Norway, and realized that he had stumbled across a genetic disease.

We know now that PKU is caused by a mutation in the gene for phenylalanine hydroxylase, an enzyme that converts phenylalanine to another amino acid, tyrosine. It is a rare disorder, affecting about 1 in 10,000 people in North America, and shows a recessive pattern of inheritance: you must have two mutated copies of the gene, one from each parent, to develop PKU. In the children affected, who lack a functioning enzyme, phenylalanine accumulates in the blood, impairing brain development and leading to severe mental handicap. Prevention is simple: PKU children raised from birth on a diet low in phenylalanine—with minimal protein and no artificially sweetened soft drinks, the two principal sources—grow up normal. Nutrition alone can make the difference between normal brain development and profound disability. Clearly it is important to know a child's PKU status as soon after birth as possible. Robert Guthrie devised a simple diagnostic test for blood levels of phenylalanine and tirelessly promoted its use until it became standard neonatal practice. Since 1966, a heel-prick blood sample has been taken from every newborn and analyzed for phenylalanine levels. Thus without ever examining a single base pair of DNA, the Guthrie test screens for a genetic disease in millions of babies every year. Prior to this testing program, as much as 1 percent of mental retardation in the United States was attributable to PKU; now there are only a handful of cases a year.

The 1950s saw the development of cytogenetics, the study of chromosomes through the microscope. Employed diagnostically, this approach soon revealed that abnormalities in chromosome number—usually one too many or one too few—invariably cause profound dysfunction. These problems stem from an imbalance in the number of genes, a departure from the norm of two of each. Such conditions do not run in families like Duchenne muscular dystrophy (DMD) or cystic fibrosis (CF), but they are still very much *genetic;* they arise spontaneously through accidents in the cell divisions leading to the generation of sperm and egg cells.

The best known is Down syndrome, named for John Langdon Down, who in 1866, as the medical superintendent of a home for the retarded, was first to describe its characteristic clinical features. He noted that 10 percent of the res-

The karyotype—the full set of chromosomes—from a male with Down syndrome. Note the extra copy of chromosome 21.

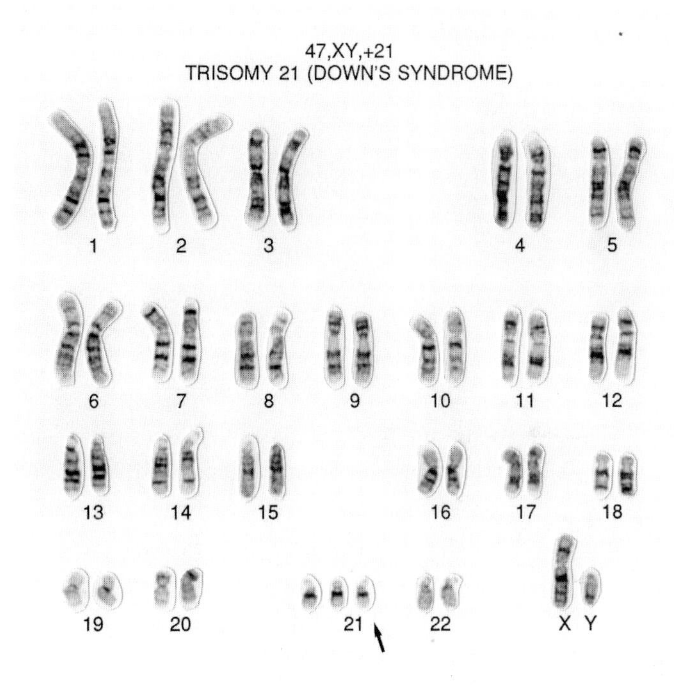

idents of his institution resembled one another: "So marked is this, that when placed side to side, it is difficult to believe that the specimens compared are not children of the same parents." But the first insight into the condition's biological basis did not come until ninety years later, when the French physician Jérôme Lejeune found that children with Down syndrome have three copies of one chromosome, later shown to be chromosome 21. The normal condition, two copies of a chromosome, is called "disomy"; so Down syndrome is known in genetic parlance as "trisomy 21."

The incidence of Down syndrome increases with the age of the mother. At age 20, a woman's chance of producing a Down baby is about 1 in 1,700; but at 35, it jumps to 1 in 400; and at 45 shoots to 1 in just 30. For this reason, many older pregnant women choose to have prenatal diagnosis performed on their fetus to determine whether it might possess 21 in triplicate. The test was first done in 1968, and today it is routinely offered to all pregnant women over 35.

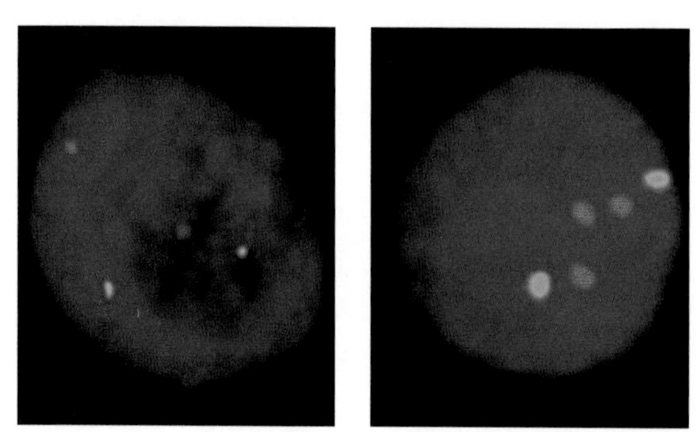

Fluorescent staining for chromosome number. A cell nucleus (dark blue) is probed for chromosome 10 (light blue) and chromosome 21 (pink). The image on the far left shows a normal karyotype with two copies of each chromosome; in the other, we see a Down karyotype, which has an extra copy of chromosome 21.

Because the developing fetus must be big enough to withstand safely the extraction of a tissue sample, such diagnosis cannot be performed in the very early stages of pregnancy. Typically it is done in the fifteenth to eighteenth week using amniocentesis, a procedure that entails drawing some amniotic fluid (which naturally contains cells from the fetus). An alternative test, which may be done as early as the tenth week, gathers cells from the chorionic villus, the part of the placenta that attaches to the uterine wall, but this method is less reliable. Because both procedures are mildly risky—amniocentesis results in a 1 percent rate of miscarriage, and chorionic villus sampling in a 2 percent rate—younger women are usually advised to avoid them: the probability that their fetus has a genetic defect is actually lower than the probability that it will be damaged by the procedure. At one time, extracted fetal cells had to be grown in petri dishes before being processed for chromosome analysis. Nowadays, a more rapid diagnosis can be done using fluorescence in situ hybridization (FISH); in this method, a small fluorescent molecule is attached to a stretch of DNA sequence specific to chromosome 21 and introduced to the sample, where it binds to the fetal chromosome 21 DNA. If two fluorescent patches appear in the nucleus of a cell, the fetus is normal; if three, the fetus has Down syndrome.

In Britain, 30 percent of Down pregnancies are detected by routinely testing the oldest 5 percent of women bearing children. This method boasts a clear efficiency in simple terms of detections per pound spent (Britain's National Health Service has been subject to such a calculus ever since Mrs. Thatcher's assault on health spending), but what of the remaining 70 percent of Down cases? Down is rarer in the babies of younger mothers, but these women account for the vast majority of all pregnancies. Since the standard tests are statistically not worth their attendant risks, there have been attempts to find alternative, noninvasive, indicators. It turns out that substances detectable in the mother's blood yield useful information. Low levels of alpha-fetoprotein and high levels of chorionic gonadotropin correlate to a significant degree with Down (though they are by no means ironclad indicators of trisomy). Modern practice, then, is to offer younger women the blood test, and, if it suggests the possibility of Down, they are then counseled to undergo amniocentesis or chorionic villus sampling for a definitive diagnosis.

Sadly, today a woman who learns that her fetus has Down syndrome has only two choices: to become the mother of a Down baby or to abort the fetus. It is a painful decision that is made no easier by the variable severity of the affliction. People with Down all share the characteristic facial features identified by Dr. Down—a broad flat face, a small nose, and narrow slanting eyelids*—but they range considerably in IQ, scoring between 20 and 85 (i.e., from severely handicapped to low normal). They are especially prone to a range of ailments, including heart disease (which claims about 15 percent in the first year of life), gastrointestinal anomalies, leukemia, and, with increasing age, cataracts and Alzheimer; but it's also perfectly possible that an individual will have relatively few health problems. With improved care, and better knowledge of medical hazards posed by possession of that extra chromosome, life expectancy has increased substantially: 50 percent of affected individuals today survive into their fifties. Despite typically acquiring over time

JD, six, with his father. JD has Down syndrome.

*Dr. Down originally called the disorder "mongolism" on the basis of these characteristics, entitling his 1866 paper "Observations of an Ethnic Classification of Idiots." Subscribing to the racist evolutionary views of his day, he believed that Down represented an evolutionary step backward from the exalted Caucasian state to the "inferior" Mongoloid one. To give him his due, though, he concluded that what he called "retrogression" undermined the claims of those who refused to accept that Caucasians and non-Caucasians were members of the same species.

what most would consider a depressing familiarity with the insides of hospitals, people with Down generally enjoy life, and have brightened many a family. The condition is perhaps tougher on their parents, who must adjust to caring for someone with special medical needs, as well as to the knowledge that their child will, in many ways, never really grow up.

In general, women who learn that they are carrying a Down fetus choose to terminate the pregnancy.* As a result, in countries with routine prenatal screening, the number of Down babies born is declining. Statistically, however, this claim is more complicated than it sounds: the trend toward deferring motherhood—often for professional reasons—has actually increased the ranks of women at risk for a Down pregnancy. In Britain, therefore, the efficacy of screening programs is measured relative to the *expected* number of Down babies given the ages of the women having children that year. We are seeing an ever-decreasing proportion of Down babies; in 1994, for instance, screening programs reduced the incidence of Down by about 40 percent.

Trisomies can also occur for other chromosomes, but these result in abnormalities so severe that the pregnancies abort spontaneously in all cases except trisomies of chromosomes 13 and 18. But children with trisomy 13 seldom live more than a few weeks, and those with trisomy 18 usually die before their first birthday. Chromosomal abnormalities, trisomies included, are probably very common. While many are lethal—a current estimate is that as many as 30 percent of conceptions end in spontaneous abortion, and in about half of these there is some form of chromosomal aberration—some have little or no effect. Alterations may be far less drastic than the loss or gain of an entire chromosome, involving the rearrangement of segments within a chromosome or the transfer of part of one chromosome to another. If there has been a net loss or gain of genetic material, then, as in the case of a whole extra chromosome, the resulting imbalance will usually prove deleterious. Unfortunately, standard cytological analysis of fetal chromosomes can detect only gross imbalances, and yet even minor ones can have disastrous effects.

*In the United Kingdom, 92 percent of fetuses diagnosed with Down are currently aborted. Typically only women who are willing to consider an abortion undergo prenatal testing (there's no point subjecting a fetus to the risks associated with testing if the mother intends to carry the pregnancy to term whatever the result), so we would expect this figure to be high.

After struggling to become pregnant for the first time at thirty-seven, Kathleen McAuliffe was relieved to learn that only two chromosome 21s had shown up in her amniocentesis. But what she had not realized was that the test could reveal other chromosomal abnormalities as well. The cytogeneticist had spotted an inversion in the fetus's chromosome 2: it was as though a segment had been popped out of the chromosome, flipped, and reinserted the other way around. The information was not accompanied by any useful advice: there was a chance that the inversion might create a problem—it might, for example, have resulted in a genetic imbalance—but then again it might have no effect. One way to find out more was to look at McAuliffe's own second chromosome, and that of her husband. If either parent had the inversion (i.e., if it was not a spontaneous alteration in their child), one could infer it would have little or no impact since both parents were normal. But neither McAuliffe nor her husband had an inverted chromosome 2, implying that it had arisen de novo in the sperm or egg. What would the inversion do to the baby? McAuliffe suddenly found herself confronting a life or death decision. After agonizing at length, she decided that the uncertainty was too great and she terminated the pregnancy. Despite a specific request that she not be informed of the autopsy results—she was sad and guilt-ridden over the loss of her fetus—by some administrative gaffe the report was sent to her home and she discovered that the fetus had indeed been profoundly abnormal. But this was cold comfort, and McAuliffe still keeps the ultrasound image tucked away in a drawer. Happily, subsequent pregnancies have met with no such complications, and McAuliffe is now blessed with two young children in, as she puts it, "ear-piercing good health."

Genetic knowledge creates ethical dilemmas. McAuliffe had never been warned that her amniocentesis might detect problems other than trisomy 21; perhaps the cytogeneticist overstepped the bounds of duty and should have reported only the results for the test that had been ordered. Certainly there would have been no choice had the clinician used the FISH method, which reveals *only* the number of 21s present. As it grows more sophisticated, genetic testing becomes a Pandora's box, its consequences going far beyond the original issues motivating the test, sometimes spreading to lives beyond those of the tested individuals. Nowhere is this more evident than in genetic testing conducted among families with histories of an inherited condition like DMD,

Huntington disease, or cystic fibrosis. In these cases, diagnosis is carried out not by a cytogeneticist but by a molecular biologist, who analyzes not chunks of chromosomes but specified stretches of DNA. DNA is extracted from a sample of tissue, which is obtained from a fetus by amniocentesis, or from a child or adult by taking blood or harvesting cheek cells from inside the mouth with the scrape of a spatula. These days tests usually involve PCR amplification of the critical region—the suspect gene—from the DNA sample, followed by sequence analysis to determine whether or not it carries the mutation. And the test results for any individual may tell us something about the genetic status of his or her relatives.

Let's take, for example, a test for Huntington disease. In a recent case, a man in his twenties came into a genetics clinic to request that he be tested for Huntington. His paternal grandfather had died of the disease, and his father, in his forties, had decided not to be tested, preferring, like Nancy Wexler, to live with 50-50 uncertainty over knowing for sure. Because Huntington strikes relatively late in life, it was possible that the father was in fact carrying the mutation even though symptoms had not yet appeared. The young man knew that the probability of his having the mutation—and therefore the disease in his future—was 1 in 4.* But he wanted to know for sure. The problem is this: if he found out that he did indeed have the mutation, then he must have received it from his father, meaning that his father, too, would definitely develop the disease. The son's quest for genetic knowledge would directly contravene the father's desire to avoid it. A family feud developed, and in the end, only intervention by the young man's mother prevented him from proceeding with the test. His desire to know, she argued, surely paled beside her husband's right to be shielded from what may be a devastating death sentence. This dramatic example illustrates the difference between genetic diagnosis and any other kind. What I might learn about my genes has implications for my biological relatives, whether they care to know or not.

Sometimes the implications may not bear on the present generation but rather on generations to come. Fragile X is the commonest form of inherited

*There is a 1 in 2 chance that the father had received the mutation from the grandfather, and then, if the father has it, another 1 in 2 chance of his having passed it on to the son. The probability for the son is the product of these independent events, a 1 in 4 or 25 percent chance.

mental retardation. (Down syndrome is more frequent, but, occurring spontaneously, it is not usually inherited.) In addition to a low IQ, symptoms typically include a notably long face with an outsize jaw and ears, and a hyperactive, occasionally irritable temperament. Like DMD, it is a sex-linked disorder (the gene responsible is on the X chromosome), but, unlike DMD, it affects females as well as males. One normal copy of the gene is evidently not enough to render the effect of a mutated one negligible; still, women tend to suffer less severe symptoms, and their incidence is 1 in 8,300, compared with 1 in 5,000 for males. Fragile X is caused by a mutation similar to the one responsible for Huntington disease: a DNA triplet, CGG, is repeated over and over again. Normal individuals have about 30 while carriers of fragile X have at least 50 and sometimes as many as 90. For reasons we do not fully understand, the number of repeats tends to increase with each generation; and once there are about 230 CGG triplets, the gene can no longer make mRNA and therefore ceases to function. The condition gets its name from a discernible structural weakness in the X chromosome caused by all these repeats.

As the number of repeats increases from one generation to the next, so the severity of the condition increases, and the age of onset decreases in each family line. The latest descendants in a fragile X pedigree have the largest numbers of repeats and are typically affected earlier in life and more severely than those from whom they inherited the mutation. Geneticists may therefore identify individuals carrying a "premutation"—too few repeats to cause problems at present, but sufficient to result in fragile X in subsequent generations, given the likely expansion next time around. We do not yet know exactly what the protein produced by the affected gene does, but it seems to bind to messenger RNA molecules in the connections—synapses—between nerve cells.

Like ongoing research into Huntington, DMD, and many other genetic afflictions, studies of fragile X have been galvanized by those most directly affected: the families and loved ones of sufferers. FRAXA, the Fragile X Association, has been hugely effective in raising money and in inducing Congress to support fragile X research. Though some scientists may cynically view such groups merely as agencies that offer individuals in dire straits the comforting illusion that they are not entirely powerless, experience shows that dedicated, resourceful, and, above all, motivated organizations like FRAXA sometimes do hold the key to cracking these diseases against the long odds. To those who take the biggest gambles—financial and scientific—sometimes, with luck, go the biggest rewards.

Many women reading this may be asking themselves the question: Why wasn't I tested during my pregnancy for cystic fibrosis, or fragile X, or DMD? Sadly, some of them may even have children with one of these afflictions. In the wake of the genetic revolution that has transformed medical technology, one notes a singularly depressing and senseless fact: the uncoupling of scientific progress and patient care. Actually, it might be more accurate to say that due attention never was paid to coupling them properly in the first instance. In any event, many women are simply not informed of their options, and tests now readily available are hugely underused.

As a head of the Human Genome Project, I made sure to fund efforts to promote understanding of how the knowledge that would soon be pouring out of the sequencing machines would affect, for good or ill, the lives of countless people. Having set aside initially 3 percent of our total budget (and later 5 percent) for this purpose, I appointed Nancy Wexler, the Huntington expert, to run a panel called ELSI charged with exploring the ethical, legal, and social implications of our research. One of ELSI's major initiatives was a series of pilot studies of genetic screening. At a time when every newborn was screened for PKU, it was necessary to ask whether medicine could responsibly fail to offer at least the option of screening for cystic fibrosis, DMD, fragile X, and every other grave human ill that was within the power of science to predict. That was in the early nineties. Today, things have scarcely advanced beyond the pilot stage: small-scale studies are still carried out here or there. The reasons for this paralysis are varied, ranging from dollars-and-cents practicality to profound philosophical disagreements about the essence of human life and dignity. In short, they encompass the gamut of social phenomena, from jockeying for funds to collective soul-searching, that have attended the genetic revolution.

Testing for DMD and Huntington is ordinarily done only in families that already have an affected member. The rationale for the limitation is that these disorders are rare and the tests are costly. This social calculus is debatable, but the same reasoning does not hold in the case of cystic fibrosis, for which testing is nevertheless also limited. Cystic fibrosis, remember, affects about 1 in 2,500 people, making it one of the most prevalent genetic disorders. It is especially common among people of northern European descent. The high incidence seems all the more remarkable when we consider that the underlying defect, which occurs in a gene on chromosome 7, follows a recessive pattern of inheritance, meaning that in order to develop cystic fibrosis one must receive two mutated copies. People with just one are unaffected but being carriers they can pass the mutation on to their children. Epidemiological surveys and calculations tell us that 1 in 25 Americans of European ancestry are carriers.

One difficulty with cystic fibrosis testing is technical, having to do with variability in the underlying defect. One specific form of mutation accounts for about 70 percent of cases: a deletion called ΔF508, which eliminates three bases, CTT.* If just a few other mutations accounted for the remaining 30 percent, then general population screening for CF carriers would not be impractical. But most of the other causative mutations occur in just a single family line and more than a thousand different CF-causing mutations have been discovered to date. What does this mean for population screening? In practice any test could screen for at most twenty-five different mutations, but those twenty-five most common forms would still account for only about 85 percent of all cases. As a result, we'd be missing about one in six mutations—not a very good batting average for a diagnostic. Now, say we have a couple, both of whom have tested negative for CF mutations according to our highly imperfect screen. We could hardly tell them with any confidence that there was no danger of their producing a child with cystic fibrosis. Why bother, the argument goes, with an inconclusive test that typically costs $300?

But, despite the technical difficulties, prenatal screening for cystic fibrosis can still identify a large proportion of all affected fetuses. Why isn't it more widely adopted? Paradoxically, cystic fibrosis advocacy groups have played a major part in limiting CF testing to families already affected. Broader testing, it

*The 70 percent figure applies to people of northern European ancestry, the population in which cystic fibrosis is most common. However, ΔF508 only accounts for around 35 percent of CF mutations in both the African American and Ashkenazi Jewish populations. Ancestral differences like these complicate the design of screening programs.

is feared, would divert limited resources away from the ultimate goal of finding a cure. That concern is understandable, particularly at this moment. An estimated thirty thousand Americans have cystic fibrosis. Treatment advances have already extended life expectancies considerably, and it is conceivable that a cure will be in hand before long. Having said that, it would be irresponsible to suggest that a cure is around the corner: babies born today with cystic fibrosis still face the prospect of a lifelong struggle against a debilitating disease. Though curing cystic fibrosis is definitely a top priority, there should still be room to permit an expectant mother to have access, if she wants, to testing. Then, fully informed about the status of her fetus, she has the freedom to make whatever choices she sees fit.

Broader testing is also resisted for less material reasons. There are those who view screening as an admission of defeat, the wrong manner of solution. Advocacy groups are in the business of ensuring that people with the disease feel that they belong to a community and are valued by society—how does one reconcile that mission with testing, which, to put it in bluntest possible terms, is promoting the abortion of affected fetuses?

Cystic fibrosis advocates are anxious that people with CF do not become stigmatized, and they worry that indirectly testing does just that. In fact, there is an unfortunate precedent in the history of genetic testing that haunts all patient advocacy groups. Long before the advent of DNA screening, one of the earliest diagnostics for a genetic disease was developed to detect sickle-cell anemia, which in the United States affects primarily African Americans. As we saw in chapter 3, those with two copies of the mutated "sickle" hemoglobin gene will suffer painful, debilitating symptoms, while those with only a single copy—carriers—will notice little effect.

Following the development of simple blood tests in the 1960s, screening programs were hastily established across the country. Despite the best of intentions they did more harm than good. Screeners generally failed to counsel test subjects properly as to the significance of the test or its results. Many diagnosed as carriers mistakenly assumed that they themselves had the disease; some were even denied jobs or health insurance on the basis of the test; and couples who were at risk of producing sickle-cell children were advised rather heavy-handedly to think twice. The tests—some programs were mandatory—were coercive in effect, suggesting to some that the United States was entering a renaissance of racist eugenics, stigmatizing all who tested positive. The sad irony is that, from a purely medical point of view, the campaign was in fact sensibly conceived: despite advances in treatment, sickle-cell anemia remains a chronic, painful condition. Screening is the best remedy for a disease that is far more easily avoided than confronted, but the first mechanisms designed to eradicate it were so badly managed as to have rightly angered many intended beneficiaries.

Fortunately in 1972 new federal guidelines redesigned the sickle-cell screening program, allowing it to be effective without the widespread concern raised by the initial effort. Harder to repair has been the trust of advocacy groups for genetic diseases in general; the experience of the community of those affected by sickle-cell disease has left them ever leery of screening programs, and the fear of stigma persists, sometimes, alas, at the expense of better public health.

In so many ways, genetic testing, despite its incontrovertible usefulness, proves to be flypaper for controversy. Randi Hagerman, then at Denver Children's Hospital, decided to apply a DNA test for fragile X to children in special education classes in Denver. The reasoning was simple: children whose learning was impaired by that disorder would be better served if they were identified, whereupon their schooling could be tailored to their particular needs. Of 439 students tested, 5 with fragile X mutations were discovered. (A more extensive survey of schools in Holland had found 11 previously undiagnosed cases of fragile X in a group of 1,531 students.)

Perhaps the most interesting part of the Denver study was the response of the parents and guardians to Hagerman's offer. Most recognized the benefit of a diagnosis, both for the potential to improve their child's education and for the identification of the presence of the disorder in the family line. But fully a third refused the test, citing either a certainty that *their* children did not have fragile X or a concern that their children might find the test too stressful. Hagerman has been criticized for her efforts; it was a field day for those who insist upon seeing the menace of a totalitarian genetic future in every attempt at harnessing DNA to address a social problem.

The issue is indeed social as well as personal. The high incidence of the fragile X premutation—it is on perhaps as many as 1 in 200 X chromosomes—may warrant population screening. In the United States, it is estimated that just one reasonably severe case will cost, over a lifetime of nonwork and institutionalization, some $2 million in current dollar value. The ever-increasing challenge of providing affordable health care should itself suggest a potent argument for giving every mother the opportunity to be tested. The logic of hard realities is not lost on smaller countries, where the margins for policy error are not as great. A pilot study in Israel screened 14,334 women; 207 were found to have a premutation. Prenatal diagnosis was made available upon request, identifying five fetuses with extended CGG repeats. The fate of those pregnancies was rightly the choice of the expectant mother: a free society should no more require a woman to abort a fetus with a genetic disorder than it should require her to carry it to term. But not every woman is prepared to raise a disabled child, nor is every woman prepared to terminate a pregnancy on account of the child's foreseeable quality of life. Whatever the individual choice, however, the fact remains that screening can only reduce the incidence of affliction, and that is an unambiguous social good.

Despite the frustrating reluctance to take advantage of genetic screening on a broad scale, the short history of the practice hasn't been entirely one of small-scale pilot studies and damning controversy. There are some happy and illuminating stories to tell about the triumphs of screening programs for genetic disorders in high-risk populations.

Hemoglobinopathies are diseases caused by some malfunction in the hemoglobin molecule. Including the various thalassemias and sickle-cell disease, they are thought to make up the most common class of genetic disorders, with about 4.5 percent of the world's population carrying a mutation for one of them. As we have seen, the sickle-cell gene carried with it antimalarial properties, and so was promoted by natural selection in areas where malaria was prevalent. As a result, the mutation was originally at high frequency only in such parts of the world. The same adaptive advantage accounts for the similar distribution pattern of other hemoglobinopathies as well. Medicine has for some time understood that certain mutations therefore tend to be much more common in some ethnic groups than others, wherever the individuals may now find themselves.

Among the population of Greek Cypriot immigrants in London, thalassemia carriers represent a remarkable 17 percent. In its severe form the condition is the most pernicious of the hemoglobinopathies, resulting in misshapen and sometimes nucleated red blood cells that cause enlargement of the liver and spleen, often leading to death before adulthood. A systematic screening program begun in 1974 by Bernadette Modell of the Royal Free Medical School was welcomed enthusiastically by London's Cypriots, who were only too well aware of the seriousness of the disorder that had long blighted their community. A similar program in Sardinia, also begun in 1974, has dramatically reduced the incidence of thalassemia from 1 in 250 to 1 in 4,000.

Ashkenazi Jews are another group with a bitter awareness of what a deadly mutation can do to a genetically isolated population. Tay-Sachs (TS) is a ghastly disease 100 times more common in this group than in most non-Jewish ones. TS babies are born apparently healthy, but gradually their development slows and they begin to go blind. By about two, they are stricken with seizures. Deterioration continues until they die usually by the age of four, blind and paralyzed. Unlike hemoglobinopathies, whose relative commonness in certain populations can usually be explained by the concomitant adaptive protection against malaria, the high frequency of TS among the Ashkenazim remains a mystery. Perhaps a genetic bottleneck is to blame: the mutation may have been present among the relatively small segment who branched off to become the Ashkenazim during the second Diaspora. A similar phenomenon might also account for why the mutation is also anomalously common among the French Canadians of southwest Quebec as well as the Cajuns in Louisiana: the chance presence of an unfortunate mutation in the small founding populations. An alternative explanation holds that being a carrier of this recessive gene (having one copy of the TS mutation) may confer some resistance to tuberculosis, an advantage perhaps for European Jews who historically tended to live in densely populated urban centers.

The cause of Tay-Sachs was discovered in 1968 when it was recognized that the red blood cells of patients were overloaded with ganglioside GM2. This chemical is an essential component of the cell membrane, and in normal individuals any excess is broken down into related compounds by a key enzyme, which is lacking in TS sufferers. In 1985, Rachel Myerowitz and her colleagues at NIH isolated the gene coding for that enzyme and showed that it was indeed mutated in Tay-Sachs patients.

Thereafter we had the basis for a foolproof prenatal test and a well-defined target population—conditions tailor-made for the implementation of a successful screening program. But prenatal screening effectively offers only one remedy in the event of a positive diagnosis: abortion, which, at least among the observant Orthodox segment of the Ashkenazim, is forbidden. Fortunately, it is also possible to screen prospective parents and so the solution morally acceptable to the devout was a program aimed at couples. Rabbi Yosef Eckstein of New York saw four of his ten children die from Tay-Sachs. In 1985, he established Dor Yeshorim, the "generation of the righteous," a program to carry out TS testing in the local Orthodox Jewish community. Young people are encouraged to take advantage of free-testing days at high schools and colleges. An unusual aspect of this program is its extreme confidentiality: not even those tested are informed whether they are carriers; instead, each is given a code number. Later, when two people are contemplating marriage, they each phone Dor Yeshorim and give their numbers. Only in the event of both being carriers is the status of either partner revealed, together with an offer of counseling. This disclosure on a need-to-know basis is intended to avoid stigmatization of carriers, while still countering the threat of Tay-Sachs.

To date, the Dor Yeshorim program has tested more than seventy thousand individuals and detected more than a hundred couples at risk. Steadily reducing the incidence of Tay-Sachs, it would appear an unqualified success, yet there are those within the Jewish community who find fault with it. Some see coercion in the program's call for all young people to be tested, and intimidation in its strong recommendation that some individuals reconsider their decision to marry. Opponents have labeled Rabbi Eckstein's crusade "eugenics" (a word whose resonance is nowhere more painful than in the Jewish community), but such demagoguery hardly alters the central fact of the matter: the program clearly enjoys strong support within the community it serves, a community that understands the horrors of Tay-Sachs. Indeed, Dor Yeshorim has demonstrated that a screening program can be both effective and culturally responsive, working even in a situation where social mores and religious precepts seem to be at odds with genetic testing in principle.

Prenatal screening offers a stark choice for any woman carrying a fetus that has tested positive for a genetic disorder: to terminate or not to terminate the pregnancy. The fact that amniocentesis cannot be performed until a fetus is at least fifteen weeks old makes the option of termination only more traumatic. At this stage an abortion does not eliminate a featureless ball of cells, but a tiny being—real enough that a parental bond may have already formed with the developing fetus, thanks to the power of ultrasound imaging. Most parents—at least those who do not oppose all abortion on principle—would infinitely prefer to make hard choices presented by genetic testing at an earlier stage of development. Such was the inspiration for the invention of preimplantation diagnosis.

Robert Winston at the Hammersmith Hospital in London is a leading gynecological microsurgeon, an expert in such procedures as the correction of Fallopian tube defects that prevent a woman from conceiving. He has also become one of British television's leading popularizers of science and biomedical research and even finds time, as Lord Winston of Hammersmith, to sit in Parliament advising the government on such matters. By combining two state-of-the-art technologies—in vitro fertilization (IVF)* and PCR-based DNA diagnosis—Winston pioneered a method for checking the genetic status of an embryo before it is implanted in a woman's uterus and begins to develop. After in vitro fertilization, the several conceptuses are grown in the laboratory until each fertilized egg has divided three or four times to produce a ball of eight to sixteen cells. One or two cells are carefully removed from each for DNA extraction and then PCR is used to amplify the relevant sequences to determine in each case whether or not a mutation is present. It is the astonishing capacity of PCR to amplify even the tiniest quantities of target DNA that makes possible

*IVF is a method of assisted reproduction in which sperm and egg are fused in a laboratory dish. The resulting embryo—rather ominously called a "test-tube baby" in the early days of the technology—is then transferred to the uterus to develop naturally.

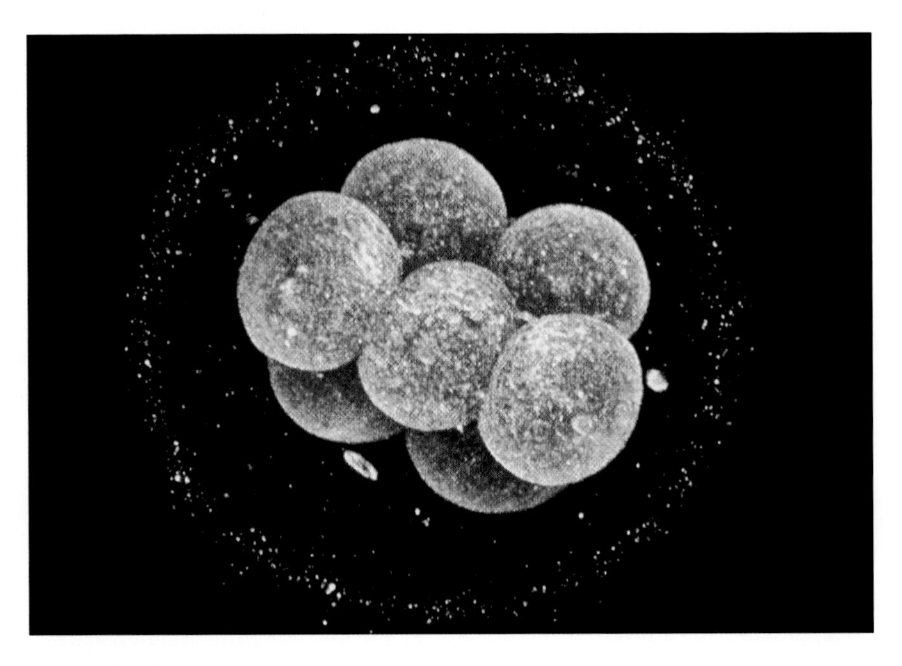

Eight-cell embryo

this method of ultra-early diagnosis. Parents are then free to implant only those embryos that test negative for genetic disease.

The first preimplantation tests performed in 1989 screened for the sex of the fetus—important information when the risk is for a sex-linked disorder such as DMD. A mother who is a carrier may select only female embryos on the premise that they will not be affected by the disorder, though they may be carriers. It was Winston's colleague, Alan Handyside, and others who subsequently extended preimplantation diagnosis beyond simple sex determination to the detection of specific mutations: in 1992, they first applied the technique to screen for cystic fibrosis, which is not sex-linked.

As we have seen, despite being sex-linked, fragile X can affect both males and females. The disorder is therefore a natural target for gene-specific preimplantation diagnosis, but it still took impassioned parents familiar with the difficulties of raising a fragile X child to mobilize doctors to do it. Debbie Stevenson, a CNBC television news reporter, has a son, Taylor, whose fragile X was only diagnosed after the birth of a second son, James. Though James fortunately beat the 50-50 odds of being affected, the Stevensons were unwilling to trust their third child to fate. They decided to seek preimplantation diagnosis: "Some people think it's unethical to select for healthy embryos," says Debbie Stevenson, "but I think it's better than having to make the wrenching decision about whether to terminate or continue a pregnancy after learning that your baby has a significant disorder." In 2000, at the end of the family's frustrating yearlong search for a lab willing to carry out the procedure, the newest member of the Stevenson family was conceived and, just days later, tested for fragile X. Like James, Samantha is free of Taylor's debilitating disorder.

Debbie Stevenson's family. The oldest boy, Taylor, has fragile X. Preimplantation diagnosis ensured that Samantha, the baby, was free of the disorder.

In our culture human reproductive biology seems an inexhaustible source of controversy, and a procedure involving the manipulation of human embryos for any purpose is sure to become a lightning rod. Preimplantation diagnosis has been no exception. Ethical considerations aside, however, the procedure still has two major drawbacks: it requires a huge commitment on the part of the couple undertaking it, and, like all forms of IVF, it is very expensive. But the method is so powerful in principle, and so much less traumatic than abortion, one can only hope that with time we shall see improved techniques, and with them diminished cost—the usual pattern with developing technology. Preimplantation diagnosis has the potential to become an extremely important weapon in our war on genetic disease.

The disorders we have discussed so far are all "simple" in the genetic sense: they are caused by a mutation in a single gene, and environment has no bearing on whether or not you will get one of these diseases. The situation is rather more complicated in the case of illnesses like cancer, which, as we've seen, may be triggered by a combination of hereditary and environmental influences. But even with cancer, there are some individual genes that have a major effect, regardless of the environment. Though BRCA1, one of the genes implicated in breast cancer, only accounts for about 5 percent of all cases, women with mutations in the gene have an estimated 90 percent chance of developing the disease by the age of sixty.

In the early 1990s, Francis Collins, then at the University of Michigan, joined forces with Mary-Claire King at UC Berkeley in the hunt for BRCA1. They took the standard approach: collecting families, preparing DNA samples, and testing markers, all with a view to homing in on the gene. One family of more than fifty members included multiple cases of breast cancer—a clear instance of an inherited predisposition to the disease. In September 1992, one member of that family—I'll call her Anne—revealed to Barbara Weber, an associate of Collins's, that she had scheduled a bilateral mastectomy the following week, even though there was no sign that she had cancer. Anne had decided that she could no longer tolerate the uncertainty, the question mark hanging over her future, and preferred to take this drastic preemptive step. Weber, however, had concluded from the DNA analysis that Anne was actually not in particular danger: her risk of breast cancer was no greater than that of a woman without a family history of the disease. But this inference was made in the context of a research project and it had been agreed long in advance that as a rule such preliminary data should not be used for clinical diagnosis.

Weber and Collins, however, decided that Anne's plight outweighed the rule book: they informed Anne that her risk was low, and with great relief she canceled the surgery. But having disclosed their findings to one member of the family, the researchers felt obliged to offer the same benefit to others who asked for it; and so Weber and Collins set up an ad hoc breast cancer genetic counseling program. One family member who also proved to be at no special risk had already undergone a prophylactic bilateral mastectomy five years before. She received the belated diagnosis philosophically: the surgery, she figured, had bought her five years' peace of mind. But had she tested positive for the mutation, the radical course might in fact have bought her more than peace of mind. For years, prophylactic mastectomy had been recommended by clinicians, even though no surgery can feasibly remove all the breast tissue and there were no solid data showing that the measure was saving lives. Today, however, there is proof that the extreme approach does indeed reduce mortality rates among women at high risk; in one group of 639 who had the surgery, only 2, rather than the 20 to 40 statistically expected, actually died of breast cancer. Similarly, removal of the ovaries before age forty (but after a woman has finished having children) reduces the risk of both ovarian and breast cancers. Genetic analysis can give women the power to make decisions that can literally make the difference between life and death.

But the keyhole view into the future that DNA analysis permits can also create opportunities to defeat breast cancer by less extreme means, as another story from the Michigan study reveals. A cousin of Anne's was told that she was in all likelihood carrying the BRCA1 mutation that was devastating her family. Since she had not had a mammogram in years—a fear-based negligence ironically not uncommon in high-risk families—she panicked. Weber scheduled one for later that day and a tiny incipient tumor was found; it was easily removed but would almost certainly have been missed in a routine examination. Self-examination and regular mammography have doubtless saved many lives, but the campaign to universalize these procedures may have had the unintended consequence of creating a false sense of security in some cases. Screening for genetic risk allows us to find those individuals whose imaging examinations merit extra-high scrutiny. Greater risk demands greater surveillance. And in the long run more needles will be found the smaller we make the haystack.

Nancy Wexler, as a member of a Huntington family, and Anne, as a member of a breast cancer family, are both part of a new generation for whom newly available screening tests can provide glimpses of genetic destiny. And as we learn more about the genetic basis of relatively common adult afflictions, from diabetes to heart disease, the biological crystal ball will become ever more powerful, telling the genetic fortunes relevant to us all.

In the past decade, few diseases have struck terror in as many hearts as Alzheimer, which each year draws ever greater numbers into its grip of awful mental and physical debilitation—the disease affects more than 4 million Americans. Family and friends of sufferers may notice first some minor lapses of memory—trouble recalling recent events or finding the right word—which they might hopefully attribute to the ordinary effects of getting on in years. The afflicted may then begin to show evidence of mood swings, also not altogether unnatural among the elderly. But as the disease progresses, the symptoms become more pronounced and unmistakable; the memory loss soon grows unnaturally severe, making the familiar challenges at work and even simple household tasks unmanageable. Speech becomes more labored; sentences go unfinished as the victim loses the train of thought. And the person's awareness of these changes may lead to depression, which in turn intensifies the effect of other increasingly distressing changes in personality. Advanced Alzheimer patients do not know who or where they are; they cannot recognize even their closest kin. With the inexorable erosion of memory and personality, their very essence as individuals is gradually destroyed.

Alzheimer typically first appears at around age sixty, but a rarer form, accounting for about 5 percent of all cases, strikes individuals in their forties. This early-onset form of the disease puts families through the same kind of hell as Huntington disease does, seizing its victims in the prime of life and gradually, relentlessly destroying them. One family with multiple affected members over several generations was described as having been struck by its own "biological Holocaust." Following the argument first advanced by Mary-Claire King in her breakthrough study of breast cancer, that any early-onset version of a disease is likelier than the ordinary form to have a clear genetic basis, most initial Alzheimer research focused on the early-onset form. By 1995 three genes had been found, all of them involved in some way with the processing of amyloid protein deposits, whose accumulation in the brains of patients was noted as early as 1906, in Dr. Alois Alzheimer's original description of the disease. Early-onset Alzheimer is, then, clearly inherited. But what of the more common variety?

Allen Roses at Duke University preferred to ignore the wisdom of the majority, and set out straightaway to tackle the much more familiar late-onset form, which only occasionally runs in families. Ronald Reagan, for instance, who announced his affliction in 1994, lost his brother Neil to late-onset Alzheimer two years later. Their mother had died of it as well.

With training as a neurologist and a background in muscle disorders like DMD, Roses began his search in 1984. His claim in 1990 that a gene on chromosome 19 appeared to correlate with the disease was met by skepticism.

Nothing, however, gives Roses more pleasure than an opportunity to prove everyone else wrong. Two years later, he had actually identified the critical gene. It turned out to code for apolipoprotein E (APOE), a protein involved in processing cholesterol. The gene comes in three forms (alleles), APOEε2, APOEε3, and APOEε4, but it was APOEε4 that proved the crucial one: a single copy of that variant increased fourfold one's risk of developing Alzheimer. And individuals with two copies were at a risk ten times greater than that of persons with no APOEε4 allele. Roses found that 55 percent of those with two copies of APOEε4 will have developed Alzheimer by age eighty. Could this correlation be the basis for a genetic test? Probably not. Despite being correlated with the disease, the APOEε4 allele is common and is not a good enough predictor of Alzheimer for testing purposes: though their risk is higher, plenty of people with two APOEε4 alleles never develop Alzheimer. But the use of APOEε4 screens in conjunction with clinical evaluations does improve the accuracy of Alzheimer diagnosis. And perhaps once we understand the correlation in causal terms, the genetic analysis can be refined. Recent work that induced Alzheimer-like symptoms in mice has suggested APOE is involved in the metabolism of the protein that causes nerve cell death in human Alzheimer sufferers.

What of treatment? Most genetic diseases present us with much the same heartrending frustration that comes with Huntington: we know enough to diagnose them, perhaps to evade them, but not to treat them. Happily, there are a few cases in which our genetic understanding has taken us the rest of the way, providing therapies that work. Unfortunately, few of these remedies are as simple and effective as that for PKU, from which a normal life can be retrieved through a few dietary restrictions.

Too often genetic disorders result in the cell-by-cell decimation of particular tissues: muscles in DMD, nerve cells in Huntington and Alzheimer. There is no quick fix to this kind of insidious decay. But though these are early days yet, I think there is a realistic chance that we will eventually be able to treat diseases like these using stem cells. Most cells in the body are capable only of reproducing themselves—a liver cell, for instance, produces only liver cells—but stem cells can generate a variety of specialized cell types. In the simplest case, a newly fertilized egg—the stem cell with maximum potential—will ultimately give rise to every one of the 216 recognized human cell types. Stem cells are accordingly most readily derived from embryos; they can also be found in adults, but such cells tend to lack that embryonic ability to differentiate into *any* cell type. We are beginning to learn how to induce stem cells to produce particular cell types, and someday, I hope, we will be able to replace the lost brain cells in people with Huntington and Alzheimer with new healthy cells. But I caution that we have a long way to go before we have a thorough understanding of the molecular triggers that cause a cell to develop in one direction rather than another. It will take ten years or so of grappling with this fundamental problem in developmental biology before we are in a position to explore properly the therapeutic value of stem cells. I think it would be a tragedy for science, and for all people who may eventually benefit from stem-cell therapy, if research is hindered by religious considerations. Polls consistently show that the majority of Americans favor research using embryonic stem cells, and yet politicians continue to pander to the outspoken religious minority that opposes it. The result is restrictive legislation in the United States that is hampering efforts to develop this potentially valuable technology.

For now, treating genetic disorders does not extend to the wholesale replacement of cells à la stem-cell therapy, but it may involve replacement of a missing protein. Striking 1 in 40,000 individuals, Gaucher disease is a rare condition resulting from a mutation in the gene for glucocerebrosidase, an enzyme that helps break down a particular kind of fat molecule, which otherwise accumulates harmfully in the body's cells. The disorder can be devastating, with a suite of symptoms including bone pain and anemia. Initial attempts to supply the missing enzyme directly were made as early as 1974. The results were promising but the logistics were nightmarish: the replacement enzyme had to be extracted from human placentas, and twenty thousand placentas were needed to furnish a year's supply for a single patient. A big breakthrough came in the early 1990s, when researchers synthesized a modified form of the enzyme, one that was taken up more efficiently by the cells that most needed it. In 1994, Genzyme, a biotech company, started to produce the modified form using recombinant methods. Treatment of Gaucher does not combat the genetic root of the disorder but rather the effect of the mutation: it provides the patient with the vital protein that the faulty gene cannot.

Righting the genetic abnormality via this biochemical route is evidently feasible and effective. But even given the remarkable efficiency of recombinant methods, the treatment is expensive—$175,000 per year—and the need for continual infusions imposes a burden on patients. Naturally, therefore, geneticists have long dreamed of a practical way to fix the cause of the problem rather than compensate for its effects. The ideal treatment for genetic disorders would be a form of genetic alteration, a correction of the genes that cause the problem. And the benefit of such gene therapy would last the patient's whole life; once fixed, it's fixed for good. There are, at least in principle, two approaches: somatic gene therapy, by which we change the genes within a patient's body cells; or germ-line therapy, by which we alter the genes in a patient's sperm or egg cells, preventing the transmission of the harmful mutation to the next generation.

Such solutions to the ravages of genetic defects might be obvious, but the idea of gene therapy has not met with the warmest of professional or public receptions. Such reactions are not altogether surprising: a culture wary of genetic modification in a corn plant might be expected to be averse to transgenic people—GM humans, if you prefer—however great the potential benefit. And more vociferous objections are made, also not unexpectedly, to the germ-line approach, because of the risk of causing genetic damage when manipulating the DNA. In somatic gene therapy, such damage may be limited in its effect; in germ-line therapy the possibility exists of accidentally producing impaired people. Even its proponents—of whom I count myself one—would never suggest that such a procedure should be carried out until our techniques are good enough for us to be confident that we will not inadvertently cause damage. Many scientists are convinced, though, that we should never attempt germ-line gene therapy. Whether based in ethics or unfounded fears of the unknown, such arguments are ultimately not compelling in my judgment. Germ-line therapy is in principle simply putting right what chance has put horribly wrong. But for now the controversy is academic: germ-line therapy is still way beyond our technical powers. Until it is in reach, we should concentrate our efforts on making somatic gene therapy a powerful tool in its own right.

The first apparently successful gene therapy was carried out by French Anderson, Michael Blaese, and Ken Culver at the National Institutes of Health in 1990. They chose a very rare disorder called adenosine deaminase deficiency (ADA), in which the lack of an enzyme disables the immune system, leaving one as defenseless as David Vetter, the boy in the plastic bubble. The experimental subjects were two young girls, four-year-old Ashanti DeSilva and nine-year-old Cindy Cutshall.

How do you shoot a new gene into a patient? At that time, retroviruses suggested themselves as the logical weapon of choice. In general, viruses are efficient genetic vectors; they make their living by injecting DNA into other cells. Retroviruses, a special group, have RNA, rather than DNA, as their genetic material. But while most viruses infect a cell, reproduce, and then kill the host cell as the "daughter" viruses escape to infect others, retroviruses are typically kinder and gentler, at least to the host cell: new viral copies are dispatched without destroying it. This does not necessarily mean that a retrovirus is any easier on the host organism; sometimes it is quite the contrary, as demonstrated by the effects of HIV, perhaps the best-known retrovirus. But it does mean that the viral genes—and any extra gene the virus may be induced to ferry—will become a permanent part of the undestroyed cell's genome. Genetic engineering has produced retroviruses that are as safe as possible for gene therapy; stripped of all the viral genes that aren't essential for invading the host cell's genome—and their means for accomplishing this purpose are formidable—retroviruses become the ideal gene vector.

But we are still left with the problem of how to target only the cells affected by the mutation, the ones that need the replacement gene. Today this remains the greatest challenge facing gene therapy: how do you get the good gene into muscle cells to treat DMD, lung cells to treat cystic fibrosis, or brain cells to treat Huntington disease? The choice of the obscure disease ADA was therefore a very sensible one for the first gene therapy trial: the target cells for ADA are readily available—immune-system cells circulating in the blood. Anderson's team was able to extract millions and millions of immune cells from the girls' blood and grow them in petri dishes, where they could be infected with a retrovirus carrying a functional copy of the gene. Once the cell's natural DNA had

Cindy Cutshall, pioneering gene-therapy patient. After a visit to Cold Spring Harbor, she sent me her portrait of me in action.

incorporated the viral genome carrying the replacement gene, the cells were ready to be fed back into the patients' blood.

In September 1990, Ashanti DeSilva was the first to undergo the procedure; Cindy Cutshall's therapy followed four months later. Each received infusions of genetically doctored immune cells every few months. At the same time, each was continued on the non-gene therapy of enzyme replacement, the same manner in which Gaucher disease patients are treated, but in lower doses. This precaution was required by the NIH's Human Gene Therapy Subcommittee, which argued reasonably that it was too dangerous to expose the girls to a new therapy without a safety net. The experiment, though not a perfectly controlled one, did seem to work: the immune systems of both girls improved, and they were better able to fight off minor infections. I personally can attest that Cutshall seemed a very healthy eleven-year-old when she and her family visited Cold Spring Harbor in 1992. Eleven years later, however, the results are not entirely conclusive. DeSilva's immune system is now approaching normal function, but only about one-quarter of her T cells are derived from the gene therapy. Cutshall's blood contains an even smaller proportion of gene-therapy T cells, though her immune system is working well. It is, however, difficult to say exactly how much of the girls' improvement is due to the gene therapy and how much to the continuing enzyme treatment. The result is therefore too ambiguous to be understood without reservation as a clear gene-therapy success.

The Cutshall/DeSilva trials were not the first time the NIH had thrown its weight around in the world of gene therapy. In fact, the Human Gene Therapy Subcommittee was formed at the NIH in 1980 in response to the first gene-therapy experiment ever performed. The trial was a failure and stirred up such a controversy that the government almost moved to strangle the newborn enterprise in its cradle. By all accounts, the man at the center of the storm, Martin Cline, was a clever, ambitious clinician, devoted to the relief of his patients' woes. His special interest was beta-thalassemia, the hemoglobinopathy Bernadette Modell had screened for among London's Cypriot community. After successful animal experiments, Cline applied to the review board of the University of California Los Angeles, where he worked, for permission to try the gene therapy on humans using nonrecombinant DNA. While the application was still being reviewed, an overzealous Cline had arranged to treat two women outside the United States, one in Israel and one in Italy, but he used recombinant genes, whose use was still prohibited under NIH guidelines. On returning to Los Angeles, Cline found that his application had been rejected; the review board ruled that more animal data would have to be presented before an attempt with humans could be sanctioned. Cline had broken just about every rule in the book: not only had he proceeded to treat human subjects without authorization, but he had also used an unequivocally prohibited method. Cline suffered the consequences: he lost his federal funding and was forced to resign as chairman of his department. Gene therapy had lost its first practitioner.

The Cline episode was by no means the last time that scientists attempting gene therapy found themselves in hot water with regulators. Tragically, it took the death of a patient in a gene-therapy trial to bring home the sobering message: gene therapy—that complicated cocktail of viruses, growth factors, and patients—is dangerous. But the message was more than that: because there are so many unknowns in the gene-therapy equation, strict oversight of all procedures involving humans is absolutely necessary. Jesse Gelsinger died both because we do not know enough to predict with complete confidence an individual's response to gene therapy and because scientists took inexcusable shortcuts.

In 1999, Gelsinger, an Arizona teenager, heard about an experiment being conducted by James Wilson, director of the Institute for Human Gene Therapy at the University of Pennsylvania. Gelsinger was suffering from ornithine transcarbamylase deficiency (OTC), a hereditary impairment of the liver's ability to process urea, a natural product of protein metabolism. Untreated, the disease can be lethal, and though, like PKU, it can be managed with simple medication and an appropriate diet, OTC does leave its victims particularly vulnerable to other ailments. The eighteen-year-old Gelsinger had only a mild case, but a childhood brush with death precipitated by his condition emboldened him to volunteer in the hope of helping to find a cure for himself and others like him. The Pennsylvania therapy aimed to use an adenovirus (a member of the group that causes the common cold) as the vector of the corrected gene. But a few hours after the viruses carrying a normal version of the OTC gene had been injected into his liver, Gelsinger developed a fever. A rampant infection followed, accompanied by blood clots and liver hemorrhaging. Three days after the injection, Jesse Gelsinger was dead.

The teenager's death was a shock not only to his family but to the research community as well. A detailed investigation revealed serious procedural lapses. Most glaring perhaps was this: although two patients had shown signs of liver toxicity earlier in the same study, the cases had gone unreported to any regulatory authority and were never disclosed to the volunteers in the study. Had the Gelsingers been informed, Jesse would likely not have been so quick to volunteer, and he might well be alive today. The tragedy dealt a serious blow to the progress of gene therapy. For a time, the FDA halted all such experiments at the university and at several other programs across the country. Bill Frist of Tennessee, the Senate's sole physician, conducted an investigation into reporting procedures in human trials; President Clinton called for improvement in standards of "informed consent," championing the right of experimental subjects to be apprised of all potential risks. If any good has come from Jesse Gelsinger's death, it is that federal oversight of human trials has been tightened.

The gene-therapy community was still reeling from the shock wave caused by Gelsinger's death when heartening news of a success story came from France. The disorder targeted was SCID, the immune deficiency that condemned David Vetter to life in a bubble. Though bone marrow transplants can effect a cure—the recipient of the first transplant, done in 1968, is still healthy today—the success rate is only about 40 percent and even successful transplants frequently lead to grim complications, as in Vetter's sad case. In 2000 a team under Alain Fischer at the Necker Hospital in Paris carried out gene therapy on two infants who, like David, had been kept in sterile isolation since birth. As with the treatment for ADA, a retrovirus was used to ferry the needed gene into cells extracted from the babies; the cells were then reintroduced. But, in a notable innovation, the French group harvested the cells to be modified from the infants' bone marrow. By using the marrow's immune stem cells rather than ordinary T cells found in the blood, the method, if successful, promised to furnish a self-perpetuating genetic fix. When stem cells reproduce, they increase not only their own numbers but also the numbers of the specialized somatic cells into which they naturally differentiate. Therefore any T cells produced

When all the news was good: Alain Fischer and Marina Cavazzana-Calvo announce their breakthrough gene-therapy triumph in April 2000.

from altered stem cells would also carry the inserted gene, making the repeated infusions of modified cells unnecessary.

And that is exactly what happened: ten months later, T cells containing a working copy of the missing gene were found in both patients, and their immune systems were performing as well as those of any normal children. Fischer's method has since been applied to other SCID children. After a long and not altogether auspicious start, gene therapy had finally notched an unequivocal success. But the champagne celebration did not last long. In October 2002, doctors found that one of the two original patients was suffering from leukemia, a cancer of the bone marrow in which certain types of cell are overproduced. Though it has not been established for sure that the genetic procedure was responsible, the circumstantial evidence is mighty strong. Gene therapy seems to have cured the baby's SCID but caused leukemia as a side effect.

Side effects have always dogged medicine. Drugs may affect more than just their intended target, or surgical procedures may end up causing complications. Though in many ways a departure from conventional medicine, genetic medicine, we now know, is also subject to the same law of unintended consequences. Fischer's SCID treatment probably inadvertently created new problems in the process of fixing the original one. After all, any treatment requiring the insertion of viral DNA into the DNA of patients' cells is inherently risky because the foreign DNA may by chance disrupt the functioning of a critical gene. Because typically the cell in which this has occurred will die, such events usually have no impact. But it is possible that the disrupted gene is one whose elimination does not kill the cell, but rather unlocks its capacity to multiply unchecked: the viral insertion can cause cancer. This seems to be what happened to the SCID baby.

Gene therapy may yet be a long way from delivering the miracles foreseen at the dawn of the genetic revolution. Jesse Gelsinger's death was a severe setback. The leukemia side effect of the SCID treatment, however, is even more damaging. In Gelsinger's case, it seems that unpardonable mismanagement was largely responsible—a problem that has hopefully been fixed by tighter regulation. But there is no ready solution to the side-effect problem. Probably we will have to rely on the depressing calculus that applies in this case: that gene therapy has at least cured a condition, SCID, worse than the one it has caused, leukemia. The good news is that the baby boy in question is apparently responding well to the chemotherapy used to treat the leukemia. However, between them the Gelsinger and leukemia incidents have crystallized many of the difficult issues that need yet to be resolved if somatic gene therapy is to enter the medical mainstream. And I am not so naive as to deny that future trials will likely uncover yet more difficulties. It may be some time before we can claim beyond doubt to have neutralized every conceivable danger, but I nevertheless believe that the potential of this technology to lift the curse of genetic disease is simply too great for medicine to turn away from it.

Your DNA can tell someone a lot about you. As we've seen, if Huntington runs in your family, your DNA can literally reveal your future; soon, depending on whether you possess a particular variant of a gene (or combination of genes), DNA may also speak to your relative risk of succumbing to common killers like heart disease. What version you have of the APOE gene can already serve as a predictor of Alzheimer. But should you worry that this profoundly personal information might be used against you? Not surprisingly, for many Americans the greatest concern is that genetic profiling may one day lead to their being denied health-care insurance.

In 2000 the *American Journal of Human Genetics* published the results of a survey that had asked health insurers whether they would adjust rates to take account of genetic information if it were available to them. Would they, in principle, be prepared to charge more for a customer in perfect health who carried a mutation predisposing him or her to a disorder? About two-thirds admitted they would. The other third were in all likelihood lying. Insurance companies are not philanthropies, but businesses, with shareholders to please. There is no reason to suppose that, left to their own devices, they wouldn't do what they have always done: maximize the premiums of those at risk, and where possible avoid altogether those customers most likely to collect. The same report described a case where an insurance company raised an individual's rates based on a suspicion of a genetic disorder, merely because this individual had requested a diagnostic test for Huntington disease.

As we come to know ourselves at the molecular level, is it inevitable that those of us who have drawn the short straws in the genetic lottery will be made to pay a price, in this way and others? And why presume such abuse would begin and end with insurers? My DNA profile might show that I am likely to have a heart attack or stroke, become an alcoholic, or suffer clinical depression. Might such information cause a prospective employer to think twice about hiring me?

Such questions suggest that *Brave New World* may be upon us well before the twenty-fifth century of Huxley's imagination. DNA is a potent fact of twenty-first-century life—a genie that will never be put back in the bottle. What we allow to be done with it, however, is something we must decide as a democratic society. Unfortunately, in such societies, laws tend to lag behind the need for them: a traffic light isn't typically installed at a dangerous intersection until after a few accidents have occurred. It may require a few horror stories of gross injustice, of individuals made the victims of their own genomes, to motivate the passage of appropriate legislation. What should it look like? Genetic privacy should be a touchstone, but not necessarily the ultimate objective. Balances will need to be struck with society's other priorities, not least the fight against disease, an effort whose progress will more and more depend upon giving medical researchers access to as much genetic data as can be collected from the general population. While legislation ought not defeat our ambition to exploit the full potential of DNA to alleviate human suffering, or to tell us about ourselves and our origins, or to identify those among us guilty of crimes, it must minimally ensure that no citizen be deprived of civil or human rights on the basis of what might be inscribed in his or her genes.

Meanwhile, it may be a comfort to know that despite a wealth of genetic information already at the industry's disposal and regardless of what they tell pollsters, insurance companies have, on the whole, shown little impulse toward factoring genetic considerations into their calculations for setting rates. The wretched pale skin I inherited has already proved its susceptibility to cancer, but the last time I looked, I still wasn't being charged a higher premium for it. Again, the rationale is business, not charity. Insurers have traditionally set rates using actuarial tables that estimate overall health and longevity mainly on the basis of how we live. I suspect that even if genetic data were universally available, insurers would still find such lifestyle factors—whether one smokes or not, whether one works in a coal mine as opposed to a flower shop—vastly more predictive of one's health risk than the overwhelming majority of relatively subtle differences determined by genetic variation from one person to another. It's indisputable that those whose DNA reveals an unavoidable destiny of debilitation need special protection under the law, but the propensity toward ailments like heart disease and cancer are certain to prove so widespread and complicated as to make them an impractical basis for cost-cutting discriminations. The essential premise of insurance, that payments of the happy many who never have cause to collect will underwrite the relief of the unfortunate few, is not likely to be abolished owing to the accumulation of any amount of genetic information.

But even if our individual rights can be secured against what our DNA may disclose, our peace of mind may not be so easily restored; as Nancy Wexler well knows, genetic knowledge can be a scary prospect. And I agree with her: there is no point in knowing about something that we are powerless to remedy or ameliorate. Alzheimer is a major concern of people my age, but in the absence of proven treatment possibilities, I have no desire to be tested for the presence of the APOEε4 allele. Craig Venter, incidentally, does indubitably have one copy of it. We know this because he insisted on disclosing that the genome sequenced by Celera was his own. And this allele, through its role in cholesterol processing, is associated not only with a higher risk of Alzheimer but of heart disease as well. (APOEε4 is not an asset in any respect.) Stuck with this genetic self-awareness, Venter is wisely trying to respond prophylactically as best he can: he is taking drugs called statins, which lower cholesterol levels and may retard or prevent the onset of Alzheimer. Even without knowledge of my APOE alleles, I'm also taking statins, figuring that a little preemptive medication can't do any harm. If statins are as effective as some claim them to be, we can look forward to many more years of Venter-generated (and, hopefully, Watson-generated) controversy.

Genetic knowledge will remain frightening so long as we remain in the present intermediate stage, possessing in general the power to diagnose but not to cure. But ours is not an unprecedented medical predicament. Think back to the early years of the twentieth century: a diagnosis of infant diabetes was a death sentence. Today, with insulin therapy, such a child can expect to live to a ripe old age. The hope of our research efforts is that one day soon a diagnosis of diseases like Huntington will be transformed in the same way—from death sentence to prescription.

Already, we are in a far stronger position to deal with bad rolls of the genetic dice than we were even twenty years ago; the ever-increasing life expectancies of people with, for example, Down syndrome and cystic fibrosis attest to this progress. But for now our most powerful weapons are diagnostics. The choice of whether to be tested is one best left to each individual or parent, those who will most directly bear the burden of genetic knowledge. In the case of prenatal diagnosis, it is the prospective mother who should make the decisions. That is not to say that others shouldn't participate, but ultimately the choices should lie with the woman: not only is she the one having the baby, but, like it or not, our world is still one in which women are expected to bear the brunt of the day-to-day care of children. Regardless of the specifics of decision making, however, one thing is completely clear to me: over the ages genetic disorders have rained unthinkable misery upon countless families such as Carol Carr's, which has been ravaged by Huntington disease. Testing holds the power to reduce misery by preventing it. Having developed the tests, it unconscionable not to make their existence known to those who might want to use them, inexcusable not to make them universally available.

Twins
Twins

CHAPTER THIRTEEN

WHO WE ARE: NATURE VS. NURTURE

Growing up, I worried quietly about my Irish heritage, my mother's side of the family. My ambition was to be the smartest kid in the class, and yet the Irish were the butt of all those jokes. Moreover I was told that in the old days signs announcing the availability of jobs often ended with "No Irish Need Apply." I wasn't yet equipped to understand that such discrimination might have to do with more than an honest assessment of Irish aptitudes. I knew only that though I myself possessed lots of Irish genes there was no evidence that I was slow-witted. So I figured that the Irish intellect, and the shortcomings for which it was known, must have been shaped by the Irish environment, not by those genes: nurture, not nature, was to blame. Now, knowing some Irish history, I can see that my juvenile conclusion was not far from the truth. The Irish aren't in the least stupid, but the British tried mightily to make them so.

Oliver Cromwell's conquest of Ireland surely ranks high among history's most brutal episodes. It culminated in banishing the native Irish population to the country's undeveloped and inhospitable western regions like Connaught while the spoils of the more salubrious east were divided up among the lord protector's supporters, who would start to Anglicize the vanquished province. With the incoming Protestants believing the heresies of Catholicism to be a one-way ticket to perdition, Cromwell duly proclaimed in 1654 that the Irish had a choice: they could "Go to Hell or Connaught." At the time, it probably wasn't clear which was worse. Seeing Catholicism as the root of the "Irish Problem,"

Identical twins attending an annual convention held at Twinsburg, Ohio the British took draconian measures to suppress the religion, and with it, they hoped, Irish culture and Irish national identity. The ensuing period of Irish history was thus characterized by a form of apartheid every bit as severe as that so infamously practiced in South Africa, with the principal difference being the basis of discrimination: religion rather than skin color.

Among the "Penal Laws" passed to "Prevent the Further Growth of Popery," education was a particular target. One statute of 1709 included the following provisions:

> Whatever person of the popish religion who shall publickly teach school or instruct youth in learning in any private house, or as an usher or assistant to any protestant schoolmaster, shall be prosecuted.
>
> For discovering, so to lead to the apprehension and conviction of any popish archbishop, bishop, vicar general, Jesuit, monk, or other person exercising foreign ecclesiastical jurisdiction, a reward of 50 pounds, and 20 pounds for each regular clergyman or non-registered secular clergyman so discovered, and 10 pounds for each popish schoolmaster, usher or assistant; said reward to be levied on the popish inhabitants of the county where found.

The British hoped that the Irish young attending British-sponsored Protestant schools would wean themselves off Catholicism. But they hoped in vain: it would take more than oppression or even bounty to prise apart the Irish and their religion. The result was a spontaneous underground educational movement, the "hedge schools," with itinerant Catholic teachers leading secret classes in ever-changing outdoor locations. Often conditions were appalling, as a visitor noticed in 1776: "They might as well be termed ditch schools, for I have seen many a ditch full of scholars." But by 1826, of the entire student body of 550,000 an estimated 403,000 were enrolled in hedge schools. Increasingly a romantic symbol of Irish resistance, the schools inspired the poet John O'Hagan to write:

> Still crouching 'neath the sheltering hedge,
> Or stretched on mountain fern,
> The teacher and his pupils met feloniously to learn.

But if the British had failed in their goal to enforce religious conversion, they had, despite the heroic efforts of the hedge schoolteachers, successfully impaired the quality of education for generations of Irish. The resulting archetype of the "stupid" Irishman would have been more aptly identified as the "ignorant" Irishman, a direct legacy of the anti-Catholic policies of Cromwell and his successors.

And in this way my boyish conclusion was not off the mark: the so-called curse of the Irish was indeed the result of nurture—development in an environment of substandard educational opportunities—rather than nature—Irish genes. Today, of course, nobody, not even the most bigoted Englishman, can legitimately claim that the Irish aren't as smart as other people. Ireland's modern education system has more than undone the damage of the hedge school era: the Irish population is today one of the best educated on the planet. My youthful reasoning on the subject, however absurdly misinformed, nevertheless taught me a very valuable lesson: the danger of assuming that genes are responsible for differences we see among individuals or groups. We can err mightily unless we can be confident that environmental factors have not played the more decisive role.

This tendency to prefer explanations grounded in "nurture" over ones rooted in "nature" has served a useful social purpose in redressing generations of bigotry. Unfortunately, we have now cultivated too much of a good thing. The current epidemic of political correctness has delivered us to a moment when even the possibility of a genetic basis for difference is a hot potato: there is a fundamentally dishonest resistance to admitting the role our genes almost surely play in setting one individual apart from another.

Science and politics are to a degree inseparable. The connection is obvious in countries like the United States, where a considerable proportion of the scientific research budget depends on the appropriations of the democratically elected government. But politics intrudes upon the pursuit of knowledge in more subtle ways as well. The scientific agenda reflects society's preoccupations, and all too often social and political considerations end up outweighing purely scientific ones. The rise of eugenics, a response on the part of some geneticists to prevailing social concerns of the era, is a case in point. With a scientific basis weak to the point of vanishing, the movement progressed mainly as a pseudoscientific vehicle for the notably unscientific prejudices of men like Madison Grant and Harry Laughlin.

Modern genetics has taken to heart the lessons of the eugenics experience. Scientists are typically careful to avoid questions with overtly political implications and even those whose potential as political fodder is less clear. We have seen, for instance, how such an obvious human trait as skin color has been neglected by geneticists. It's hard to blame them: after all, with any number of interesting questions available for investigation, why choose one that might land you in hot water with the popular press or, worse, earn you an honorable mention in white supremacist propaganda? But the aversion to controversy has an even more practical—and more insidious—political dimension. It happens that scientists, like most academics, tend to be liberal and vote Democratic. While no one can tell how much of this affiliation is principled and how much is pragmatic, it's certainly the case that Democratic administrations are assumed to be invariably more generous toward research than Republican ones.* And so having signed on to the liberal end of the political spectrum, and finding themselves in a climate intolerant of truths that don't conform to ideology, most scientists carefully steer clear of research that might uncover such truths. The fact that they duly hew to the prevailing line of liberal orthodoxy—which seeks to honor and entitle difference while shunning any consideration of its biochemical basis—is, I think, bad for science, for a democratic society, and ultimately for human welfare.

Knowledge, even that which may unsettle us, is surely to be preferred to ignorance, however blissful in the short term the latter may be. All too often, however, political anxiousness favors ignorance and its apparent safety: we had better not learn about the genetics of skin color, goes the unspoken fear, lest such information be marshaled somehow by hatemongers opposed to mixing among the races. But that same genetic knowledge may actually be vitally useful to people like me, whose Irish-Scots complexion is vulnerable to skin cancer in climes sunnier than Tipperary and the Isle of Skye, where my mother's ancestors hailed from. Similarly, research into the genetics of difference in mental

*Wrongly, it turns out. The stingiest science budget in recent history was Jimmy Carter's.

ability among people may raise awkward questions, but that knowledge would also be a boon to educators, allowing them to develop an individual's educational experience with his or her strengths in mind. The tendency is to focus on the worst-case scenario and to shy away from potentially controversial science; it is time, I think, we looked instead at the benefits.

There is no legitimate rationale for modern genetics to avoid certain questions simply because they were of interest to the discredited eugenics movement. The critical difference is this: Davenport and his like simply had no scientific tools with which to uncover a genetic basis for any of the behavioral traits they studied. Their science was not equipped to reveal any material realities that would have confirmed or refuted their speculations. As a consequence, all they "saw" was what they wished to see—a practice that really doesn't merit the name science—and often they came to conclusions manifestly at odds with the truth: for instance that "feeblemindedness" is transmitted as an autosomal recessive. Whatever the implications of modern genetics may be, they simply bear no relation to this manner of reasoning. Now, if we find a certain mutation in the gene associated with Huntington disease, we can be sure that its possessor will develop the disease. Human genetics has moved from speculation to fact. Differences in DNA sequence are unambiguous; they're not open to interpretation.

It is ironic that those who worry most about what unchecked genetics might reveal should lead the way in politicizing the field's most basic insights. Take, for example, the discovery that the history of our species implies that there are no major genetic differences among the groups traditionally distinguished as "races": it has been suggested that as a matter of general practice our society should accordingly cease to recognize the category "race" in any context, eliminating it, for instance, from medical records. The theory here goes that the quality of treatment you receive in a hospital may vary depending on how you identify your ethnicity on the admission form. Racism can surely be found in the ranks of any profession, medicine included. But it's not altogether apparent how much protection your ethnic anonymity on the form will confer once you are face-to-face with a doctor who is a bigot. What is more apparent, however, is the danger of withholding information that might be diagnostically important. It is a fact that some diseases have higher rates of incidence within certain ethnic groups as compared with the human population as a whole: Native Americans of the Pima tribe have a particular propensity to Type II diabetes; African Americans are much more likely to suffer from sickle-cell anemia than Irish Americans; cystic fibrosis affects mainly people of northern European origin; Tay-Sachs is much more common among Ashkenazi Jews than others. This is not fascism, racism, or the unwelcome intrusion of Big Brother. This is simply a matter of making the best possible use of whatever information is available.

For such a young science, genetics has played a central role in a remarkable number of notably ugly political episodes. Eugenics, as we have seen, was partly of the geneticists' own making. The pseudoscience known as Lysenkoism, which flourished in the Soviet Union in the middle of the twentieth century, however, was visited upon genetics from on high—literally: Stalin had plenty to say on the matter. Lysenkoism represents the most egregious incursion of politics into science since the Inquisition.

In the late 1920s, the Soviet Union was still finding its feet. Stalin had won the battle of succession after Lenin's death and was consolidating power. The collectivization of agriculture was under way. And in an obscure agricultural research station in distant Azerbaijan, an uneducated but ambitious peasant was making a name for himself. Trofim Lysenko, born in the Ukraine in 1898, appeared an unlikely choice to oversee Stalin's agricultural revolution. Barely literate, he was working as a minor technician at Gandzha at the Ordzhonikidze Central Plant-Breeding Experiment Station when, in 1927, he was catapulted from obscurity by a visiting *Pravda* correspondent who, perhaps at a loss for good copy, was inspired by the sight of Lysenko: here was the "barefoot professor" solving agricultural problems so that the local "Turkic peasant can live through the winter without trembling at the thought of the morrow." Critically, the article painted Lysenko as a problem solver, not a highfalutin academic: "He didn't study the hairy legs of [fruit] flies, but went to the root of things."

The image of the barefoot professor was irresistible to Soviet apparatchiks. Here was a son of the soil, the true flowering of the Soviet man, of the rural peasant class; his agricultural intuition was surely worth more than all the book learning of the shiftless intellectuals. Not to disappoint, Lysenko was quick to capitalize on his newfound prominence by proposing that winter wheat be "vernalized." Winter wheat is normally planted in the fall; it overwinters as a shoot, with some of the crop perishing, the rest maturing during the spring. Through "vernalization," Lysenko suggested, the losses of winter could be avoided. He claimed that you could fool the wheat seeds into germinating in the spring simply by chilling and wetting them, and that increased yields would be achieved in the bargain. The definitive experimental demonstration of the method was carried out by none other than Lysenko's father in his own fields. Indeed, the yield was some three times greater than that of conventional unvernalized wheat planted in the same district.

Vernalization did not in fact originate with Lysenko; wherever he may have picked it up, the procedure dates back to the preceding century at least, appearing, for example, in the Ohio agricultural literature of the 1850s. But here Lysenko's lack of education (and therefore ignorance of what had been accomplished elsewhere) stood him in good stead when it came to claiming originality. The same, however, could not be said for every further attempt to apply the method, whose results can vary a good deal depending on local conditions—something the Ohio farmers knew but the barefoot professor apparently did not.

Within a couple of years, beset by failures, Lysenko stopped advocating the vernalization of winter wheat and was pushing instead the vernalization of spring wheat—a ploy worthy of the sharpest Soviet satire, considering that the crop is indeed named after the season in which it is normally planted. Later, his wheat yield policy did another U-turn when Lysenko called for *warming* (instead of cooling) the seed prior to planting. Wheat vernalization was but one of many agricultural nostrums that Lysenko peddled, but it illustrates well his overall strategy. A complete disregard for expert knowledge was de rigueur, as was a refusal to conduct consistent and rigorous tests. Essentially, any idea intuitively appealing to Lysenko was good enough to be implemented. What scientific method he did espouse almost seems inspired by theological reasoning, odd coming from the tool of a godless Communist state: "In order to obtain a certain result, you must want to obtain precisely that result; if you want to obtain a certain result, you will obtain it."

Much more astutely reasoned was Lysenko's careful manipulation of the media. His original brush with fame in *Pravda* had taught him that the state-

Trofim Lysenko measuring wheat plants in a rare burst of empiricism on a collective farm near Odessa, Ukraine

controlled press was a better venue for scientific self-promotion than the dusty pages of academic or trade journals. In 1929 *Pravda* twice featured the barefoot professor's success with vernalization, each time reporting in loving detail the down-home contribution of Lysenko Senior.

At that point the Soviet Union needed a Lysenko. The "agricultural reorganization," as Stalin preferred to call the collectivization of farms, was proving a catastrophe. Even the official estimates, notorious for their rosy overstatements, painted a grim picture of rural productivity during this period. Lysenko's intuitive quick fixes made him the man of the hour, even if they wound up doing more harm than good the morning after. He embodied an important Bolshevist ideal, *deistvennost*—"action quality." No messing around with grand theories or arcane academic concerns, Lysenko, the can-do barefoot professor, was all about action and solving practical problems.

Lysenko learned quickly how to play the Soviet system. His lectures made no pretense to being scientific in any sense that we would recognize: they consisted instead of ideological rants peppered with all the Marxist-Leninist jargon du jour. Small wonder Stalin was a fan, leading the standing ovation (at a Congress of Collective Farm Shock-Brigade Workers) with cries of "Bravo, Comrade Lysenko!" In return, Lysenko astutely named his latest big idea, a variety of branched wheat, for Stalin. Happily accepting the honor, the generalissimo fortunately never found out that the branched wheat was another bust: though inherently higher in yield, it requires such low-density planting as to more than offset the advantage of its multiple seed heads.

Sucking all of Soviet agriculture into a vast experiment each time he introduced another hopelessly impractical new scheme, Lysenko was ultimately responsible for the starvation of millions. But since Soviet records of the era—especially those kept by Lysenko himself—are woefully self-serving, we will probably never know the actual number of lives sacrificed on the altar of Lysenko's career. Suffice it to say that at the time of Stalin's death in 1953, the availability of meat and vegetables has been estimated by more objective analysts to have been no greater than in the darkest feudal days of Tsar Nicholas II. Lysenko's pernicious influence, however, was not limited to agriculture.

If Soviet farming had its homegrown principles, Soviet science too needed a scientific credo of its own; it pained Lysenko and his followers to imagine the new Soviet man following so meekly in the footsteps of "bourgeois" Western scientists. And Lysenko's wild theories of agricultural development had boiled down to the idea that you could transform any crop so long as you subjected it to the right environment: winter wheat could *become* spring wheat through a simple environmental manipulation. And it was no one-season fix, for, according to Lysenko, such changes would then breed true—acquired traits would be passed on to the next generation. Eventually Lysenko became a full-fledged Lamarckist.* In an unusual fit of experimental enthusiasm, he even commissioned experiments to "disprove" Mendelism—the basis of genetics in the decadent Western tradition. In his mathematical incompetence, Lysenko actually became convinced that the results refuted Mendel's ratios, even when a reanalysis of the data by a distinguished Soviet mathematician showed that in fact the ratios fitted Mendel's predictions exactly. Lysenko, then, was not above doing an occasional experiment, but was never one to countenance a result contradicting even the most outlandish hypothesis.

The late 1930s saw a series of debates between Lysenko, backed by what have

*In 1801 Jean-Baptiste Lamarck first published his theory of inheritance of acquired characteristics, erroneously suggesting that traits acquired during an individual's life could be passed on to its offspring. Flawed though his idea was, Lamarck, unlike Lysenko, was at least trying to base his inferences on observation.

been described as his "hard core of militant ignoramuses," and the Soviet genetics community—a distinguished group by the standards of the international science of the day. H. J. Muller, one of T. H. Morgan's students (and my professor in graduate school at Indiana University), went to Russia to participate in the great social experiment of Communism and found himself instead embroiled in bizarre, largely stage-managed public discussions about Larmarckian inheritance. In this era of Stalin's purges, political truths carried much more weight than mere scientific ones. To what extent Lysenko directly contributed to the "repression"—to use the preferred Soviet euphemism for Stalin's purges—of the geneticists who spoke against him will probably never be known but, no matter who gave the orders, the fact remains that much of the opposition to Lysenko's Lamarckian ideas simply disappeared as the 1930s drew to a close. Some geneticists heroically stood their ground as outspoken critics. Muller was forced to flee for his life. The doyen of Soviet genetics (and an ardent Soviet patriot), Nikolai Vavilov, was arrested in 1940. He died in prison of malnutrition.

In 1948, it was officially decreed that the debate was over: Mendelism was out, Lysenkoism was in—an absurd and tragic outcome particularly when one considers that it came four years *after* Avery's landmark experiment showing DNA to be the transforming (Mendelian) factor. The Lysenkoite response to the discovery of the double helix, incidentally, was characteristically obscurantist: "It deals with the doubling, but not the division of a single thing into opposites, that is, with repetition, with increase, but not with development." I've no idea what this means, but it seems to be consistent (in its meaninglessness) with Lysenko's other writings on heredity:

> In our conception, the entire organism consists only of the ordinary body that everyone knows. There is in an organism no special substance apart from the ordinary body. But any little particle, figuratively speaking, any granule, any droplet of a living body, once it is alive, necessarily possesses the property of heredity, that is, the requirement of appropriate conditions for its life, growth and development.

Darwin was next in line for the Lysenko treatment. The out-of-control peasant-made-good denied the cardinal precept of Darwinism—competition among individuals within a species for access to limited resources—and postulated, as perhaps a good Communist should, that individuals do not compete, but cooperate. He went further, combining his anti-Mendelian and anti-Darwinian views in a bizarre unified theory of the origin of species: given that organisms are molded by their environment, it should be possible, with the right environmental conditions, to transform any one species into any other. Change a warbler's diet to caterpillars, to take his favorite example, and you can produce a cuckoo. Ardent Lysenkoists from around the country were soon writing in with reports of their own transformational successes: viruses turned into bacteria, a rabbit into a chicken. Soviet biology was itself undergoing a transformation of sorts: from science to joke.

Lysenko's rejection of Darwin eventually put him in a position so awkward as to tax even his own formidable skills of political survival. Stalin's final years witnessed the "Great Stalin Plan for the Transformation of Nature." In part, this involved planting a lot of trees to protect the steppes from the vicious east winds, thus moderating the climate in general. It was not a bad idea in principle, but as one might expect, Lysenko had his notions about the best way to grow trees: plant them in a cluster, he argued, and the individual seedlings will *not* compete with each other for sunlight and nutrients but will rather cooperate for the good of the community. In the late forties, armies of peasants fanned out across the steppe planting oak trees in clusters in accordance with the Lysenko method. The result? Intense competition among individual trees, which enfeebled all members of each cluster. By 1956 only 4 percent of the oaks planted were thriving; only 15 percent had even survived. The Ministry of Agriculture withdrew its endorsement of the Lysenko planting protocol, but only after a sum estimated at over 1 billion rubles had been squandered.*

It was a stunning setback, but so entrenched was Lysenko's authority, and so crowded with his protégés were the ranks of Soviet biology, that it wasn't until 1964 that the Kremlin turned its back on him for good. The barefoot professor

*The modern dollar equivalent is difficult to compute because the official exchange rates of that era generally reflected Communist wishful thinking rather than financial reality. To put the sum in context, however, it can be noted that in 1956 one-sixth of the Soviet workforce earned an annual wage of around three thousand rubles.

had managed to persuade Stalin's successor that he was still the man to create a Soviet agricultural miracle. Indeed, when Khrushchev was bundled ignominiously out of office by the Soviet Central Committee (and replaced with Brezhnev), it was rumored that one important reason for the intervention was a general frustration with Khrushchev's continued reliance on Comrade Lysenko. Lysenko himself died in 1976. His family requested that he be buried in the most prestigious Russian national cemetery at the Novo-Devichi convent. The request was denied.

I would not for a moment wish to suggest, by telling the parable of Lysenko, that the fate of Soviet science under that fool's sway is remotely comparable to the state of contemporary Western research in even the most politically overpowering university setting. But the extreme instance should suffice to demonstrate that ideology—of any kind—and science are at best inappropriate bedfellows. Science may indeed uncover unpleasant truths, but the critical thing is that they are *truths.* Any effort, whether wicked or well-meaning, to conceal truth or impede its disclosure is destructive. Too often in our free society, scientists willing to take on questions with political ramifications have been made to pay an unjust price. When in 1975 E. O. Wilson of Harvard published *Sociobiology,* a monumental analysis of the evolutionary factors underlying the behavior of animals ranging from ants—his own particular subject of expertise—to humans, he faced a firestorm of rebuke in the professional literature as well as the popular media. An anti-Wilson book published in 1984 bore a title that said it all: *Not in Our Genes.* Wilson was even attacked physically, when protesters objecting to the genetic determinism they perceived in his work dumped a jug of water on him during a public meeting. Similarly, Robert Plomin, whose work on the genetics of human intelligence we shall presently address, found the American academy so hostile that he decamped from Penn State to England.

Passions inevitably run high when science threatens to unsettle or redefine our assumptions about human society and our sense of ourselves—our identity as a species, and our identities as individuals. What could be a more radical question than this: Does the way I am owe more to a sequence of As, Ts, Gs, and Cs inherited from my parents, or to the experiences I've had ever since my father's sperm and mother's egg fused together many long years ago? It was

Francis Galton, the father of eugenics, who was the first to frame the question as one of nature versus nurture. And the implications spill over into less philosophical, more practical areas. Are all math students born equal, for instance? If the answer is no, it may well be a waste of time and money trying to force differential equations down the throats of people like me who are simply not wired to take the stuff on board. In a society built on an egalitarian ideal, the notion that all men are *not* born equal is an anathema to many people. And not only is there a lot at stake, but the issues are very difficult to resolve. An individual is a product of both genes and environment: how can we disentangle the two factors to determine the extent of each one's contribution? If we were dealing with laboratory rats, we could conduct a set of simple experiments, involving breeding and rearing under specified uniform conditions. But, happily, humans are not rats, so illuminating data are hard to come by. This combination of the debate's importance and the near impossibility of satisfactorily resolving it makes for perennially lively argument. But a free society should not shrink from honest questions honestly asked. And what is critical is that the truths we discover are then applied only in ethical ways.

With a lack of reliable data, the nature/nurture debate was entirely subject to the shifting winds of social change. Early in the twentieth century, during the heyday of the eugenics movement, nature was king. But when the fallacies of eugenics became apparent, culminating in its horrific applications by the Nazis and others, nurture began to gain the upper hand. In 1924 John Watson (no relation), the American father of an influential school of psychology called "behaviorism," summarized his nurture-ist perspective as follows:

> Give me a dozen healthy infants, well-formed, and my own specified world to bring them up in and I'll guarantee to take any one at random and train him to become any type of specialist I might select: doctor, lawyer, artist, merchant-chief, and yes, even beggar-man and thief, regardless of his talents, penchants, tendencies, abilities, vocations, and race of his ancestors.

The notion of the child as tabula rasa—a blank slate upon which experience and education can write any future—dovetailed nicely with the liberal agenda that grew out of the sixties. Genes (and the determinism they stood for) were out. Discounting inheritance, psychiatrists preached that mental illness was caused by varieties of environmental stress, an assertion that inspired endless guilt and paranoia among the parents of the afflicted: where did we go wrong? they asked. The tabula rasa remains the paradigm of choice among the politically entrenched defenders of some increasingly untenable views of human development. Among some unregenerate hardliners of the women's movement, for example, the notion of biological—genetic—differences in cognitive aptitudes between the sexes is simply unspeakable: men and women are equally capable of learning any task, period. The fact that men are more common in some fields and women in others is, these theorists would have it, purely a result of divergent social pressures: the male slate is inscribed with one destiny, the female with another, and it begins with our laying that pink blanket on the baby girl and the blue one on the baby boy.

Today we are seeing a swing away from the extreme nurture-ist position embodied by the other Watson. And it is no coincidence that this drift away from behaviorism is coinciding with our first glimpses of the genetics underpinning behavior. As we saw in chapter 11, for years the genetics of humans has lagged behind that of fruit flies and other creatures owing to a lack of genetic markers and the impossibility of doing breeding experiments on people. But since the introduction in 1980 of DNA-based genetic markers the analysis of human traits through the mapping of related genes has advanced by leaps and bounds. Most of the effort has understandably been expended on meeting the most urgent human need genetics can address: diagnosis and treatment of inherited disease. Nevertheless, some efforts have been directed toward non-medical questions. Robert Plomin, for example, has used this approach to hunt up genes influencing IQ, taking advantage of an annual gathering in Iowa of superbright schoolchildren from across the nation. With an average IQ of 160, these slightly scary kids were an obvious place to start to look for genes that might affect IQ. Plomin compared their DNA to that of a sample of "normal" kids with average range IQs like yours and mine*—and indeed he found a weak

*Mine is a respectable, but definitely not stellar, 122. I discovered it by sneaking a quick look at a list on a teacher's desk when I was eleven.

association between a genetic marker on chromosome 6 and stratospheric IQ. Here was reason to suppose that a gene or several genes in that region might in some way contribute to IQ. Of course, any mechanism governing such a complex trait is apt to involve many genes.

In chapter 11, we discussed the difficulty of mapping polygenic traits, like heart disease, that are produced by multiple genes, each with a small individual effect, each mediated by the environment. Behavioral traits generally fall in this category. So far as we know, having the appropriate variant on chromosome 6 would not by itself a genius make: there are doubtless necessary variants in genes as yet undiscovered. And even a solid genetic basis might not get you there unless you were also reared in an environment in which learning and thinking were honored over Nickleodeon. But the discovery, and the acknowledgment, of any molecular basis for intelligence is a breakthrough such as only the genetic revolution could foster.

Before DNA markers were available, the meat and potatoes of behavioral genetics was twin studies. Twins come in two varieties: dizygotic (DZ), meaning two individuals develop from two separate eggs, each fertilized by a different sperm; and monozygotic (MZ), meaning that both come from a single fertilized egg, which in early development—usually the 8- or 16-cell stage—splits into two balls of cells. DZ twins are no more genetically similar than any two siblings, but MZ twins are genetically identical. MZ twins are therefore always the same sex, while DZ twins may or may not be. Surprisingly, it hasn't been very long since we first understood this fundamental difference between twin types. In 1876, when Francis Galton first suggested that twins might be useful in determining the relative contributions of heredity and upbringing, he was unaware of the difference (whose basis had been worked out just two years earlier), and assumed wrongly that it was possible for different-sex twins to be derived from a single fertilized egg. From his later publications, however, it's clear that the message eventually got through.

MZ twinning occurs globally in about four of every thousand pregnancies and seems to be nothing more than a random accident. DZ twinning, on the other hand, may run in families, and varies from population to population: a group in Nigeria tops the list with DZ twins accounting for forty pregnancies per thousand, while there are only three per thousand in Japan.

The basic premise of the standard form of twin study is that both members of a same-sex pair of twins, whether DZ or MZ, are raised the same way (i.e., receive similar "nurtures"). Suppose we are interested in a simply measurable characteristic like height. If DZ1 and DZ2 were both raised on the same diet of food, love, and so on, any difference in height between them would be attributable to some combined effect of genetic differences and whatever subtle differences of nurture may have crept in (for instance, DZ1 always finishes her milk; DZ2 never does). But if we follow the same program with MZ1 and MZ2, the fact that these twins are genetically identical eliminates genetic variation as a factor; any differences in height must be a function of *only* those subtle environmental differences. All things being equal, MZ twins will then tend to be more similar in height than DZ twins, and the extent to which this is true gives us a measure of how much genetic factors influence height. Similarly, the extent to which MZ twins have more similar IQs than DZ twins reflects the effect of genetic variation on IQ.

This kind of analysis is also applicable to the inheritance of genetic diseases. We say that twins are "concordant" if they both have the disease. An increase in concordance when we look from DZ twins to MZ twins would support the claim of a strong genetic basis to the disease: for example, DZ twins are 25 percent concordant for late-onset diabetes (if one twin has it, then there is a one in four chance that the other does too), whereas MZ twins are 95 percent concordant for the disease (if one twin has it, then nineteen out of twenty times the other does too). The conclusion: late-onset diabetes has a strong genetic component. Even here, though, the environment plays an obvious role: if it were not so, we would see 100 percent concordance in MZ twins.

A long-standing criticism of this kind of twin study addresses methodology: MZ twins tend to be treated more alike by their parents than DZ twins. Parents sometimes make a virtual fetish of identicalness: often, for instance, MZ twins are even dressed exactly alike, a habit some weirdly carry into adulthood. This is a legitimate criticism insofar as the more pronounced similarity of MZ twins (as compared with DZs) is interpreted as evidence of genetic influence when in fact it could simply be a reflection of the more precisely similar nurtures shared by MZs. And here is a further wrinkle in the same problem: how do we tell whether a pair of same-sex twins is DZ or MZ? "It's easy," you say. "Just look at them." Wrong. In a small but significant proportion of cases, parents mistake their same-sex DZ twins for MZ twins (and thus tend to subject them to the supersimilar nurture routine—the same frilly pink frock for each); and conversely a small proportion of parents with MZ twins wrongly take them for DZs (dressing one in frilly pink and the other in bright green). Fortunately, DNA fingerprinting techniques have rescued twin studies from this comedy of errors. The test can determine for sure whether the pair are indeed as they were supposed to be, whether DZ or MZ. The mistaken-identity groups then serve as the perfect experimental control in the analysis: for example, height difference in DZ twins cannot be put down to differences in nurture if the parents were raising them as MZs.

Perhaps no form of twin study holds more popular appeal than analysis of MZs separated at birth. In such cases, the rearing environments are often very different, and so marked similarities are attributable to what the twins have in common: genes. It makes good copy: you see reports of MZs separated at birth who, it turns out, both have red velvet sofas and dogs called Ernest. Striking though these similarities may be, however, chances are they are mere coincidences. There is almost assuredly no gene coding for the red velvet upholstery preference, or for impulses in dog naming. Statistically, if you list a thousand attributes—make and model of car, favorite TV show, etc.—of any two people, you will inevitably find ones that overlap, but in the press these are the ones that get reported, usually in the Believe It or Not column. My coauthor and I both drive Volvo station wagons and appreciate a cocktail or two, but we are most definitely not related.

Popular or no, twin studies have had a checkered history. Part of the ill repute stems from the controversy surrounding Sir Cyril Burt, the distinguished British psychologist who did much to establish the use of twins in studies of the genetics of IQ. After his death in 1971, a detailed examination of his work suggested that some of it was fraudulent—Sir Cyril, some alleged, was not above inventing a few twins from time to time if he needed to bolster sample sizes. The truth of these charges is still debated, but one thing is undeniable: the episode cast a shadow of suspicion over not only twin studies but all attempts to understand the genetic basis of intelligence. In fact, the combination of the Burt affair and hair-trigger political sensitivity to the topic have in effect stifled research by cutting the flow of grant money. No money, no research. Tom Bouchard at the University of Minnesota, a distinguished scientist whose massive 1990 survey of twins reared apart redefined twin studies, had such difficulty raising funds that he was forced to go cap in hand to a right-wing organization that supports behavioral genetics to further its own dubious political agenda. Founded in 1937, the Pioneer Fund counts among its early luminaries Harry Laughlin, the chicken geneticist we encountered in chapter 1 who turned his attentions to humans and entered the vanguard of American scientific racism. The fund's charter was "race betterment with special reference to the people of the United States." That legitimate researchers like Bouchard should be faced with a choice between seeking such a tainted sponsor or seeing their work perish represents a staggering indictment of the federal agencies that fund scientific research. Tax dollars are being allocated according to political rather than scientific merit.

Bouchard's Minnesota twins study revealed that a host battery of personality traits—as measured using standardized psychological tests—were substantially affected by genes. In fact, more than 50 percent of the variability observed in a range of characteristics—the tendency to be religious, to name one—was typically caused by underlying variation in genes. Bouchard concluded that one's upbringing has surprisingly little effect upon personality: "On multiple measures of personality and temperament, occupational and leisure-time interests, and social attitudes, MZ twins reared apart are about as similar as MZ twins reared together." In other words, when it comes to measurable components of personality, nature seems to trump nurture. This lack of impact of upbringing on personality development has even Bouchard scratching his head. Upbringing has little effect, and yet the data still show the environment's considerable effect: MZ twins raised apart are as similar to each other as those raised together, but there are nevertheless differences in both cases between members of a pair. Could there be an aspect of environment distinguishable from upbringing? One suggestion is that variation in prenatal experience, the life of a fetus in utero, may be important; even small differences at this early developmental stage—when, after all, the brain is being assembled—may have a significant impact on who we become. Even MZ twins may find themselves in very different uterine settings courtesy of the natural whims of implantation—the lodging of the embryo in the wall of the uterus—and the development of the placenta. The popular belief that all MZ twins share a single placenta (and therefore have similar uterine environments) is wrong: 25 percent of MZ pairs have separate placentas. Studies have shown that such twins differ more from each other than do pairs who have shared a placenta.

The elephant in the living room of all twin studies is the genetics of intelligence. How much of our smarts is determined by our genes? Everyday experience suffices to prove there is a lot of variation out there. While teaching at Harvard, I became intimately acquainted with the familiar pattern: in any population, there are a few who really aren't too bright, a few who are alarmingly smart, and a vast majority who are middling. The fact that the setting was Harvard, where the population had been preselected in favor of intelligence, makes no difference: the same proportions hold whatever the group. This "bell curve" distribution of course can describe just about any trait that varies in humans: most of us are medium tall, but there are a few super-tall and a few super-short among us. But when used to describe variations in human intelligence the bell curve has demonstrated powers to raise a dust storm of objection. The reason is that in a land of equal opportunity, where we are each free to advance as far as our wits will carry us, intelligence is a trait with profound socioeconomic implications: the measure of it is predictive of how one will fare in life. And so in this matter the nature/nurture debate becomes entangled with the noble aspirations of our meritocratic society. But given the complex interplay of the two factors, how can we reliably judge their respective weights? Smart parents not only pass on smart genes; they also tend to rear their children in ways that foster intellectual growth, thus confounding the effects of genes and environment. This is the reason careful twin studies are so valuable in permitting us to analyze the constituents of intelligence.

Bouchard's study and earlier ones as well have found that as much as 70 percent of the variation in IQ is attributable to corresponding genetic variation: a strong argument for the primacy of nature over nurture. But does this really mean that our intellectual fate is largely sealed by our genes—that education (even our own free will) has little to do with who we are? Not at all. As with all traits, it is nice to be blessed with favorable genes, but there is much that nurture can do to influence the standing of any individual, at least in the bell curve's vast midland, where variations in social circumstance are mainly determined.

Take the case of one group within Japan, the Buraku. They are the descendants of Japanese who by feudal custom had once been condemned to perform society's "unclean" tasks, like slaughtering animals. Despite the modernization of Japanese society, the Buraku remain impoverished and marginalized outsiders, scoring on average ten to fifteen IQ points lower than the national Japanese mean. Are they genetically inferior, or is their IQ simply a reflection of their lowly status in Japan? It would seem to be the latter: Buraku who have immigrated to the United States, where they are indistinguishable from other Japanese Americans, have shown an increase in IQ and over time the fifteen-point gap with their fellows in the homeland disappears. Education matters.

In 1994, Charles Murray and Richard Herrnstein published *The Bell Curve,* arguing that, despite the well-established effect of education, the discrepancies in the average IQ scores of different races may themselves be attributable to genes. It was a profoundly controversial claim, but not as simpleminded as many have supposed. Murray and Herrnstein understood that the combined observations of a genetic basis to IQ and of differences in average IQ among groups do *not* lead directly to the conclusion that genes are responsible for the intergroup differences. Imagine sowing the seeds of a particular plant species in which height varies genetically. Put one set of seeds in a tray with high-grade soil, and another in a tray of poor soil: in both trays we see variation in height; some individuals are taller than others—as expected, given genetic variation. But we also observe that the average height for plants in the tray of poor soil is less than the average for those in the tray of rich earth. The environment, in the form of soil quality, has affected the plants. While genetics is the dominant factor in determining height differences among plants within a tray—all other factors being equal—genetics has nothing to do with the differences seen between the two trays.

Does this same argument apply to African Americans, who lag behind other Americans in measures of IQ? Since poverty rates among African Americans are relatively high, with a large proportion of individuals finding themselves rooted in the relatively poor educational soil of the inner city, environment surely does contribute to their underperformance on IQ tests. Murray and Herrnstein's point, however, was that the discrepancy was so great that environment likely couldn't explain it all. Similarly, environmental factors alone may not account for why, globally, Asians, have on average higher IQs than other racial groups. The idea of measurable variations in average intelligence among ethnic groups is not one, I admit, I want to live with. But though *The Bell Curve*'s claims remain questionable, we should not allow political anxieties to keep us from looking into them further.

There is perhaps no more heartening proof of the role of environment in human intelligence than the Flynn effect, the worldwide phenomenon of upwardly trending IQ, named for the New Zealand psychologist who first described it. Since the early years of the twentieth century, gains have ranged between nine and twenty points per generation in the United States, Britain, and other industrialized nations for which reliable data-sets are available. With our knowledge of evolutionary processes, we can be sure of one thing: we are *not* seeing wholesale genetic change in the global population. No, these changes must be recognized as largely the fruits of improvement in overall standards both of education and of health and nutrition. Other factors as yet not understood doubtless play a role, but the Flynn effect serves nicely to make the point that even a trait whose variation is largely determined by genetic differences is in the end significantly malleable. We are not mere puppets upon whose strings our genes alone tug.

The finding that there is a substantial genetic component to our behavior should not surprise us; indeed, it would be far more surprising if this were not the case. We are products of evolution: among our ancestors, natural selection indubitably exerted a strong influence over all traits that have figured in our survival. The human hand, with its marvelous opposable thumb, is the product of natural selection. In the past, therefore, there must have been varying forms of the hand, with natural selection favoring the version we have today by promoting the spread of the genetic variants underlying it; in this way, evolution ensured that every member of our species would be endowed with this supremely valuable asset.

Behavior, too, has been critical to human survival, and therefore sternly governed by natural selection. Presumably our enthusiasm for fatty and sweet foods evolved this way. Our ancestors were ever pressed to meet their nutritional requirements; therefore the propensity to take full advantage of all energy-rich foods whenever any became available was of huge benefit. Natural selection would have favored any genetic variations that ensured a sweet tooth since those with it survived better. Today those same genes are the scourge of everyone who struggles to keep off the weight in parts of the world with abundant food sources: what was adaptive in our ancestors is now maladaptive.

Ours is a strikingly social species; it is logical, therefore, to infer that natural selection once favored genetic adaptations facilitating social interaction. Not only would gestures, like smiling, have evolved as a means of signaling one's state of mind to other members of the group, but presumably there would also have been strong selective pressures in favor of psychological adaptations permitting one to judge the intentions of others. Social groups are prey to parasitism; there are always individuals who seek to benefit from membership without contributing to the general good. The capacity to detect such freeloaders is vital to the success of a cooperative social dynamic. And though we are no longer hovering in small groups around one fire roasting the communal supper, our gifts for sensing one another's moods and motivations may nevertheless come from those early phases of our development as a social species.

A Victorian illustration of a baby doing what comes naturally

Since the publication of E. O. Wilson's *Sociobiology* in 1975, evolutionary approaches to understanding human behavior have themselves evolved, giving rise to the modern discipline of evolutionary psychology. In this field the search is for the common denominators of our behavior—human nature, the characteristics shared by all of us, whether New Guinea Highlander or Parisienne—which we seek to understand, trait by trait in relation to some past adaptive advantage conferred by each. Some such correlations are simple and relatively uncontroversial: the grasping reflex of

The cultural wonder that is Homo sapiens. *Two contrasting notions of chic: Paris 1950s and the highlands of Papua New Guinea. Evolutionary psychology seeks the common denominators underlying all our widely divergent behavior.*

a newborn, for instance, strong enough that a baby can use its hands and feet to suspend its full body weight, is presumably a legacy from the time when the ability to cling to a hirsute mother was important for infant survival.

Evolutionary psychology does not, however, limit its scope to such mundane faculties. Is the relatively low representation of women in the mathematical sciences worldwide a universal fact of culture, or might eons of evolution have selected male and female brains for different purposes? Can we understand in strictly Darwinian terms the tendency of older men to marry younger women? With a teenager likely to produce more children than a thirty-five-year-old, might such men be seen as succumbing to the power of evolutionary hardwiring that urges each of them to maximize the number of his offspring? Similarly, do younger women go for wealthy older men because natural selection has operated in the past to favor such a preference: a powerful male with plenty of resources? For now any answers to these questions are mainly conjectural. As we discover more of the genes underpinning behavior, however, I am confident that evolutionary psychology will migrate from its current position on the fringes of anthropology to the very heart of the discipline.

For now the power of genes to affect behavior is more evident in other species, whose nature we can actually manipulate using genetic tricks. One of the oldest, and most effective, of those tricks is artificial selection, which farmers have long used to increase milk yield in cows or wool quality in sheep. But its applications have not been limited to agriculturally valuable traits like these. Dogs are derived from wolves—possibly from wolf individuals that tended to hang around human settlements looking for a scrap and thereby conveniently assisting in garbage disposal. It is thought that they first laid claim to the title of "man's best friend" 10,000 years ago, roughly coinciding with the origin of agriculture. In the brief time since, the anatomical and behavioral diversity engendered by dog breeders has become literally a thing to behold. Dog shows are celebrations of the power of genes, with each breed effectively a genetic isolate—a freeze frame in the spectacular feature film of canine genetic diversity. Of course, the morphological differences are the most striking and fun to consider: the fluff ball called Pekingese; the enormous shaggy English mastiff, which sometimes weighs over 300 pounds; the stretched-out dachshund; the face-flattened bulldog. But it is the behavioral differences that I find most impressive.

Of course, not all dogs within a breed behave alike (or look alike), but typically individuals in the same breed have much more in common with one another than with specimens from other breeds. The Labrador retriever is affectionate and pliant; the greyhound is twitchy; the border collie will round up anything available if a flock of sheep is not to be had; the pit bull, as news reports occasionally remind us, is the canine embodiment of aggression. Some dog behaviors are so engrained as to have become stereotypes. Think of a pointer's elaborate "pointing" stance: that is no "stupid pet trick" taught each individual dog, but a hardwired part of the breed's genetic makeup. Despite their diversity, all modern dogs remain members of one species—meaning that in principle even the most apparently dissimilar pair can produce offspring, as one doughty male dachshund demonstrated in 1972 when he managed to inseminate a sleeping Great Dane bitch. Thirteen "Great Dachshunds" resulted.

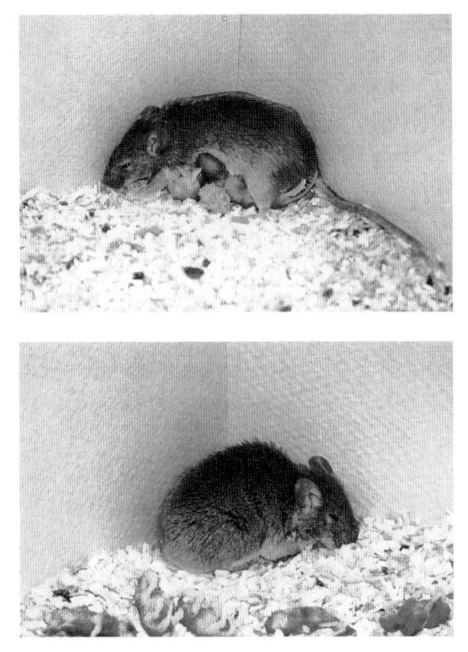

The impact of just one gene. At top, a normal mouse mother is highly attentive to her offspring. The mother below, lacking a functional fos-B *gene, ignores her newborns.*

While the basis of most behavior is surely polygenic—affected by many genes—a number of simple genetic manipulations in mice reveal that changing even a single gene can have a major behavioral effect. In 1999, Princeton neurologist Joe Tsien used sophisticated recombinant DNA techniques to create a "smart mouse" with extra copies of a gene producing a protein that acts as a receptor for chemical signals in the nervous system. Tsien's transgenic rodent performed better than normal mice in a battery of tests of learning and memory; for example, it was better at figuring out mazes and at retaining that knowledge. Tsien named the mouse strain "Doogie" after the precocious medic in the television show *Doogie Howser, M.D.* In 2002 Catherine Dulac at Harvard discovered that by deleting a gene from a mouse of her own she could affect the processing of chemical information contained in pheromones, the odors mice use to communicate. Whereas male mice will typically attack other males and attempt to mate with females, Dulac's doctored males failed to distinguish between males and females and attempted to mate with any mouse they encountered. Nurturing behavior in mice is also subject to a sex-specific manipulation of a single gene. Females instinctively look after their newborns, but Jennifer Brown and Mike Greenberg at Harvard Medical School found a way to short-circuit this native sense by knocking out the function of a gene called *fos*-B. Otherwise entirely normal, a mouse so altered simply ignores her offspring.

Rodents have even given us a glimpse of the mechanistic basis of what in humans we call "love" (in rodents, it's less romantically termed "pair-bonding").

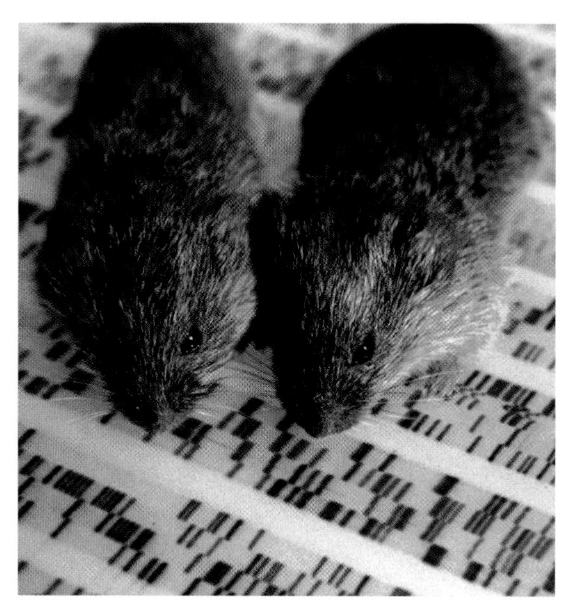

A recently pair-bonded prairie vole couple ponders the DNA sequence of the gene that makes them so lovey-dovey.

Mouselike voles are common throughout North America. Although they all look pretty much alike, different species have wildly different approaches to life. The prairie vole is monogamous, meaning that couples bond for life, but its close relative, the montane vole, is promiscuous: a male mates and moves on, and over the course of her life the female typically produces litters by several different males. What differences could underlie such widely divergent sexual strategies? Hormones provide the first half of the answer. In all mammals, oxytocin figures in many aspects of mothering: it stimulates both the contractions in labor and the production of breast milk for the newborn; thus it also plays a part in creating the nurturing bond between a mother and her young. Could the same hormone also engender another kind of bond, one between members of a prairie vole couple? In fact it does, together with another garden-variety mammalian hormone, vasopressin, which is primarily known for controlling urine production. But why is the montane vole, which also produces both hormones, such a randy little critter compared with its prairie cousin? The key, it turns out, is in the hormone receptors—the molecules that bind to circulating hormones, initiating a cell's response to the hormone signal.

Focusing on the vasopressin receptor, Tom Insel, a psychiatrist at Emory University, found a major difference between the vole species, not in the receptor gene itself but in an adjacent DNA region that determines when and where the gene is turned on. As a result the distribution of vasopressin receptors in the brain is very different in the prairie vole and the montane vole. But does this difference in gene regulation alone explain why one species is, in human terms, loving and the other cavalier? Apparently it does. Insel and his colleague Larry

Young inserted the prairie vole vasopressin gene, complete with its next-door regulatory region, into a regular lab mouse (a species that is typically promiscuous, like the montane vole). Though the transgenic mice did not instantly become romantic pair-bonders, Insel and Young observed a marked change in their behavior. Rather than doing the typical mouse thing of mating with a female and then hightailing it unceremoniously, the transgenic male appeared in contrast tenderly solicitous of the female—in short, the addition of the gene, while not a guarantee for everlasting love, did seem to make the affected mouse less of a rat.

We must not forget that human brain function remains a million miles removed from that of a mouse. No rodent, whether from the mountains or the prairies, has yet produced a major work of art. It nevertheless remains worthwhile to bear in mind that most sobering lesson of the Human Genome Project: our genome and the mouse's are strikingly similar. The basic genetic software governing both mice and men has not changed much over the 75 million years of evolution since our lineages separated.

Unable to target specific genes for inactivation or enhancement as mouse geneticists can, human geneticists must rely on what might be called "natural experiments"—spontaneous genetic changes that affect brain function. Many of the best characterized genetic disorders affect mental performance. Down syndrome, caused by having an extra copy of chromosome 21, results in lowered IQ and, in many cases, a disarmingly sunny disposition. People with Williams syndrome, caused by the loss of a small portion of chromosome 7, also have low IQs but are often preternaturally talented as musicians.

But these are instances in which the mental aspects of a given disorder are effectively by-products of systemwide dysfunction. Thus they tell us relatively little about the particular genetics of behavior. It's a bit like discovering that your computer doesn't work during a power outage. OK, so you know now that it requires electricity, but you've learned very little about the specifics of computer function. To understand the genetics of behavior we need to look at disorders that affect the mind directly.

Among the mental scourges that have attracted close attention from gene mappers, two of the most formidable are bipolar (or manic-depressive) disease (BPD) and schizophrenia. Both diseases have strong genetic components (identical-twin concordance for BPD is as high as 80 percent; for schizophrenia it is close to 50 percent), and both take a devastating toll on mental health worldwide. One in every hundred people is schizophrenic, and the figure is about the same for BPD.

As we have seen, mapping polygenic traits is difficult because each individual gene has only a small incremental effect, and the trait as a whole is often mediated strongly by the environment, as with both these afflictions. But the overall difficulty has also bred a bad habit among researchers: they tend to publish only positive results, failing to notify the field of eliminated possibilities. Compounding the problem is the converse: the understandable but ultimately counterproductive impulse to publish any correlation that appears, after countless other genetic markers have proved a dead end. Ideally, spotting a correlation should only be the beginning of more in-depth analysis to sort out meaningful results from statistical coincidence—after all, if we try enough markers, we should expect occasionally to see, even in the absence of genetic linkage, a correlation produced by chance alone. Too often the pressure to get results leads to premature pronouncements, which must later be withdrawn sheepishly after other groups fail to replicate the finding.

There are additional impediments to the gene hunt when mental illness is the target. Diagnosis, however psychiatric manuals may try to standardize it, is often more an art than a science. Cases may be identified on the basis of ambiguous symptoms, and so a portion of the individuals in a pedigree may be misdiagnosed; these false positives wreak havoc in the mapping analysis. Another complicating factor is that the disorders are defined and diagnosed according to their symptoms, and yet it is likely that a number of genetic causes result in a similar set of symptoms. Thus the genes underlying schizophrenia may differ from one case to the next. Even apparently clear differences between syndromes can prove messy when viewed through the microscope of genetics. Since 1957 we have known that BPD and unipolar disease (the condition characterized by depression alone) are genetically distinct syndromes, but, confusingly, there is some genetic overlap between the two: unipolar depression is much more common among relatives of BPD patients than in the overall population.

Partly for such reasons, the genetic culprits of mental illness have so far proved particularly elusive. A recent study reveals that as many as twelve chromosomes—half the total—have been shown through mapping analysis to contain genes contributing to schizophrenia. It's the same story for BPD, in which genes on ten chromosomes have been implicated. One interesting finding is that there does seem to be some overlap between the gene regions identified in the separate mapping studies of the two maladies. Perhaps, then, there are some genes responsible for the overall organization and structure of our brains. Malfunctions in these genes may be the cause of the delusional or hallucinatory episodes common to both BPD and schizophrenia. The history of this research is full of high hopes brought low. A study will identify a strong correlation in one pedigree, and then subsequent research will fail to generalize the result in other populations. Such was the case in 1987, with a much-publicized study of BPD in the Amish: a promising connection with chromosome 11 did not fare well in follow-up studies. Stanford geneticists Neil Risch and David Botstein have aptly articulated these disappointments:

> In no field has the difficulty [of mapping disease genes] been more frustrating than in the field of psychiatric genetics. Manic depression (bipolar illness) provides a typical case in point. Indeed, one might argue that the recent history of genetic linkage studies for this disease is rivaled only by the course of the illness itself. The euphoria of linkage findings being replaced by the dysphoria of non-replication [in other populations] has become a regular pattern, creating a roller coaster-type existence for many psychiatric genetics practitioners as well as their interested observers.

Without denying these difficulties I am extremely hopeful that we are right now entering an era of genetic analysis that will soon take us beyond this irritating game of "now we have it, now we don't." Two innovations hold the key. First, the "candidate gene" approach to finding the genes. With both the complete human genome sequence and a rudimentary functional understanding of many genes finally in hand, we can narrow our search as never before, homing in on genes with functions related to a given disorder. In the case of BPD, for example, a condition apparently connected with a fault in the mechanism by which the brain regulates its concentration of certain chemical neurotransmitters like serotonin and dopamine, we might choose to concentrate on genes that produce neurotransmitters or their receptors. Having chosen our candidate gene, we simply compare its sequence in affected and unaffected individuals to determine whether or not a particular variant might correlate with the disorder. In 2002, Eric Lander's team at MIT's Whitehead Institute surveyed seventy-six BPD candidate genes. Only one—a gene encoding the brain-specific nerve growth factor, the neurochemical tested as a possible treatment for Lou Gehrig's disease (see chapter 5)—proved to correlate with the disorder. But one truly relevant gene can be extremely valuable. The one found resides on chromosome 11, apparently vindicating the original Amish study, which long ago implicated the same region of the chromosome in BPD.

Technological improvements underlie the other reason for my optimism about the hunt for these elusive genes. To detect the subtle effect of a particular gene, we need extra-sensitive statistical analyses, which themselves require very large data-sets. Only with the advent of high-throughput sequencing and genetic-typing technologies have we had the capacity to collect appropriate data for huge numbers of markers from huge numbers of people. Not surprisingly, such industrial-scale genetic analysis is beyond the reach of most academic labs, so we will see biotech companies bankrolled by the pharmaceutical industry come to play an increasingly prominent role in this area. In 2002, two such companies, Genset in France and deCODE in Iceland, identified separate genes implicated in schizophrenia. These discoveries are a major step forward: because we have now fingered actual genes—as opposed to merely mapping an effect to a region of a chromosome—we can study gene function to learn about the biochemical basis of the disorder. Strikingly, both genes are involved in regulating the function of a particular neurotransmitter, glutamate.

With these new approaches—candidate genes and super-powerful genetic mapping—I am confident that we will soon uncover the major genes contributing to BPD and schizophrenia. Hopefully that will lead to improved treatments, as well as to a better understanding of how genes govern the workings of our brain.

For traits about whose neurochemical basis we have no clue, however, the roller-coaster ride of euphoric expectations dysphorically dashed is likely to continue. This has often been the case in studies of nonpathological behavior. Dean Hamer's 1993 analysis of the genetics of male homosexuality provides a case in point. It caused quite a stir when he found a particular region on the X chromosome that seemed to correlate with being gay. If being gay were proven to be as much a function of genes as is, say, skin color, then perhaps antidiscrimination legislation applicable to skin color was equally applicable to gays. Hamer's finding, however, has not withstood the test of time. Nevertheless, I suspect that as we develop more statistically powerful means of analysis (and learn to recognize and discount weaker correlations), we will indeed eventually identify some genetic factors that predispose us to our respective sexual orientations. But this should not be taken as purely determinist conjecture; environment is never to be discounted and a predisposition does not a predetermination make. My pasty complexion may predispose me to skin cancer but, absent the effects of ultraviolet input from the environment, my genes are merely a matter of potential.

Hamer's other high-profile discovery looks more robust. He looked into the genetics underlying the urge for novelty, one of five key "personality dimensions" identified by psychologists. Do you cower in a corner when your routine gets disrupted? Or do you go out of your way to avoid a rut, subjecting yourself to an ever-changing kaleidoscope of new adventures? These, of course, are the extremes. Hamer's evidence pointed to a slight but significant effect of variation in a gene underlying a receptor for the brain signal molecule dopamine. Some attempts to replicate this result have failed, but others have extended it, finding the same gene implicated in particular types of novelty seeking, including drug abuse.

Violence, too, can be viewed through the lens of genetics. Some people are more violent than others. That's a fact. And violent behavior may be governed by a single gene interacting with environmental factors. This does not, of course, mean that we all carry a "violence gene" (though it's likely that most violent individuals do possess a Y chromosome), but we have identified at least one simple genetic change that can lead to violent outbursts. In 1978, Dr. Hans Brunner, a clinical geneticist at University Hospital in Nijmegen, Holland, learned of a family whose men tended to border on mental retardation and were prone to aggressive episodes. Thirty years earlier, in an attempt to document this "curse," a relative had compiled an extensive dossier on the family's woes. Brunner brought the survey up-to-date. He found eight men in the clan who, despite coming from different nuclear families, evinced similar patterns of violence. One had raped his sister and subsequently stabbed a prison guard; another used his car to hit his boss after being mildly reprimanded for laziness; two others were arsonists.

That only men were affected suggested sex linkage. The inheritance pattern was consistent with a gene likely on the X chromosome, and recessive, meaning that it was typically unexpressed in women, in whom the other (normal) copy on their second X would mask the faulty one's effect. In men, with their single X, the recessive variant was automatically expressed. By comparing the DNA of affected and unaffected members of the family, Brunner and his team duly mapped the gene to the long arm of the X chromosome. In collaboration with Xandra Breakfield at Massachusetts General Hospital, he found that the eight violent men all had a mutated—and nonfunctional—copy of a gene coding for monoamine oxidase. This protein, found in the brain, regulates levels of a class of neurotransmitters called "monoamines," which include adrenaline and serotonin.

The monoamine oxidase story does not end with the eight violent Dutchmen. It turns out to provide an illuminating glimpse of the interaction between genes and the environment, the complex duet of nature and nurture that informs all our behavior. In 2002, Avshalom Caspi and others at London's Institute of Psychiatry examined why some boys from abusive homes grow up normal while others end up antisocial (in the technical sense of having a history of behavioral problems—not in the sense of preferring to keep company with Web pages over people or of tending to be found picking at the canapés in a lonely corner at parties). The survey revealed a genetic predictor of development: the presence or absence of a mutation in the region adjacent to the monoamine oxidase gene, the switch regulating the amount of the enzyme produced. Maltreated boys with high levels of the enzyme were less likely to become antisocial than those with low levels. In the latter case, genes and the environment conspired to predispose the boys to lives punctuated by brushes with the law. Girls are less likely to be affected because, with the gene located on the X chromosome, they must inherit *two* copies of the low-level version rather than one. Girls who do have two copies, however, are likely to have antisocial tendencies similar to those of affected boys. But again the causal relation is nowhere near 100 percent in either boys or girls: growing up abused and having low monoamine oxidase levels in no way guarantees a career in crime.

Among the most surprising discoveries of a monogenic (single-gene) impact on a complex form of human behavior is what the press have dubbed the "grammar gene." As we discussed in the context of human evolution (chapter 9), in 2001 mutations detected by Tony Monaco at Oxford in the FOXP2 gene were found to impair the ability to use and process language. Not only do those so affected have difficulty articulating, but they are stymied by simple grammatical reasoning that poses no trouble for the typical four-year-old: "Every day I wug; yesterday I ____." FOXP2, remember, encodes a transcription factor—a genetic switch—that apparently plays a crucial role in brain development. Rather than exerting a simple direct behavioral impact (like that of monoamine oxidase), FOXP2 affects behavior by shaping the very organ at the center of it all. FOXP2 will prove, I believe, a model for momentous discoveries yet to be made; if I am right, many of the most important genes governing behavior will indeed turn out to be those involved in constructing that most extraordinary of organs, that still supremely inscrutable mass of matter, the human brain. These genes influence us by how they build the exquisite piece of hardware that mediates all we do.

We are as yet in the early days of our attempts to understand the genetic underpinnings of our behavior, both that which we all have in common—human nature—and that which sets us apart, one person from another. But this is a fast-moving area of research; I'm sure that what I've written will be out-of-date by the time this book is published. The future promises a detailed genetic dissection of personality, and it is hard to imagine that what we discover will not tip the scales of the nature/nurture debate more and more in the direction of nature—a frightening thought for some, but only if we persist in being held hostage to a static, ultimately meaningless dichotomy. To find that any trait, even one with formidable political implications, has a mainly genetic basis is not to find something set immutably in stone. It is merely to understand the nature upon which nurture is ever acting, and those things we, as a society and as individuals, need to do if we are better to assist the process. Let us not allow transient political considerations to set the scientific agenda. Yes, we may uncover truths that make us uneasy in the light of our present circumstances, but it is those circumstances, not nature's truth, to which policy makers ought to address themselves. As those Irish children who packed the hedge schools understood very well, knowledge, however awkwardly acquired, is still preferable to ignorance.

CODA

. . .

OUR GENES AND OUR FUTURE

"The event on which this fiction is founded has been supposed, by Dr. Darwin, and some of the physiological writers of Germany, as not of impossible occurrence."

So begins Percy Bysshe Shelley's anonymous preface to his wife Mary Shelley's novel *Frankenstein,* a story whose grip on the modern imagination has exceeded by far that of anything the poet himself ever wrote. Perhaps no work since *Frankenstein* has so hauntingly captured the terrifying thrill of science at the point of discovering the secret of life. And probably none has dealt so profoundly with the social consequences of having appropriated such godlike power.

The idea of animating the inanimate, and improving upon life as it occurs naturally on earth, had captured the human imagination long before the publication of Mary Shelley's work in 1818. Greek mythology tells of the sculptor Pygmalion, who successfully petitioned Aphrodite, goddess of love, to breathe life into the statue of the beautiful woman he had carved from ivory. But it was during the feverish burst of scientific progress following the Enlightenment that it first dawned upon scientists that the secret of life might be within human reach. Indeed, the Dr. Darwin to whom the preface refers is not the familiar Charles but rather his grandfather Erasmus, whose experimental use of electricity to spark life back into dead body parts fascinated his acquaintance Shelley. In retrospect we know that Dr. Darwin's exploration of what was called "galvanism" was a red herring; the secret of life remained a secret until 1953.

Only with the discovery of the double helix and the ensuing genetic revolution have we had grounds for thinking that the powers held traditionally to be the exclusive property of the gods might one day be ours. Life, we now know, is nothing but a vast array of coordinated chemical reactions. The "secret" to that coordination is the breathtakingly complex set of instructions inscribed, again chemically, in our DNA.

But we still have a long way to go on our journey toward a full understanding of how DNA does its work. In the study of human consciousness, for example, our knowledge is so rudimentary that arguments incorporating some element of vitalism persist, even as these notions have been debunked elsewhere. Nevertheless, both our understanding of life and our demonstrated ability to manipulate it are facts of our culture. Not surprisingly, then, Mary Shelley has many would-be successors: artists and scientists alike have been keen to explore the ramifications of our newfound genetic knowledge.

Many of these efforts are shallow and betray their creators' ignorance of what is and is not biologically feasible. But one in particular stands out in my mind as raising important questions, and doing so in a stylish and compelling way. Andrew Niccols's 1997 film *Gattaca* carries to the present limits of our imagination the implications of a society obsessed with genetic perfection. In a future world two types of humans exist—a genetically enhanced ruling class and an underclass that lives with the imperfect genetic endowments of today's humans. Supersensitive DNA analyses ensure that the plum jobs go to the genetic elite while "in-valids" are discriminated against at every turn. Gattaca's hero is the "in-valid" Vincent (Ethan Hawke), conceived in the heat of reckless passion by a couple in the back of a car. Vincent's younger brother, Anton, is later properly engineered in the laboratory and so endowed with all the finest genetic attributes. As the two grow up, Vincent is reminded of his own inferiority every time he tries, fruitlessly, to best his little brother in swim races. Genetic discrimination eventually forces Vincent to accept a menial job as a porter with the Gattaca Corporation.

At Gattaca, Vincent nurtures an impossible dream: to travel into space. But to qualify for the manned mission to Titan he must conceal his "in-valid" status. He therefore assumes the identity of the genetically elite Jerome (Jude Law), a one-time athlete, who, crippled in an accident, needs Vincent's help. Vincent buys samples of Jerome's hair and urine and uses them to secure illicit admission into the flight-training program. All seems to be going well when he encounters the statuesque Irene (Uma Thurman) and falls in love. But a week before he is to fly off into space, disaster strikes: the mission director is murdered and in the ensuing police investigation the hair of an "in-valid" is discovered at the crime scene. An eyelash Vincent has lost threatens not only to dash his desperate dream but to unjustly implicate him by DNA evidence as the director's murderer. Vincent's unmaking seems foreordained, but he evades a nightmarish genetic dragnet until another of Gattaca's directors is found to be the actual murderer. The film's ending is only semi-happy: Vincent will fly off into space but without Irene, who is found to carry certain genetic imperfections incompatible with long space missions. In real life, the two actors who play Vincent and Irene have their futures more under their personal control. Ethan Hawke and Uma Thurman later married and now live in New York City.

Few, if any, of us would wish to imagine our descendants living under the sort of genetic tyranny suggested by *Gattaca.* Setting aside the question of whether the scenario foreseen is technologically feasible, we must address the central issue raised by the film: Does DNA knowledge make a genetic caste system inevitable? A world of congenital haves and have-nots? The most pessimistic commentators foresee an even worse scenario: Might we one day go so far as to breed a race of clones, condemned to servile lives mandated by their DNA? Rather than strive to fortify the weak, would we aim to make the descendants of the strong ever stronger? Most fundamentally, should we manipulate human genes at all? The answers to these questions depend very much on our views of human nature.

Today much of the public paranoia surrounding the dangers of human genetic manipulation is inspired by a legitimate recognition of our selfish side—that aspect of our nature that evolution has hardwired to promote our own survival, if necessary at the expense of others. Critics envision a world in which genetic knowledge would be used solely to widen the gap between the privileged (those best positioned to press genetics into their own service) and the downtrodden (those whom genetics can only put at greater disadvantage). But such a view recognizes only one side of our humanity.

If I see the consequences of our increasing genetic understanding and know-how rather differently, it is because I acknowledge the other side as well. Disposed though we might be to competition, humans are also profoundly social.

Compassion for others in need or distress is as much a genetic element of our nature as the tendency to smile when we're happy. Even if some contemporary moral theorists are content to ascribe our unselfish impulses to ultimately selfish considerations—kindness to others seen as simply a conditioned way of promoting the same benefit in return—the fact remains: ours is a uniquely social species. Ever since our ancestors first teamed up to hunt a mammoth for dinner, cooperation among individuals has been at the heart of the human success story. Given the powerful evolutionary advantage of acting collectively in this way, natural selection itself has likely endowed each of us with a desire to see others (and therefore our society) do well rather than fail.

Even those who accept that the urge to improve the lot of others is part of human nature disagree on the best way to go about it. It is a perennial subject of social and political debate. The prevailing orthodoxy holds that the best way we can help our fellow citizens is by addressing problems with their nurture. Underfed, unloved, and uneducated human beings have diminished potential to lead productive lives. But as we have seen, nurture, while greatly influential, has its limits, which reveal themselves most dramatically in cases of profound genetic disadvantage. Even with the most perfectly devised nutrition and schooling, boys with severe fragile X disease will still never be able to take care of themselves. Nor will all the extra tutoring in the world ever grant naturally slow learners a chance to get to the head of the class. If, therefore, we are serious about improving education, we cannot in good conscience ultimately limit ourselves to seeking remedies in nurture. My suspicion, however, is that education policies are too often set by politicians to whom the glib slogan "leave no child behind" appeals precisely because it is so completely unobjectionable. But children *will* get left behind if we continue to insist that each one has the same potential for learning.

We do not as yet understand why some children learn faster than others, and I don't know when we will. But if we consider how many commonplace biological insights, unimaginable fifty years ago, have been made possible through the genetic revolution, the question becomes pointless. The issue rather is this: Are we prepared to embrace the undeniably vast potential of genetics to improve the human condition, individually and collectively? Most immediate, would we want the guidance of genetic information to design learning best suited to our children's individual needs? Would we in time want a pill that would allow fragile X boys to go to school with other children, or one that would allow naturally slow learners to keep pace in class with naturally fast ones? And what about the even more distant prospect of viable germ-line gene therapy? Having identified the relevant genes, would we want to exercise a future power to transform slow learners into fast ones before they are even born? We are not dealing in science fiction here: we can already give mice better memories. Is there a reason why our goal shouldn't be to do the same for humans?

One wonders what our visceral response to such possibilities might be had human history never known the dark passage of the eugenics movement. Would we still shudder at the term "genetic enhancement"? The reality is that the idea of improving on the genes that nature has given us alarms people. When discussing our genes, we seem ready to commit what philosophers call the "naturalistic fallacy," assuming that the way nature intended it is best. By centrally heating our homes and taking antibiotics when we have an infection, we carefully steer clear of the fallacy in our daily lives, but mentions of genetic improvement have us rushing to run the "nature knows best" flag up the mast. For this reason, I think that the acceptance of genetic enhancement will most likely come about through efforts to prevent disease.

Germ-line gene therapy has the potential for making humans resistant to the ravages of HIV. The recombinant DNA procedures that have let plant molecular geneticists breed potatoes resistant to potato viruses could equally well make humans resistant to AIDS. But should this be pursued? There are those who would argue that rather than altering people's genes, we should concentrate our efforts on treating those we can and impressing upon everyone else the dangers of promiscuous sex. But I find such a moralistic response to be profoundly *immoral*. Education has proven a powerful but hopelessly insufficient weapon in our war. As I write, we are entering the third decade of the worldwide AIDS crisis; our best scientific minds have been bamboozled by the virus's remarkable capacity for eluding attempts to control it. And while the spread of the disease has been slowed for the moment in the developed world, huge swaths of the planet tick away as demographic time bombs. I am filled with dread for the future of those regions, populated largely by people who are neither wealthy nor educated enough to mount an effective response. We may wishfully expect that powerful antiviral drugs or effective HIV vaccines will be produced economically enough for them to be available to everyone everywhere. But given our record in developing therapies to date, the odds against such dramatic progress occurring are high. And yet those who propose to use germ-line gene modifications to fight AIDS may, sadly, need to wait until such conventional hopes turn to despair—and global catastrophe—before being given clearance to proceed.

All over the world government regulations now forbid scientists from adding DNA to human germ cells. Support for these prohibitions comes from a variety of constituencies. Religious groups—who believe that to tamper with the human germ line is in effect to play God—account for much of the strong knee-jerk opposition among the general public. For their part, secular critics, as we have seen, fear a nightmarish social transformation such as that suggested in *Gattaca*—with natural human inequalities grotesquely amplified and any vestige of an egalitarian society erased. But though this premise makes for a good script, to me it seems no less fanciful than the notion that genetics will pave the way to utopia.

But even if we allow hypothetically that gene enhancement *could*—like any powerful technology—be applied to nefarious social ends, that only strengthens the case for our developing it. Considering the near impossibility of repressing technological progress, and the fact that much of what is now prohibited is well on its way to becoming practicable, do we dare restrain our own research community and risk allowing some culture that does not share our values to gain the upper hand? From the time the first of our ancestors fashioned a stick into a spear, the outcomes of conflicts throughout history have been dictated by technology. Hitler, we mustn't forget, was desperately pressing the physicists of the Third Reich to develop nuclear weapons. Perhaps one day, the struggle against a latter-day Hitler will hinge on our mastery of genetic technologies.

I see only one truly rational argument for delay in the advance of human genetic enhancement. Most scientists share this uncertainty: can germ-line gene therapy ever be carried out safely? The case of Jesse Gelsinger has cast a long shadow on gene therapy in general. It's worth pointing out, though, that contrary to appearances, germ-line gene therapy should in principle be easier to accomplish safely than somatic cell therapy. In the latter case, we are introducing genes into billions of cells, and there is always a chance, as in the recent SCID case in France, that a crucial gene or genes will be damaged in one of those cells, resulting in the nightmarish side effect of cancer. With germ-line gene therapy, in contrast, we are inserting DNA into a single cell, and the whole process can accordingly be much more tightly monitored. But the stakes are even higher in germ-line therapy: a failed germ-line experiment would be an unthinkable catastrophe—a human being born flawed, perhaps unimaginably so, owing to our manipulation of his or her genes. The consequences would be tragic. Not only would the affected family suffer, but all of humankind would lose because science would be set back.

When gene therapy experiments in mice run aground, no career is aborted, no funding withdrawn. But should gene improvement protocols ever lead to children with diminished rather than improved potential for life, the quest to harness the power of DNA would surely be delayed for years. We should attempt human experimentation only after we have perfected methods to introduce functional genes into our close primate relatives. But even when monkeys and chimpanzees (an even closer match) can be safely gene enhanced, the start of human experimentation will require resolute courage; the promise of enormous benefit won't be fulfilled except through experiments that will ultimately put lives at some risk. As it is, conventional medical procedures, especially new ones, require similar courage: brain surgery too may go awry, and yet patients will undergo it if its potential positives outweigh the dangers.

My view is that, despite the risks, we should give serious consideration to germ-line gene therapy. I only hope that the many biologists who share my opinion will stand tall in the debates to come and not be intimidated by the inevitable criticism. Some of us already know the pain of being tarred with the brush once reserved for eugenicists. But that is ultimately a small price to pay to redress genetic injustice. If such work be called eugenics, then I am a eugenicist.

Over my career since the discovery of the double helix, my awe at the majesty of what evolution has installed in our every cell has been rivaled only by anguish at the cruel arbitrariness of genetic disadvantage and defect, particularly as it blights the lives of children. In the past it was the remit of natural selection—a process that is at once marvelously efficient and woefully brutal—to eliminate those deleterious genetic mutations. Today, natural selection still often holds sway: a child born with Tay-Sachs who dies within a few years is—from a dispassionate biological perspective—a victim of selection against the Tay-Sachs mutation. But now, having identified many of those mutations that have caused so much misery over the years, it is in our power to sidestep natural selection. Surely, given some form of preemptive diagnosis, anyone would think twice before choosing to bring a child with Tay-Sachs into the world. The baby faces the prospect of three or four long years of suffering before death comes as a merciful release. And so if there is a paramount ethical issue attending the vast new genetic knowledge created by the Human Genome Project, in my view it is the slow pace at which what we now know is being deployed to diminish human suffering. Leaving aside the uncertainties of gene therapy, I find the lag in embracing even the most unambiguous benefits to be utterly unconscionable. That in our medically advanced society almost no women are screened for the fragile X mutation a full decade after its discovery can attest only to ignorance or intransigence. Any woman reading these words should realize that one of the important things she can do as a potential or actual parent is to gather information on the genetic dangers facing her unborn children—by looking for deleterious genes in her family line and her partner's, or, directly, in the embryo of a child she has conceived. And let no one suggest that a woman is not entitled to this knowledge. Access to it is her right, as it is her right to act upon it. She is the one who will bear the immediate consequences.

Two years ago, my views on this subject received a very cold reception in Germany. The publication of my essay, "Ethical Implications of the Human Genome Project," in the highly respected newspaper *Frankfurter Allgemeine Zeitung* (FAZ), provoked a storm of criticism. Perhaps this was the editors' intent: Without my knowledge, let alone consent, the paper had given my essay a new title devised by the translator as "The Ethic of the Genome—Why We Should Not Leave the Future of the Human Race to God." While I subscribe to no religion and make no secret of my secular views, I would never have framed my position as a provocation to those who do. A surprisingly hostile response came from a man of science, the president of the German Federal Chamber of Medical Doctors, who accused me of "following the logic of the Nazis who differentiate between a life worth living and a life not worth living." A day later, an editorial entitled "Unethical Offer" appeared in the same paper that had published mine. The writer, Henning Ritter, argued with self-righteous conviction that in Germany the decision to end the lives of genetically damaged fetuses would never become a private matter. In fact, his grandstanding displayed a simple ignorance of the nation's law; in Germany today, it is solely the right of a pregnant woman, upon receipt of medical advice, to decide whether to carry her fetus to term.

The more honorable critics were those who argued openly from personal beliefs, rather than exploiting the terrifying specter of the German past. The respected German president, Johannes Rau, countered my views with an assertion that "value and sense are not solely based on knowledge." As a practicing Protestant, he finds truths in religious revelation while I, a scientist, depend only on observation and experimentation. I therefore must evaluate actions on the basis of my moral intuition. And I see only needless harm in denying women access to prenatal diagnosis until, as some would have it, cures exist for the defects in question. In a less measured comment, the Protestant theologian Dietmar Mieth called my essay the "Ethics of Horror," taking issue with my assertion that greater knowledge will furnish humans better answers to ethical dilemmas. But the existence of a dilemma implies a choice to be made, and choice to my mind is better than no choice. A woman who learns that her fetus has Tay-Sachs now faces a dilemma about what to do, but at least she has a choice, where before she had none. Though I am sure that many German scientists agree with me, too many seem to be cowed by the political past and the religious present: except for my longtime valued friend Benno Müller-Hill, whose brave book on Nazi eugenics, *Murderous Science* (*Tödliche Wissenschaft*), still rankles the German academic establishment, no German scientist saw reason to rise to my defense.

I do not dispute the right of individuals to look to religion for a private moral compass, but I do object to the assumption of too many religious people that atheists live in a moral vacuum. Those of us who feel no need for a moral code written down in an ancient tome have, in my opinion, recourse to an innate moral intuition long ago shaped by natural selection promoting social cohesion in groups of our ancestors.

The rift between tradition and secularism first opened by the Enlightenment has, in more or less its present form, dictated biology's place in society since the Victorian period. There are those who will continue to believe humans are creations of God, whose will we must serve, while others will continue to embrace the empirical evidence indicating that humans are the product of many millions of generations of evolutionary change. John Scopes, the Tennessee high school teacher famously convicted in 1925 of teaching evolution, continues to be symbolically retried in the twenty-first century; religious fundamentalists, having their say in designing public school curricula, continue to demand that a religious story be taught as a serious alternative to Darwinism. With its direct contradiction of religious accounts of creation, evolution represents science's most direct incursion into the religious domain and accordingly provokes the acute defensiveness that characterizes creationism. It could be that as genetic knowledge grows in centuries to come, with ever more individuals coming to understand themselves as products of random throws of the genetic dice—chance mixtures of their parents' genes and a few equally accidental mutations—a new gnosis in fact much more ancient than today's religions will come to be sanctified. Our DNA, the instruction book of human creation, may well come to rival religious scripture as the keeper of the truth.

I may not be religious, but I still see much in scripture that is profoundly true. In the first letter to the Corinthians, for example, Paul writes:

> Though I speak with the tongues of men and of angels, but have not love, I have become sounding brass or a clanging cymbal.
>
> And though I have the gift of prophecy, and understand all mysteries and all knowledge, and though I have all faith, so that I could remove mountains, but have not love, I am nothing.

Paul has in my judgment proclaimed rightly the essence of our humanity. Love, that impulse which promotes our caring for one another, is what has permitted our survival and success on the planet. It is this impulse that I believe will safeguard our future as we venture into uncharted genetic territory. So fundamental is it to human nature that I am sure that the capacity to love is inscribed in our DNA—a secular Paul would say that love is the greatest gift of our genes to humanity. And if someday those particular genes too could be enhanced by our science, to defeat petty hatreds and violence, in what sense would our humanity be diminished?

In addition to laying out a misleadingly dismal vision of our future within the film itself, the creators of *Gattaca* concocted a promotional tag line aimed at the deepest prejudices against genetic knowledge: "There is no gene for the human spirit." It remains a dangerous blind spot in our society that so many wish this were so. If the truth revealed by DNA could be accepted without fear, we should not despair for those who follow us.

NOTES

. . .

INTRODUCTION: THE SECRET OF LIFE

xiii "Today, we are": White House Press Release, available at: http://www.ornl.gov/hgmis/project/clinton1.html

CHAPTER 1: BEGINNINGS OF GENETICS

6 "Hair, nails, veins": Anaxagoras as cited in F. Vogel and A. G. Motulsky, *Human Genetics* (Berlin, N.Y.: Springer, 1996), p. 11.

9 "very difficult for me": Mendel as cited in R. Marantz Henig, *A Monk and Two Peas* (London: Weidenfeld & Nicolson, 2000), pp. 117–18.

15 "The elephant is": Charles Darwin, *The Origin of Species* (New York: Penguin, 1985), p. 117.

17 "I profess to be": Francis Galton, *Narrative of an Explorer in Tropical South Africa* (London: Ward Lock, 1889), pp. 53–4.

19 "a devil": William Shakespeare, *The Tempest* (IV:i:188–9).

19 "I have no patience": Francis Galton, *Hereditary Genius* (London: MacMillan, 1892), p. 12.

19 "It is easy": ibid., p. 1.

20 "there is now no": George Bernard Shaw as cited in Diane B. Paul, *Controlling Human Heredity* (Atlantic Highlands, N.J.: Humanities Press, 1995), p. 75.

22 "The Wyandotte is": ibid., p. 66.

24 "family with mechanical": C. B. Davenport, *Heredity in Relation to Eugenics* (New York: Henry Holt, 1911), p. 56.

25 "broad-shouldered, dark hair": ibid., p. 245.

26 "More children": Margaret Sanger as quoted in D. M. Kennedy, *Birth Control in America* (New Haven: Yale University Press, 1970), p. 115.

27 "criminals, idiots": Harry Sharp as cited in E. A. Carlson, *The Unfit* (Cold Spring Harbor, N.Y.: Cold Spring Harbor Laboratory Press, 2001), p. 218.

27 "It is better": Oliver Wendell Holmes as cited in ibid., p. 255.

28 "Under existing": Madison Grant, *The Passing of the Great Race* (New York: Scribner, 1916), p. 49.

29 "America must": Calvin Coolidge as cited in D. Kevles, *In the Name of Eugenics* (Cambridge, Mass.: Harvard University Press, 1995), p. 97.

29 "the farseeing": Harry Laughlin as cited in S. Kühl, *The Nazi Connection* (New York: Oxford University Press, 1994), p. 88.

31 "must declare": Adolf Hitler's *Mein Kampf* as cited in Paul, p. 86.

31 "those who": Adolf Hitler, *Mein Kampf,* trans. Ralph Manheim (Boston: Houghton Mifflin Company, 1971), p. 404.

31 "law for": Benno Müller-Hill, *Murderous Science* (Cold Spring Harbor, N.Y.: Cold Spring Harbor Laboratory Press, 1998), p. 35.

31 "extra-marital": ibid.

32 "simply the meddlesome": Alfred Russel Wallace as cited in A. Berry, *Infinite Tropics* (New York: Verso, 2002), p. 214.

32 "orthodox eugenicists": Raymond Pearl as cited in D. Miklos and E. A. Carlson, "Engineering American Society: The Lesson of Eugenics," *Nature Genetics* 1 (2000): 153–58.

CHAPTER 2: THE DOUBLE HELIX

36 "Inheritance insures": Friedrich Miescher as cited in Franklin Portugal and Jack Cohen, *A Century of DNA* (Cambridge, Mass.: MIT Press, 1977), p. 107.

46 "stupid, bigoted": Rosalind Franklin as cited in Brenda Maddox, *Rosalind Franklin* (New York: HarperCollins, 2002), p. 82.

55 "Nobody told me": Linus Pauling, interview, as cited at http://www.achievement.org/autodoc/page/pau0int-1

59 "The most beautiful experiment": John Cairns as quoted in Horace Judson, *The Eighth Day of Creation* (New York: Simon & Schuster, 1979), p. 188.

CHAPTER 3: READING THE CODE

71 "That's when I saw": Sydney Brenner, *My Life in Science* (London: BioMed Central, 2001), p. 26.

74 "We're the only two": Francis Crick as quoted in Horace Judson, *The Eighth Day of Creation* (New York: Simon & Schuster, 1979), p. 485.

81 "Without giving me": François Jacob as quoted in ibid., p. 385.

CHAPTER 4: PLAYING GOD

88 "rivaled the importance": Jeremy Rifkin as quoted by Randall Rothenberg in "Robert A. Swanson: Chief Genetic Officer," *Esquire,* December 1984.

91 "from corned beef to": Stanley Cohen, http://www.accessexcellence.org/AB/WYW/cohen/

95 "She made": Paul Berg as quoted at http://www.ascb.org/profiles/9610.html

96 "scientists throughout": Paul Berg et al., "Potential Biohazards of Recombinant DNA Molecules," letter to *Science* 185 (1974): 303.

96 "until the potential": ibid.

97 "our concern": ibid.

97 "the molecular biologists had clearly reached": Michael Rogers, "The Pandora's Box Congress," *Rolling Stone* 189 (1975): 36–48.

100 "I felt": Leon Heppel as quoted in James D. Watson and J. Tooze, *The DNA Story* (San Francisco: W. H. Freeman and Co., 1981), p. 204.

102 "In his cranberry": Arthur Lubow as cited in ibid., p. 121.

102 "In today's": Alfred Vellucci as cited in ibid., p. 206.

104 "Compared to": Watson as cited in James D. Watson, *A Passion for DNA* (Cold Spring Harbor, N.Y.: Cold Spring Harbor Laboratory Press, 2001), p. 73.

108 "You get a nice medal": Fred Sanger as quoted by Anjana Ahuja, "The Double Nobel Laureate Who Began the Book of Life," *The Times* (London), 12 January 2000.

CHAPTER 5: DNA, DOLLARS, AND DRUGS

113 "to become": Herb Boyer as quoted in Stephen Hall, *Invisible Frontiers* (New York: Oxford University Press, 2002), p. 65.

121 "a live human-made": *Diamond vs. Chakrabarty et al.* as cited in Nicholas Wade, "Court Says Lab-Made Life Can Be Patented," *Science* 208 (1980): 1445.

132 "I'll *make* it an issue": Jeremy Rifkin as cited in Daniel Charles, *Lords of the Harvest* (Cambridge, Mass.: Perseus, 2001), p. 94.

CHAPTER 6: TEMPEST IN A CEREAL BOX

135 "If man": http://www.nrdc.org/health/pesticides/hcarson.asp

138 "operating outside": Mary-Dell Chilton et al. as cited in Daniel Charles, *Lords of the Harvest* (Cambridge, Mass.: Perseus, 2001), p. 16.

138 "loved the smell": Rob Horsch as quoted in ibid., p. 1.

150 "Put a molecular": Bob Meyer as quoted in ibid., p. 132.

153 "I naively": Roger Beachy, Daphne Preuss, and Dean Dellapenna, "The Genomic Revolution: Everything You Wanted to Know About Plant Genetic Engineering but Were Afraid to Ask," *Bulletin of the American Academy of Arts and Sciences,* Spring 2002, p. 31.

155 "After BSE": Friends of the Earth press release as cited in Charles, p. 214.

156 "this kind": Charles, Prince of Wales, "The Seeds of Disaster," *Daily Telegraph* (London), 8 June 1998.

161 "Where genetically": E. O. Wilson, *The Future of Life* (New York: Knopf, 2002), p. 163.

CHAPTER 7: THE HUMAN GENOME

167 "put Santa Cruz": Robert Sinsheimer as quoted in Robert Cook-Deegan, *The Gene Wars* (New York: W. W. Norton & Co., 1994), p. 79.

167 "DOE's program": David Botstein as cited in ibid., p. 98.

168 "the National Bureau": James Wyngaarden as quoted in ibid., p. 139.

168 "It means": David Botstein as quoted in ibid., p. 111.

169 "an incomparable": Walter Gilbert as cited in ibid., p. 88.

171 "from the very start": James Wyngaarden as quoted in ibid., p. 142.

174 "The revelation": Kary B. Mullis, "The Unusual Origin of the Polymerase Chain Reaction," *Scientific American* 262 (April 1990): 56–65.

174 "Mullis had": Frank McCormick as quoted in Nicholas Wade, "After the Eureka, a Nobelist Drops Out," *New York Times,* 15 September 1998.

182 "If somebody": William Haseltine as quoted in Paul Jacobs and Peter G. Gosselin, "Experts Fret Over Effect of Gene Patents on Research," *Los Angeles Times,* 28 February 2000.

182 "We'd be entitled": William Haseltine as quoted in ibid.

184 "I was": Francis Collins as quoted in interview, *Christianity Today,* 1 October 2001.

185 "little more": John Sulston and Georgina Ferry, *The Common Thread* (London: Bantam Press), p. 123.

185 "After we'd invested": Bridget Ogilvie as quoted in ibid., p. 125.

188 "Fix it": President Clinton as quoted in Kevin Davies, *Cracking the Code* (New York: The Free Press, 2001), p. 238.

188 "I'd love": Rhoda Lander as quoted in Aaron Zitner, "The DNA Detective," *Boston Globe Sunday Magazine,* 10 October 1999.

188 "an isolated": Eric Lander as quoted in ibid.

188 "I pretty much": Eric Lander as quoted in ibid.

191 "Today, we are": White House Press Release, available at: http://www.ornl.gov/hgmis/project/clinton1.html

CHAPTER 8: READING GENOMES

195 "seeing the surface": Mark Patterson as cited in Kevin Davies, *Cracking the Code* (New York: The Free Press, 2001), p. 194.

206 "Really trust": Barbara McClintock as paraphrased by Elizabeth Blackburn at http://www.cshl.edu/cgi-bin/ubb/library/ultimatebb.cgi?ubb=get_topic;f=1;t=000015

212 "had felt like an outcast": Claire Fraser as quoted in Ricki Lewis, "Exploring the Very Depths of Life," *Rennselaer Magazine,* March 2001.

212 "Well, young lady": Claire Fraser as quoted in ibid.

212 "We went to": Claire Fraser as quoted in ibid.

221 "a new kind": http://cmgm.stanford.edu/biochem/brown.html

221 "We're toddlers": Pat Brown as quoted in Dan Cray, "Gene Detective," *Time* 158 (20 August 2001): 35–36.

221 "It was like thinking": Pat Brown as quoted in ibid.

225 "Because embryos are beautiful": Eric Wieschaus as quoted in Ethan Bier, *The Coiled Spring* (Cold Spring Harbor, N.Y.: Cold Spring Harbor Laboratory Press, 2000), p. 64

CHAPTER 9: OUT OF AFRICA

231 "It was like": Ralf Schmitz as quoted in Steve Olson, *Mapping Human History* (Boston: Houghton Mifflin, 2002), p. 80.

232 "I can't": Matthias Krings as quoted in Patricia Kahn and Ann Gibbons, "DNA from an extinct human," *Science* 277 (1997): 176–78.

232 "That's when": Matthias Krings as quoted in ibid.

235 "If everyone": Allan Wilson as quoted by Mary-Claire King at http://www.chemheritage.org/EducationalServices/pharm/chemo/readings/king.htm

240 "It turned out": Luigi Luca Cavalli-Sforza as quoted in Olson, p. 164.

249 "If we": Charles Darwin, *The Origin of Species* (New York, Penguin, 1985), p. 406.

CHAPTER 10: GENETIC FINGERPRINTING

261 "having a white woman": Brooke A. Masters, "For Trucker, the High Road to DNA Victory," *Washington Post,* Saturday, 8 December 2001, p. B01.

262 "unwelcome precedent": the Director of the Virginia Department of Criminal Justice as quoted at http://www.innocenceproject.org/case/display_profile.php?id=99

263 "profile provides": Alec Jeffreys, Victoria Wilson, and Swee Lay Thein, "Hypervariable 'minisatellite' regions in human DNA," *Nature* 314 (1985): 67–73.

263 "In theory": Alec Jeffreys as quoted at http://www.dist.gov.au/events/ausprize/ap98/jeffreys.html

266 "must be sufficiently": *Frye vs. United States,* 293 F.2d 1013, at 104.

268 "The implementation": Eric Lander, "Population genetic considerations in the forensic use of DNA typing," in Jack Ballantyne, et al., *DNA Technology and Forensic Science* (Cold Spring Harbor, N.Y.: Cold Spring Harbor Laboratory Press, 1989), p. 153.

269 "mountain of evidence": Johnnie Cochran as quoted at http://simpson.walraven.org/sep27.html

272 "Where is": Barry Scheck as quoted at http://simpson.walraven.org/apr11.html

275 "State of Wisconsin": Geraldine Sealey, "DNA Profile Charged in Rape," http://abcnews.go.com/sections/us/DailyNews/dna991007.html

275 "unknown male": Case Number 00F06871, *The People of the State of California vs. John Doe,* Aug. 21, 2000.

276 "DNA appears": Judge Tani G. Cantil-Sakauye, Case Number 00F06871, *The People of the State of California vs. Paul Robinson,* Motion to Dismiss, reporter's transcript, p. 136, Feb. 23, 2001.

281 "My son": Jean Blassie as quoted in Pat McKenna, "Unknown, No More," http://www.af.mil/news/airman/0998/unknown.htm

289 "The DNA evidence": Lord Woolf, Case Number 199902010 S2, "Regina and James Hanratty," Judgment, May 10, 2002, paragraph 211.

289 "In hindsight": "DNA testing also proves guilt," editorial, *St. Petersburg Times,* 30 May 2002.

291 "DNA testing": Barry Scheck et al., *Actual Innocence* (New York: Doubleday, 2000), p. xv.

CHAPTER 11: GENE HUNTING

294 "Fifty-fifty": Milton Wexler as quoted in Alice Wexler, *Mapping Fate* (New York: Random House, 1995), p. 43.

294 "Over 50 years": George Huntington as cited in Charles Stevenson "A Biography of George Huntington, M.D.," *Bulletin of the Institute of the History of Medicine* 2 (1934).

295 "gradually increase": ibid.
295 "As the disease": ibid.
295 "When either": ibid.
296 "without theater": Americo Negrette as quoted in Robert Cook-Deegan, *The Gene Wars* (New York: W. W. Norton & Co., 1994), p. 235.
298 "tends to think": ibid., p. 37.
302 "We would never": Ray White as quoted in Leslie Roberts, "Flap arises over genetic map," *Science* 239 (1987): 750–52.
309 "We own": Orrie Friedman as quoted in Richard Saltus, "Biotech Firms compete in Genetic Diagnosis," *Science* 234 (1986): 1318–20.
318 "committing": Rural Advancement Foundation International, at http://www.rafi.org/article.asp?newsid=207
319 "the creation": Althing (Icelandic Parliament), "Law on a Health Sector Database," at http://www.mannvernd.is/english/laws/law.HSD.html

CHAPTER 12: DEFYING DISEASE

327 "So marked": John Langdon Down as quoted in Elaine Johansen Mange and Arthur P. Mange, *Basic Human Genetics* (Sunderland, Mass.: Sinauer Associates, 1999), p. 267
331 "ear-piercing": Kathleen McAuliffe, "The Hardest Choice," at http://blueprint.bluecrossmn.com/topic/hardestchoice
343 "Some people": Debbie Stevenson, "The Mystery Disease No One Tests For," *Redbook,* July 2002: 137.
346 "biological Holocaust": Daniel Pollen, *Hannah's Heirs* (New York: Oxford University Press, 1993), p. 14.
354 "informed consent": U.S. Department of Health and Human Services Press Release "New Initiatives to Protect Participants in Gene Therapy Trials," 7 March 2000. Available at http://www.fda.gov/bbs/topics/NEWS/NEW00717.html

CHAPTER 13: WHO WE ARE

362 "Whatever person": Penal Laws as cited at http://www.law.umn.edu/irishlaw/education.html
362 "They might as well": Arthur Young as cited in Julie Henigan, "For Want of Education: The Origins of the Hedge Schoolmaster Songs," *Ulster Folklife* 40 (1994): 27–38.
362 "Still crouching": John O'Hagan as cited at http://www.in2it.co.uk/history/2.html
366 "barefoot professor": Vitaly Fyodorovich as cited in David Joravsky, *The Lysenko Affair* (Cambridge, Mass.: Harvard University Press, 1970), p. 189.
366 "Turkic peasant": Vitaly Fyodorovich as cited in Valery N. Soyfer, *Lysenko and the Tragedy of Soviet Science* (New Brunswick, N.J.: Rutgers University Press, 1994), p. 11.
366 "He didn't": Vitaly Fyodorovich as cited in ibid., p. 11.
367 "In order": Trofim Lysenko as cited in Joravsky, p. 110.
370 "hard core": ibid., p. 226
370 "It deals": K. Iu. Kostriukova as cited in ibid., p. 247.

370 "In our": Trofim Lysenko as cited in ibid., p. 210.

373 "Give me": John B. Watson, *Behaviorism* (New York: W. W. Norton & Co., 1924), p. 104.

378 "On multiple": Thomas J. Bouchard et al., "Sources of Human Psychological Differences: The Minnesota Study of Twins Reared Apart," *Science* 250 (1990): 223–28.

389 "In no field": Neil Risch and David Botstein, "A Manic Depressive History," *Nature Genetics* 12 (1996): 351–53.

CODA

395 "The event": Percy Bysshe Shelley, introduction to Mary Wollstonecraft Shelley, *Frankenstein* (New York: Oxford University Press, 1969), p. 13.

402 "Ethical Implications": James D. Watson in *Frankfurter Allgemeine Zeitung,* September 26, 2000.

402 "following the logic": Jörg Dietrich Hoppe as cited in Benno Müller-Hill, "Speaking Out in Favor of the Right to Choose" *Frankfurter Allgemeine Zeitung,* December 5, 2000.

403 "value and sense": Johannes Rau as cited in ibid.

403 "Ethics of Horror": Dietmar Mieth in *Frankfurter Allgemeine Zeitung* as cited in ibid.

404 "Though I speak": 1 Corinthians 13: 1–2.

FURTHER READING

. . .

CHAPTER 1: BEGINNINGS OF GENETICS

Carlson, Elof Axel. *The Unfit: A History of a Bad Idea.* Cold Spring Harbor, N.Y.: Cold Spring Harbor Laboratory Press, 2002. Discussion of eugenics beginning in biblical times and ending with contemporary clinical genetics.

Gillham, Nicholas Wright. *A Life of Sir Francis Galton: From African Exploration to the Birth of Eugenics.* New York: Oxford University Press, 2001. Engaging recent study of an extraordinary but neglected figure.

Jacob, François. *The Logic of Life: A History of Heredity.* Princeton: Princeton University Press, 1993. Reflections by one of the founders of molecular genetics.

Kevles, Daniel J. *In the Name of Eugenics: Genetics and the Uses of Human Heredity.* New York: Alfred A. Knopf, 1985. Scholarly but readable account of eugenics.

Kohler, Robert E. *Lords of the Fly: Drosophila Genetics and the Experimental Life.* Chicago: University of Chicago Press, 1994. Chronicle of the early days of fruit fly genetics.

Kühl, Stefan. *The Nazi Connection: Eugenics, American Racism, and German National Socialism.* New York: Oxford University Press, 1994.

Mayr, Ernst. *This Is Biology: The Science of the Living World.* Cambridge, Mass.: Harvard University Press, 1997. Fine overview from a biologist who has just celebrated the seventy-fifth anniversary of earning his Ph.D.

Müller-Hill, Benno. *Murderous Science: Elimination by Scientific Selection of Jews, Gypsies, and Others in Germany, 1933–1945.* Translated by Todliche Wissenschaft. New York: Oxford University Press, 1988. Reveals how German scientists and physicians were implicated in Nazi policies and how they resumed their academic positions after the war.

Olby, Robert C. *Origins of Mendelism.* Chicago: University of Chicago Press, 1985.

Orel, Vítezslav. *Gregor Mendel: The First Geneticist.* New York: Oxford University Press, 1996. The most complete biography to date.

Paul, Diane B. *Controlling Human Heredity, 1865 to the Present.* Atlantic Highlands, N.J.: Humanities Press, 1995. A succinct history of eugenics.

CHAPTER 2: THE DOUBLE HELIX

Crick, Francis H. C. *What Mad Pursuit: A Personal View of Scientific Discovery.* New York: Basic Books, 1988.

Hager, Thomas. *Force of Nature: The Life of Linus Pauling*. New York: Simon & Schuster, 1995. An excellent biography of a scientific giant.

Holmes, Frederic Lawrence. *Meselson, Stahl, and the Replication of DNA: A History of "The Most Beautiful Experiment in Biology."* New Haven: Yale University Press, 2001.

McCarty, Maclyn. *The Transforming Principle: Discovering That Genes are Made of DNA*. New York: W. W. Norton & Co., 1995. Account of the experiments that showed DNA to be the hereditary material by one of the three scientists who carried them out.

Maddox, Brenda. *Rosalind Franklin: The Dark Lady of DNA*. New York: HarperCollins, 2002. Thorough biography that casts new light on Franklin.

Olby, Robert. *The Path to the Double Helix: The Discovery of DNA*. Foreword by Francis Crick. Dover Publishers, 1994. Scholarly historical perspective.

Watson, James D. *The Double Helix: A Personal Account of the Discovery of the Structure of DNA*. New York: Atheneum Press, 1968.

CHAPTER 3: READING THE CODE

Brenner, Sydney. *My Life in Science*. London: BioMed Central Limited, 2001. A rare combination: illuminating *and* funny.

Hunt, Tim, Steve Prentis, and John Tooze, ed. *DNA Makes RNA Makes Protein*. New York: Elsevier Biomedical Press, 1983. Collection of essays summarizing the state of molecular genetics in 1980.

Jacob, François. *The Statue Within: An Autobiography*. Translated by Franklin Philip. Cold Spring Harbor, N.Y.: Cold Spring Harbor Laboratory Press, 1995. Lucid and beautifully written.

Judson, Horace Freeland. *The Eighth Day of Creation: Makers of the Revolution in Biology*. Expanded edition. Cold Spring Harbor, N.Y.: Cold Spring Harbor Laboratory Press, 1996. Classic study on the origins of molecular biology.

Monod, Jacques. *Chance and Necessity: An Essay on the Natural Philosophy of Modern Biology*. Translated by Austryn Wainhouse. New York: Alfred A. Knopf, 1971. Philosophical musings by a key figure in molecular genetics.

Watson, James D. *Genes, Girls, and Gamow*. New York: Alfred A. Knopf, 2001. Sequel to *The Double Helix*.

CHAPTER 4: PLAYING GOD

Fredrickson, Donald S. *The Recombinant DNA Controversy: A Memoir: Science, Politics, and the Public Interest 1974–1981*. Washington, D.C.: American Society for Microbiology Press, 2001. Account of turbulent times in biomedical research by the then-director of the National Institutes of Health.

Krimsky, Sheldon. *Genetic Alchemy: The Social History of the Recombinant DNA Controversy*. Cambridge, Mass.: MIT Press, 1982. A critic's perspective.

Rogers, Michael. *Biohazard*. New York: Alfred A. Knopf, 1977. Expansion of Rogers's insightful account in *Rolling Stone* of the Asilomar meeting.

Watson, James D. *A Passion for DNA: Genes, Genomes, and Society.* Cold Spring Harbor, N.Y.: Cold Spring Harbor Laboratory Press, 2000. Collection of essays drawn from newspapers, magazines, talks, and Cold Spring Harbor Laboratory reports.

Watson, James D., Michael Gilman, Jan Witkowski, and Mark Zoller. *Recombinant DNA.* New York: Scientific American Books, distributed by W. H. Freeman, 1992. Now out of date but still a sound introduction to the basic science underlying genetic engineering.

Watson, James D., and John Tooze. *The DNA Story: A Documentary History of Gene Cloning.* San Francisco: W. H. Freeman and Co., 1981. The recombinant DNA debate recounted through contemporary articles and documents.

CHAPTER 5: DNA, DOLLARS, AND DRUGS

Cooke, Robert. *Dr. Folkman's War: Angiogenesis and the Struggle to Defeat Cancer.* New York: Random House, 2001.

Hall, Stephen S. *Invisible Frontiers: The Race to Synthesize a Human Gene.* New York: Atlantic Monthly Press, 1987. Tells the insulin-cloning story with verve.

Kornberg, Arthur. *The Golden Helix: Inside Biotech Ventures.* Sausalito, Calif.: University Science Books, 1995. The founder of several companies describes the rise of the biotechnology industry.

Werth, Barry. *The Billion-Dollar Molecule: One Company's Quest for the Perfect Drug.* New York: Touchstone Books/Simon & Schuster, 1995. The story of Vertex, a company typifying the biotech approach to the pharmaceutical business.

CHAPTER 6: TEMPEST IN A CEREAL BOX

Charles, Daniel. *Lords of the Harvest: Biotech, Big Money, and the Future of Food.* Cambridge, Mass.: Perseus Publishing, 2001. Fascinating account of the genetically modified food controversy, emphasizing the business side and focusing primarily on Monsanto.

McHughen, Alan. *Pandora's Picnic Basket: The Potential and Hazards of Genetically Modified Foods.* New York: Oxford University Press, 2000. Spotty introduction to some of the issues, including scientific ones, behind the controversy.

CHAPTER 7: THE HUMAN GENOME

Cook-Deegan, Robert M. *The Gene Wars: Science, Politics, and the Human Genome.* New York: W. W. Norton & Co., 1994. Brilliantly comprehensive account of the origins and early days of the Human Genome Project.

Davies, Kevin. *Cracking the Genome: Inside the Race to Unlock Human DNA.* New York: Free Press, 2001. Continuation of Cook-Deegan's story, bringing it up to the completion of the first draft of the human genome.

Sulston, John, and Georgina Ferry. *The Common Thread: A Story of Science, Politics, Ethics, and the Human Genome.* Washington, D.C.: Joseph Henry Press, 2002. Personal account of research on the worm and of the British end of the Human Genome Project. Sulston's disdain for individuals and companies profiting from the human genome sequence drives his story and his science.

CHAPTER 8: READING GENOMES

Bier, Ethan. *The Coiled Spring: How Life Begins.* Cold Spring Harbor, N.Y.: Cold Spring Harbor Laboratory Press, 2000.

Comfort, Nathaniel C. *The Tangled Field: Barbara McClintock's Search for the Patterns of Genetic Control.* Cambridge, Mass.: Harvard University Press, 2001. A scholarly but approachable account of the life and work of Barbara McClintock.

Lawrence, Peter A. *The Making of a Fly: The Genetics of Animal Design.* Boston: Blackwell Scientific Publications, 1992. Now out of date but still an excellent introduction to the excitement generated when genetics meets developmental biology.

Ridley, Matt. *Genome: The Autobiography of a Species in 23 Chapters.* New York: HarperCollins, 1999. Hugely accessible introduction to modern studies of human genetics.

CHAPTER 9: OUT OF AFRICA

Cavalli-Sforza, L. L. (Luigi Luca). *Genes, Peoples, and Languages.* Translated by Mark Seielstad. New York: North Point Press, 2000. Personal account of human-evolution studies by the field's leader.

Olson, Steve. *Mapping Human History: Discovering the Past Through Our Genes.* Boston: Houghton Mifflin, 2002. Balanced and up-to-date account of human evolution and the impact that our past has on our present.

Sykes, Bryan. *The Seven Daughters of Eve.* New York: W. W. Norton & Co., 2001.

CHAPTER 10: GENETIC FINGERPRINTING

Massie, Robert K. *The Romanovs: The Final Chapter.* New York: Random House, 1995. The story of the Romanovs' murders and of how DNA fingerprinting established the authenticity of the remains and unmasked impostors.

Scheck, Barry, Peter Neufeld, and Jim Dwyer. *Actual Innocence: Five Days to Execution and Other Dispatches from the Wrongly Convicted.* New York: Doubleday, 2000. From the horses' mouths, an examination of the power of DNA fingerprinting to exonerate the wrongfully convicted.

Wambaugh, Joseph. *The Blooding.* New York: Bantam Books, 1989. Exciting account of the first use of DNA fingerprinting to apprehend a criminal.

CHAPTER 11: GENE HUNTING

Bishop, Jerry E., and Michael Waldholz. *Genome: The Story of the Most Astonishing Scientific Adventure of Our Time—The Attempt to Map All the Genes in the Human Body.* New York:

Simon & Schuster, 1990. Still one of the best accounts of the early days of hunting human disease genes.

Gelehrter, Thomas D., Francis Collins, and David Ginsburg. *Principles of Medical Genetics.* Baltimore: Williams & Wilkins, 1998. A short and readable textbook on modern human molecular genetics.

Pollen, Daniel A. *Hannah's Heirs: The Quest for the Genetic Origins of Alzheimer's Disease.* New York: Oxford University Press, 1993. Captures the thrill of the chase and highlights the awfulness of the disease.

Wexler, Alice. *Mapping Fate: A Memoir of Family, Risk, and Genetic Research.* New York: Random House, 1995. Searingly honest testimony from Nancy Wexler's sister.

CHAPTER 12: DEFYING DISEASE

Davies, Kevin, with Michael White. *Breakthrough: The Race to Find the Breast Cancer Gene.* New York: John Wiley & Sons Inc., 1996. Story of immensely hard work, dedication, ambition, and greed.

Kitcher, Philip. *The Lives to Come: The Genetic Revolution and Human Possibilities.* New York: Simon & Schuster, 1997. Philosophical and ethical discussion about how to use what we have learned of human molecular genetics.

Lyon, Jeff, with Peter Gorner. *Altered Fates: Gene Therapy and the Retooling of Human Life.* New York: W. W. Norton & Co., 1995. Includes a good account of the treatment of the two girls with ADA deficiency.

Reilly, Philip R. *Abraham Lincoln's DNA and Other Adventures in Genetics.* Cold Spring Harbor, N.Y.: Cold Spring Harbor Laboratory, 2000. Essays on topical issues written from the unusually informed perspective of a physician-cum-lawyer.

Thompson, Larry. *Correcting the Code: Inventing the Genetic Cure for the Human Body.* New York: Simon & Schuster, 1994. Account of the development of gene therapy, including the Martin Cline episode.

CHAPTER 13: WHO WE ARE

Coppinger, Raymond, and Lorna Coppinger. *Dogs: A Startling New Understanding of Canine Origin, Behavior, and Evolution.* New York: Scribner, 2001. Overview of the enormous differences, in body and mind, among dogs.

Crick, Francis H. C. *The Astonishing Hypothesis: The Scientific Search for the Soul.* New York: Scribner, 1993. A materialist perspective on the problem of consciousness. Crick concludes that we are "no more than the behavior of a vast assembly of nerve cells and their associated molecules."

Herrnstein, Richard J., and Charles Murray. *The Bell Curve: Intelligence and Class Structure in American Life.* New York: Free Press, 1994. More talked about than read.

Jacoby, Russell, and Naomi Glauberman, ed. *The Bell Curve Debate: History, Documents, Opinions.* New York: Times Books, 1995. Collection of eighty essays about and reviews of *The Bell Curve.*

Further Reading

Lewontin, R. C., Steven Rose, and Leon J. Kamin. *Not in Our Genes: Biology, Ideology, and Human Nature.* New York: Pantheon Books, 1984. The academic left's response to Wilson's *Sociobiology.*

Mendvedev, Zhores A. *The Rise and Fall of T. D. Lysenko.* New York: Columbia University Press, 1969. Firsthand account by a scientist who suffered from the Communist Party's control of Soviet science.

Pinker, Steven. *The Blank Slate: The Modern Denial of Human Nature.* New York: Viking Penguin, 2002.

Pinker, Steven. *How the Mind Works.* New York: W. W. Norton & Co., 1997. Evolutionary psychology outlined by one its most eloquent proponents.

Ridley, Matt. *Nature via Nurture: Genes, Experience, and What Makes Us Human.* New York: HarperCollins, 2003.

Soyfer, Valery N. *Lysenko and the Tragedy of Soviet Science.* Translated by Leo Gruliow and Rebecca Gruliow. New Brunswick, N.J.: Rutgers University Press, 1994. An account from someone who knew Lysenko.

Wilson, Edward O. *Sociobiology: The New Synthesis.* Cambridge, Mass.: Belknap Press of Harvard University Press, 1975. Proposes an evolutionary explanation for much of our behavior.

ACKNOWLEDGMENTS

This book is one of several strands that together comprise a major effort to commemorate the fiftieth anniversary of the discovery of the double helix. All of the projects—this book, a five-part TV series, a multimedia educational product, and a short film for science museum audiences—are interconnected in many ways. We therefore find ourselves indebted to more than the usual slew of readers, editors, and spouses found in the acknowledgments section of a typical nonfiction book. What follows is a reflection of the size and scope of a sprawling collaborative project.

Throughout, the Alfred P. Sloan Foundation, the Howard Hughes Medical Institute, and the University of North Carolina have been phenomenally generous in their support. With wisdom and good sense, John Cleary and John Maroney oversaw the project's alarmingly complex logistics, ensuring that its many strands never became unraveled.

The television series was produced by David Dugan of Windfall Productions in London under the direction of David Glover and Carlo Masarella. To create the educational components, Max Whitby of the Red Green & Blue Company, also in London, collaborated with a team under Dave Micklos at the Cold Spring Harbor Dolan DNA Learning Center and with genius animator Drew Berry (no relation) at the Walter and Eliza Hall Institute in Melbourne, Australia.

The illustrations for the book were prepared by Keith Roberts of the John Innes Centre in Norwich, England. With his customary flair for combining design with scientific clarity, Keith has, with Nigel Orme, produced a series of illustrations that we feel massively enhance the value of the book. Robin Reardon, the assistant editor at Knopf, managed against all odds to coax us into making deadline after deadline (well, more or less) without once having to resort to physical intimidation. Designer Peter Andersen, also at Knopf, effected the miraculous marriage between the text and the images. Keith, Robin, and Peter were indispensable members of the team.

Many people read versions of the book or of chapters addressing their particular areas of expertise. The following graciously supplied detailed and insightful comments on the manuscript: Fred Ausubel, Paul Berg, David Botstein, Stanley Cohen, Francis Collins, Jonathan Eisen, Mike Hammer, Doug Hanahan, Rob Horsch, Sir Alec Jeffreys, Mary-Claire King, Eric Lander, Phil Leder, Victor McElheny, Svante Pääbo, Joe Sambrook, and Nancy Wexler.

Many others also supplied helpful information and/or images: Bruce Ames, Jay Aronson, Antonio Barbadilla, John Barranger, Jacqueline Barataud, Caroline Berry, Sam Berry, Ewan Birney, Richard Bondi, Herb Boyer, Pat Brown, Clare Bunce, Caroline Caskey, Tom Caskey, Luigi Luca Cavalli-Sforza, Shirley Chan, Francis A. Chifari, Kenneth Culver, Charles DeLisi, John Doebley, Helen Donis-Keller, Cat Eberstark, Mike Fletcher, Judah Folkman, Norm Gahn, Wally Gilbert, Janice Goldblum, Eric Green, Wayne Grody, Mike Hammer, Krista Ingram, Leemor Joshua-Tor,

Linda Pauling Kamb, David King, Robert Koenig, Teresa Kruger, Brenda Maddox, Tom Maniatis, Richard McCombie, Benno Müller-Hill, Tim Mulligan, Kary Mullis, Harry Noller, Peter Neufeld, Margaret Nance Pierce, Naomi Pierce, Tomi Pierce, Daniel Pollen, Mila Pollock, Sue Richards, Tim Reynolds, Matt Ridley, Julie Reza, Barry Scheck, Mark Seielstad, Phil Sharp, David Spector, Rick Stafford, Debbie Stevenson, Bronwyn Terrill, William C. Thompson, Lap-Chee Tsui, Peter Underhill, Elizabeth Watson, Diana Wellesley, Rick Wilson, David Witt, Jennifer Whiting, James Wyngaarden, Larry Young, Norton Zinder.

Thank you all.

All the above did their best to ensure that we got things right. Nevertheless we are wholly responsible for the errors that no doubt remain.

INDEX

. . .

Entries referring to key terms or concepts in the text are set in **bold**.
Entries referring to figures are set in *italics*.

Index

Index

Index

Index

Index

Index

Index

ILLUSTRATION CREDITS

© AFP: 355
AIP Emilio Segrè Visual Archives: 36
© Paul Almasy/CORBIS: 204
American Philosophical Society: 12
American Philosophical Society, Eugenics Record Office: 20, 24, 25, 26
Anonymous, nineteenth century © Image Select/Art Resource, N.Y.: 252
AP/Wide World Photos: 192 (top), 260
Courtesy of Michael Baden: 278 (left and top right)
© Anthony Bannister; Gallo Images/CORBIS: 244
Courtesy of Jacqueline Barataud: 178
© A. Barrington-Brown/Photo Researchers, Inc.: viii
Courtesy of Paul Berg: 95
Dr. David Becker/Wellcome Photo Library: 342
© Bettmann/CORBIS: 105 (bottom), 255 (right), 278 (bottom right), 280, 322
Courtesy of Bio-Rad Laboratories: 140
Photo by Jim Bourg © Reuters NewMedia Inc./CORBIS: 159
Drawn by Tony Bramley for TiBS/Reprinted from *Trends in Biomedical Sciences,* vol. 3, pp. N243, Szybalski: "Dangers of regulating the recombinant DNA technique" © 1978 with permission from Elsevier Science: 103
Photo by Michael Brooke: 317
Courtesy of BRT Laboratories, Inc.: 284
Pieter Brueghel the Elder (1551–1569), *The Harvesters,* 1565, Oil on Wood/The Metropolitan Museum of Art, Rogers Fund, 1919 (19.164) / Photograph © 1998 The Metropolitan Museum of Art: 157 (top left)
Courtesy of the Archives, California Institute of Technology: 60
Courtesy of Stanley Cohen: 91
© Stewart Cohen Photography: 360
Cold Spring Harbor Laboratory Archives: 76, 169, 205, 351
Cold Spring Harbor Laboratory Archives, James D. Watson Collection: 4, 45, 51, 57, 70 (top left and right)
Joseph DeRisi: 194
John Doebley: 143
© Laura Dwight/CORBIS: 329
Florida State Archives: 136 (right)
Courtesy of Igor Gamow: 70 (bottom right)
Photo by Peter Ginter/Courtesy of the Hereditary Disease Foundation: 292
Courtesy of Eric Green: 185 (top)
Michael Greenberg and Jennifer Brown, Children's Hospital, Division of Neuroscience, Harvard Medical School: 385
Photo by Fergus Greer/Courtesy of Kary Mullis: 174
David Gregory and Debbie Marshall/ Wellcome Photo Library: 94
Jeff Hansen, Yale University: 62
Photo by Don Harris © University of California, Santa Cruz: 190
© Hulton-Deutsch Collection/CORBIS: 288, 368, 383 (left)
Courtesy of Indigo® Instruments (indigo.com): 11
© Institut Pasteur: 81
© Markowitz Jeffrey/CORBIS Sygma: 133
Courtesy of Alec Jeffreys: 263
Courtesy of Alec Jeffreys/Reprinted with permission from *Nature,* vol. 317 (31 October 1985): 264

Illustration Credits

Leemor Joshua-Tor: 218
Courtesy of Linda Pauling Kamb: 44
Courtesy of Mary-Claire King: 235
King's College London Archives: 42
Jack J. Kunz © 1964 Time Inc.: 13
By Eric Lander/Courtesy of Whitehead Institute: ii
Harry H. Laughlin Archives, Truman State University: 23, 30
Lear/Carson Collection, Connecticut College: 136 (left)
Courtesy of Phil Leder: 123
Painting © Jay H. Matternes: 228 (top)
Richard McCombie and Lance Palmer: 198 (bottom)
Courtesy of mitomap.org: 246
Courtesy of Monsanto: 137, 146
Photos by Tim Mulligan, Cold Spring Harbor Laboratory: 162
Joe Munroe: 142
Reprinted with permission from *Nature,* vol. 171 (25 April 1953): 56
Reprinted with permission from *Nature,* vol. 409 (15 February 2001): 196, 197, 198 (top), 199
Newsweek, Inc. © 1988. All rights reserved. Reprinted by permission: 239
Cover of the May 8, 1995, issue of *BusinessWeek* reprinted by permission. Copyright © 1995 by The McGraw-Hill Companies: 181
NIH/NGHRI: 164
Courtesy of Harry Noller: 84
© Diego Lezama Orezzoli/CORBIS: 255 (left)
© 1972 Oxford University Press, Inc.: 7
Courtesy of Margaret Nance Pierce: 47
Courtesy of Pioneer: 141
Private collection/Bridgeman Art Library: 2
© Roger Ressmeyer/CORBIS: 157 (bottom right)
Courtesy of Alex Rich: 70 (bottom left)
Photograph by Conly L. Rieder, Division of Molecular Medicine, Wadsworth Center, Albany, New York, 12201-0509: 223
Courtesy of Keith Roberts: 90
Keith Roberts: 52, 53, 54, 58, 59, 67, 75, 77, 85, 93, 107, 109, 116, 175, 207, 237, 271, 299
Reprinted from *Darwin and After Darwin* by George J. Romanes. Open Court Publishing Co. © 1892: 382
© 1990 Bill Sanderson: 185 (bottom)
The Sanger Centre/Wellcome Photo Library: 177
© Albrecht G. Schaefer/CORBIS: 383 (right)
© 1983 The Spokesman-Review: 306
© Rick Stafford: 101
© Rick Stafford/Courtesy of Walter Gilbert: 105 (top)
Stanford News Service: 61
Andrew A. Stern/NAS: 98
Courtesy of Debbie Stevenson: 343
Photo by Rob Teteruck, The Hospital for Sick Children: 309
Courtesy of The Institute for Genomic Research: 208, 209
Time Magazine © Time Inc./Timepix: 112
F. Rudolf Turner, Indiana University: 225
© 1945 The University Press, Cambridge, England: 34
US Army/Fort Detrick photo: 86
Courtesy of Vysis Genomic Disease Management: 328
Photo by Bob Waterston, Whitehead Institute/MIT Center for Genome Research: 189
Wessex Regional Genetics Centre/Wellcome Photo Library: 327
Westminster City Archives: 17
Sherri Wick: 37
Photo by Allison Wilson: 192 (bottom)

A NOTE ABOUT THE AUTHORS

JAMES D. WATSON was director of Cold Spring Harbor Laboratory in New York from 1968 to 1993 and is now its president. He was first director of the National Center for Human Genome Research of the National Institutes of Health from 1989 to 1992. A member of the National Academy of Sciences and the Royal Society, he has received the Presidential Medal of Freedom, the National Medal of Science, and, with Francis Crick and Maurice Wilkins, the Nobel Prize for Physiology or Medicine in 1962.

ANDREW BERRY, who holds a Ph.D. in fruit fly genetics, is a research associate of Harvard University's Museum of Comparative Zoology. A writer and teacher, he is the editor of a collection of the writings of the Victorian biologist Alfred Russel Wallace, *Infinite Tropics*.

A NOTE ON THE TYPE

This book was set in Fairfield, a typeface designed by the distinguished American artist and engraver Rudolph Ruzicka (1883–1978). In its structure Fairfield displays the sober and sane qualities of the master craftsman whose talents were dedicated to clarity. Ruzicka was born in Bohemia and came to America in 1894. He designed and illustrated many books, and was the creator of a considerable list of individual prints in a variety of techniques.

Composed by North Market Street Graphics, Lancaster, Pennsylvania
Printed and bound by R. R. Donnelley & Sons, Crawfordsville, Indiana
Designed by Peter A. Andersen